T0210178

Negotiating Darwin

Medicine, Science, and Religion in Historical Context

Ronald L. Numbers, *Consulting Editor*

Negotiating Darwin

THE VATICAN CONFRONTS EVOLUTION,

1877–1902

Mariano Artigas, Thomas F. Glick,

and Rafael A. Martínez

THE JOHNS HOPKINS UNIVERSITY PRESS

BALTIMORE

2 4 6 8 9 7 5 3 1

The Johns Hopkins University Press
2715 North Charles Street
Baltimore, Maryland 21218-4363
www.press.jhu.edu

Library of Congress Cataloging-in-Publication Data
Artigas, Mariano.
Negotiating Darwin : the Vatican confronts evolution, 1877–1902 /
Mariano Artigas, Thomas F. Glick, and Rafael A. Martínez.
    p. cm. — (Medicine, science, and religion in historical context)
Includes bibliographical references and index.
ISBN 0-8018-8389-x (hardcover : alk. paper)
    1. Evolution (Biology)—Religious aspects—Catholic Church—History
of doctrines—19th century. 2. Catholic Church—Doctrines—History—
19th century. I. Glick, Thomas F. II. Martínez, Rafael A., 1957– . III. Title.
IV. Series.
BX1795.E85A78 2006
231.7'652088282—dc22   2005032623

A catalog record for this book is available
from the British Library.

CONTENTS

ACKNOWLEDGMENTS

Work in the Vatican archives has not been easy. Systematic cataloging of records is currently in progress, but for our work the only catalogs available were the old ones. At times we had to search blindly, though usually successfully, for dossiers of whose existence we were not sure, among a host of other dossiers carrying the dust of a century and with no clear indication of their contents. Of course, our work has been made possible due to the assistance and cordiality of Monsignor Alejandro Cifres, director of the Archive of the Congregation for the Doctrine of the Faith in the Vatican, and to his whole group, to whom the authors' thanks is gratefully offered. Cardinal Joseph Ratzinger, currently Pope Benedict XVI, while prefect of that congregation, made possible the opening of its archive, long desired by researchers, and appointed one of the authors (Artigas) a member of its advisory Scientific Council.

Although the main part of this research has been done in the Vatican archives, it would have been impossible without consulting other archives. A brief but productive research stage was conducted by one of us (Martínez) at Notre Dame (Indiana). To the staff of the University of Notre Dame Archives, and especially Sharon Sumpter, and to the Indiana Province Archives Center (Congregation of Holy Cross) and Jackie Dougherty, we extend our gratitude. We are also grateful for the permission to reproduce texts from those archives. Other material has been kindly provided by the General Archives of the Dominicans in Santa Sabina (Rome), and the Archives of the Dominican Province of France, Bibliotèque du Saulchoir, Paris (Michel Albaric).

An adumbration of this work, titled "New Light on Catholic Responses to Evolution: The Reception of Evolutionary Theories in Catholic Media in the Late 19th Century," was presented in July 2002 at the Workshop on Science and Human Values, hosted in Castelgandolfo by the Reverend George V. Coyne (head of the Vatican Observatory), organized by Professor John Brooke (Oxford University) and by the Humanities section of the European Science Foundation, headed by Professor William R. Shea (Galileo Chair, University of Padua). A first version of the chapter on Raffaello Caverni was published in *Scripta Theologica* (Pamplona) 36 (2004): 37–68.

A part of the work was conducted at the Pontifical University of the Holy Cross in Rome and, in the final stages, at the University of Navarra in Pamplona (summer 2003) and at Boston University (summer 2004), profiting from the facilities provided by the three universities.

Negotiating Darwin

Introduction

When on October 22, 1996, Pope John Paul II declared that the theory of evolution was considered today as more than a hypothesis, he was acknowledging the Church's inclusion in the great evolutionary consensus, a step that followed from an open and creative debate over the issue in the years after Pope Pius XII's encyclical *Humani Generis* (1950). The occasion was provided by an address to the members of the Pontifical Academy of Sciences, gathered in the Vatican for a meeting on the origins and evolution of life. As the pope recalled, "In his Encyclical *Humani Generis* (1950), my predecessor Pius XII had already stated that there was no opposition between evolution and the doctrine of the faith about man and his vocation, on condition that one did not lose sight of several indisputable points."[1]

The "indisputable points" referred mainly, as Pope John Paul II noted in the same address, to the teaching of revelation that the human being has been created in the image and likeness of God, a doctrine qualified by the pope as "pivotal to Christian thought." A large part of the address focused on this point, drawing a distinction between the scientific theory of evolution on the one hand and its philosophical interpretations on the other. A materialist interpretation denying the spiritual dimensions in the human being would not be compatible with the Christian doctrine. The pope clearly stated that the Church does not oppose the scientific theory of evolution, which is now supported by varied and independent proofs coming from diverse branches of the sciences:

> Today, almost half a century after the publication of the Encyclical, new knowledge has led to recognize that the theory of evolution is more than a hypothesis. It is indeed remarkable that this theory has been progressively accepted by researchers, following a series of discoveries in various fields of knowledge. The convergence, neither sought nor fabricated, of the results of work that was conducted independently is in itself a significant argument in favour of this theory.[2]

Of course, the 1996 address should not be considered as an official endorsement of the theory of evolution by the Catholic Church, that is, as saying that a Catholic must accept that theory.

The 1996 address and the 1950 encyclical were two big stepping-stones, but they were not the only interventions of the Vatican authorities regarding evolutionism. Other statements by John Paul II and Pius XII could be added along the same lines. In fact, a quite peaceful accord had already been achieved by the 1930s. In 1931 Ernest Messenger published a long reliable account of the subject, whose main doctrinal point centered on the evolution of the human being.[3] According to Messenger, no opposition existed between Christianity and the scientific theory of evolution, although the Vatican's contribution to the debate had not reached the status of authoritative doctrine. Of course, Messenger could quote only from available documents, which at the time were very scarce indeed. A much fuller review is now possible, thanks to the opening of the archives of the Holy Office in Rome in 1998.

The evidence reveals that the Vatican's actions with respect to evolution have been quite moderate. For many years, Catholic theology textbooks criticized evolutionism harshly, but they were able to marshal only a few authoritative arguments. Although it was known that Rome had intervened on some occasions, the exact picture was enveloped in darkness. The limited available data could not even be found in public documents. They almost always originated in a journal published by the Jesuits in Rome, *La Civiltà Cattolica*, which, without being an official publication of the Vatican, has always had a special relationship with the Holy See.

When the Vatican opened the Archives of the Congregation for the Doctrine of the Faith, which contains the archives of the old Congregation of the Holy Office and the Congregation of the Index, it became possible to gain access to information that, until that moment, had been rigorously guarded. Since 1999, we have worked in these archives with the objective of studying in detail the conduct of the Vatican authorities with respect to evolutionism in the second half of the nineteenth century, when the theory of evolution came to prominence. At some future date, when more recent documentation becomes available for consultation, it will be possible to continue the story.

Although we strive for the most accurate interpretation of these documents and their context, we recognize that other scholars might select materials according to some criteria other than our own, which is to identify both the ideological and operational stance of the Church with respect to the reception of Darwinism. We trust that the resulting narrative will appeal not only to scholars but also to a broader spectrum of readers interested in the interaction of science, culture, and religion.

Thanks to the opening of these two archival collections, we now have the opportunity to examine, with the objectivity that historical distance affords, the conduct of the Vatican authorities and their colleagues. In the archive we found reports prepared by designated experts who were already important figures in the life of the Church, or who became so later on. Thanks to the methodical, bureaucratic administrative style of the Holy See, abundant documentation offers a detailed look into a now remote period, but one that harbors the roots of current problems, many of which were also those of our predecessors, even in the face of vastly changed circumstances.

In the present study we focus on six cases that featured Catholics who tried to integrate ("harmonize" was the word of the epoch) evolution and Christianity. These same cases have been continuously mentioned whenever the relationship of the Catholic Church and evolution is discussed. But information about these cases has been fragmentary, giving rise to confusion that persists to the present, even in studies published since the opening of the archives of the Holy Office and with access to its documents.

These six authors all asserted the compatibility of evolution and Christianity. Others shared their views, but their cases had particular resonance. Based on new archival data, the present study reconstructs what really happened: who acted and for what reasons; how the events unfolded; what decisions were made, and how they were put into practice. As often happens with historical reconstructions, reality is sometime very simple, other times complex, and on occasion so extraordinarily complex that the web of decisions made for or against each of our six protagonists proves impossible to follow all the way to its conclusion.

The length of the chapters is unequal owing to the peculiar characteristics of the six cases. That of Bonomelli constitutes at best a small anecdote in the tumultuous life of this bishop. The case of Mivart is important but, contrary to what is sometimes said, evolutionism did not occupy center stage in the unfolding of his drama. Caverni's case is what we might call of regular dimensions, without complications. By contrast, those of Leroy and Zahm are by far the most complex. Leroy provides a unique opportunity to contemplate the work of the Congregation of the Index in detail, and it also had important external repercussions. Zahm's case was not initially so complex; only after the Congregation made its decision did complications arise. These are amply documented in the correspondence of Zahm and his friends and permit us to establish the tight relationship between this case and other factors that have nothing to do with evolution. The case of Hedley is very instructive, because it helps to reveal the origin of misconceptions that have persisted to the present.

We have adopted a vantage point that places our book entirely within the perspective of Catholicism, because our purpose is to provide a reliable account of the documents of the Vatican archives that are now available for the first time. We analyze the six cases placing their protagonists and the documents in their historical context. In the first chapter we provide the context and data that are necessary to capture the meaning of the Vatican documents. An interesting but quite different work would be to place these cases in a wider context, including other Christian denominations, and to examine the relationship between science and religion in general. The reader interested in these issues can easily find books covering an ample range of fields, from historical introductions on science and religion,[4] to more specialized studies on evolution and Protestantism.[5]

We have tried to avoid preconceptions and thus allow the documents to speak for themselves. When we began our research, we did not know what we might find in the archives. Of course, we knew that no official condemnation of evolution had ever been issued by the Vatican, in spite of the fact that evolutionism provoked severe tensions. We also knew something about the actions of the Vatican authorities, but very little indeed—only those few details that had already come to the public's attention. The research led us to an unexpected conclusion. Although from the outside one might well think that the Vatican adopted a careful policy toward evolutionism, to our surprise there was, in a sense, no policy at all. The actions of the authorities responded to particular circumstances, not to any carefully designed plan.

The Vatican authorities were aware of the fact that no condemnation of evolutionism had been issued, and apparently they were not anxious to provide one. They examined the writings of our protagonists when a work was denounced, analyzing each on the basis of the existing doctrine but without the guidance of any official doctrine regarding evolutionism. This explains why the reports of the experts followed no uniform pattern. Nor was there any fixed pattern that could predict the decision of the cardinals, or even of the pope. We will see that in one of the major cases the pope prevented the publication of a prohibition decided by the cardinals.

There is an obvious difference between the actions of the Vatican authorities and the fate of evolutionism within other Christian denominations. The exercise of a centralized government in the Vatican, following precise rules and procedures and saving the corresponding documents in carefully preserved archives, enables an orderly examination of the facts, reconstructing the proceedings from the first step until the last. Studies on Protestantism usually have to decide which protagonists can be taken as representative. In our case, that choice is determined by the structure of the government of the Catholic Church. We concentrate on those cases that provoked the intervention of Vatican authorities.

The decisions of the Vatican were used in textbooks and other theological studies as the authoritative reference for judging the acceptability of evolution. But they were poorly known, because only a few details were ever made public.

The prudence of the Vatican authorities when dealing with evolution can be interpreted as an effect of the Galileo affair. In the last decades of the nineteenth century, Vatican documents on the proceedings against Galileo were published for the first time, and the behavior of the authorities of the Church was submitted to close scrutiny. Comparing their case with Galileo's, the supporters of evolution argued that they would also triumph in the long run. Most likely, Vatican authorities sought to avoid another conflict with the natural sciences if possible.

A major factor in the Roman responses to evolutionism was the complete opposition showed by *La Civiltà Cattolica*, a journal that was not an official publication of the Vatican but had a close relationship with Church authorities. It is difficult to overestimate the influence of this journal in the Catholic world. References made in *La Civiltà* to the attitude of the Vatican authorities about evolutionism, even though few and at times inaccurate, provided the basis on which Catholic theologians represented the issue for decades, even till the present day. The documents of the archives enable us to clarify this important issue. They show that the attitude of the Vatican authorities was not determined by the influence of any particular group, and also that their silence cannot be interpreted as a continuing condemnation that, due to mysterious reasons, was not given the publicity it deserved. Both interpretations have been suggested after the opening of the archives, as if they were a consequence of what the archives reveal. The reality, however, was much more complex and cannot be reduced to any simple scheme.

Until recently the relationship between science and religion has been considered as an ongoing, perpetual conflict. In our times, the "complexity thesis" has gained ground, the result of an increasing awareness that particular conflicts must be placed in their historical context. Two noteworthy books in this line are *God and Nature,* a collection of essays edited by David Lindberg and Ronald Numbers,[6] and *Science and Religion* by John Hedley Brooke.[7] According to Brooke, "Serious scholarship in the history of science has revealed so extraordinarily rich and complex a relationship between science and religion in the past that general theses are difficult to sustain. The real lesson turns out to be the complexity."[8] We think that our study provides a major illustration of the complexity thesis.

The case of John Zahm can be considered paradigmatic, as it shows how theological reasons merge with motives that were in the final analysis religious but were also closely related to social and national problems in the United States and Europe. We are surprised to find that biological evolution played no role at all in the problems of Mivart, the champion of evolutionism in the Catholic orbit. Paradoxically, the only case that reached the ultimate conclusion, namely, public condemnation, was that of Raffaello Caverni, who nevertheless did not include in his defense of evolution the origin of the human being, which was the main point in contention. Therefore, our conclusions cannot be reduced to any single simple thesis. In fact, if any thesis should be highlighted, we would emphasize that the Vatican authorities did not follow any fixed agenda, a conclusion very much in accord with the complexity thesis.

Even though the concept of evolution took shape in the second half of the nineteenth century, after the publication of Charles Darwin's *Origin of Species* in 1859, its reception by the Roman Catholic Church bore some relation to the condemnation of Galileo in 1633. The Catholic authorities regarded evolutionism with suspicion but were afraid to condemn it. In Galileo's time, modern science was almost nonexistent, and the motion of the Earth was seen as absurd and contrary to "common sense." In Darwin's time, however, modern science was already one of the main components of Western civilization. Theologians liked to say that, although evolution was not scientifically proven, the Roman authorities did not want to get involved in a second "Galileo affair." When Catholic authors attempted to harmonize evolutionism with Christianity, the authorities preferred not to condemn them by a public act but rather to persuade them to retract their ideas. A short letter published in a newspaper was enough. Galileo's shadow was always present.

The New Documents

January 22, 1998, was a historic day for the Vatican and for cultural history. In the seat of the Accademia Nazionale dei Lincei, successor of the famous academy founded by Prince Federico Cesi in 1603, there took place a symposium titled "The Opening of the Archives of the Roman Holy Office." Although the whole archive has not been preserved, the collection is still ample. For the first time in history, scholars can analyze the actions of the Vatican with respect to evolution with complete freedom.

The cases of Galileo and of evolution have become emblematic of the problems between science and religion, and both have been the subject of much debate.. Although there are many studies on the relationship between evolution and Christianity, until now little was known about conflicts with Vatican authorities, and the facts were frequently distorted owing to the lack of trustworthy information. The recent declassification of documents in the Vatican archives makes it possible to clarify numerous issues and to know in detail how the Vatican reacted in the face of the problems posed by the concept of evolution.

On April 24, 1585, the Franciscan cardinal Felice Peretti had been elected pope, taking the name Sixtus V. It is said that he entered the conclave in the fullness of his sixty-four years with a sickly cast, leaning on a cane, but that at the moment of his election, he threw the cane down, ruling with great authority from that day forward. Although the story of the cane may be apocryphal, it is a fact that within two years Sixtus had done away with thousands of bandits who had scourged the Papal States, which then became the safest territory on European soil. He also reorganized the Vatican's finances and set in motion a distinctive phase in the urbanization of Rome, including a series of public works that remain an important part of the city's urban landscape. In the five years of his pontificate, he also reorganized the central administration of the Church, creating, with a bull dated February 11, 1588, the system that continues in force in our own times: the central administration, whose head is the pope, is organized around a series of "congregations," which came to function like the ministries of modern nations.

Each congregation is directed by a cardinal, and has as members several other cardinals, assisted by consultants and an administrative staff. Although several congregations have disappeared and others have been created or modified, the same administrative organization still exists today. In particular we need to know how two of them, the Holy Office and the Index, functioned, because those were the congregations that took part in the cases that we will examine here. The documents just declassified belong to the archives of these two congregations.[1]

The Holy Office

Since 1965 the Holy Office has been called the Congregation for the Doctrine of the Faith. It has a tutelary role over matters of faith and morality throughout the Catholic world. Until 1908 it was also called the Holy Roman Inquisition, because it was the tribunal in which acts held to be crimes against faith or morality were adjudicated.

The antecedents of this congregation stretch back to the Middle Ages. Permanent inquisitors were created in Europe in 1231. They were charged with converting heretics and to sentencing them in cases of obstinacy, although bishops had the same function in each diocese. The Roman Tribunal was presided over by the pope, assisted by the assessor (the master of the Sacred Palace) and the commissioner. In 1542 Pope Paul III created the modern Roman Inquisition, with the objective of slowing the tide of Protestantism as it spread through Europe and began to make inroads into Italy. The new organism, consisting of six cardinals, extended its authority to all of Christendom. It was highly centralized—a necessity in view of the dispersion of the various tribunals of the Inquisition and of its absorption by the state in Spain (in the Spanish Inquisition, created in 1478, religious and political competencies were intermixed). In the general reform of the Roman Curia in 1588, Sixtus V placed it first among all the congregations, whence it acquired the qualifier "The Supreme," which is how it was known. The current Congregation for the Doctrine of the Faith is presided over by a cardinal, just like the other congregations, but in the period that interests us here the Holy Office was chaired directly by the pope, who participated in its weekly meetings, always held on Thursday, or *feria quinta*.

One of the competencies of the Holy Office was the examination and prohibition of books. This was also the purview of the Congregation of the Index. It was not unusual for the Holy Office to decide to prohibit a book, in which case it would communicate its decision to the Congregation of the Index to have it put into practice. The cases we consider were transacted and decided almost in totality by the Congregation of the Index, although we also find some participation by the Holy Office.

'The Index of Prohibited Books'

The *Index of Prohibited Books* was a publication that listed the books whose reading, possession, or publication was prohibited for Catholics.[2] We have records of lists of books from antiquity on that Church authorities considered dangerous for faith and customs, for example, a list produced by a Roman Council in A.D. 494. This activity continued in later centuries and acquired new importance with the rise of Protestantism in the sixteenth century. Moreover, when this period coincided with the spread of printing, an avalanche of Protestant books and pamphlets provoked a reaction by the Catholic Church aimed at impeding their printing, sale, possession, and reading. The Council of Trent took up this problem, which finally was entrusted to the new Congregation of the Index.

The first *Index of Prohibited Books* was published in 1544 by the Faculty of Theology of the University of Paris. New books were then added to this *Index*, as to others published in subsequent years by civil and ecclesiastical authorities in different places: Venice (1549), Venice and Milan (1549, 1554), Anvers (1569, 1570, 1571), Louvain (1546, 1550, 1558), Spain (1551, 1554, 1559), Portugal (1547, 1551, 1561). In 1557 the first Roman *Index* was published.[3] The decisions of the Council of Trent led to the publication of a new *Index* in 1564 and to the formulation of ten general rules that remained in force for more than 300 years.

In 1571 Pope Pius V, mindful of the numerous matters that occupied the Holy Office, created the Congregation of the Index as a permanent institution in the Vatican to scrutinize publications and whose operations would later be more precisely defined by successive popes. It occupied seventh place among the congregations established in Sixtus V's 1588 reform of the Roman Curia. The Congregation published a new *Index of Prohibited Books* in 1596, and it continued to publish new editions in which books prohibited since the preceding edition were added and other emendations introduced.

In 1753 Benedict XIV established with greater precision the procedures to be followed in the examination and censorship of books and set the very norms that continued in force during the period here examined. When he received a charge against a book, the secretary of the Congregation was obliged to examine it and to name referees, called "consultors," to do likewise. Then a written report was prepared for presentation at a meeting with the consultors and, afterward, at another meeting of the full Congregation of the member cardinals, who composed a definitive resolution submitted for the pope's approval.

In 1897 Leo XIII initiated a comprehensive reform of the *Index*, simplifying the body of rules that had accumulated over the centuries and revising their content. In 1917 Benedict XV abolished the Congregation of the Index and assigned its jurisdiction to the Holy Office, which, from that time forward, assumed the tasks related to the *Index of Prohibited Books,* just as it had had in its early years. Finally, in 1965 Paul VI left the *Index* with no status in ecclesiastical law, while preserving its spirit.[4] The final edition of the *Index* was published in 1948.

The Congregation of
the Index

The Congregation of the Index examined publications that had been reported to it because they presumably contained doctrines contrary to the faith and morality, although it could also carry out such an inquiry on its own initiative. If the result of the inquiry was negative, a decree was published whereby the book was added to the *Index,* whose contents were brought up to date whenever a new edition was issued, something that did not happen with any regularity.

As was the case with the other congregations, that of the Index was headed by a cardinal prefect, aided by the master of the Sacred Palace (*maestro di Sacro Palazzo*) as his permanent assistant, an official equivalent to the current theologian of the pontifical household. The Congregation had a secretary; with the exception of the first, who was a Franciscan, the rest were always Dominicans. In the period that interests us here, the secretary directed the business of the Congregation. In addition, there was a group of cardinal members, a team of expert consultors who drew up the reports on the books examined, and various subalterns.

The Congregation of the Index reached its decisions in three phases. In the first place, the prefect, assisted by the secretary, charged one or more consultors with the examination of the work denounced. The consultor submitted his report in writing. In the period under consideration, these verdicts, with a few exceptions, were printed for distribution in the meetings of the Congregation.

The second step was a meeting called the Preparatory or Particular Congregation, in which the consultors of the Congregation of the Index met, chaired by the secretary of the Congregation and assisted by the master of the Sacred Palace.[5] This Preparatory Congregation had no power to make decisions. Its function was to prepare the cardinals for their deliberations, examining and discussing the reports that the same consultors had prepared. After their discussion, the consultants were to draw up a recommendation for each work examined, usually including a vote tally. The secretary then wrote a summary giving the result of the vote, and that proposal was transmitted to the cardinals, together with the reports of each consultor.

Several days later, the General Congregation would meet, in which the cardinals who were members of the Congregation of the Index participated. The task of the cardinals was to judge the works submitted for examination, taking into account the reports of the consultors and the deliberations of the Preparatory Congregation, and to decide what kind of punishment should be accorded to each one of them. The secretary of the Congregation and the master of the Sacred Palace also attended the meetings of the General Congregation.

The Particular Congregation and General Congregation were convened with varying frequency, generally twice or three times yearly. Attendance was not very regular. The number of cardinal members of the Index, although the total varied, was always quite high: between 1894 and 1900, it varied from around twenty to thirty. A total of forty-six different cardinals were members of the Index in those years. Many resided outside of Rome and only appeared infrequently at the General Congregation, if it coincided with their presence in Rome for other reasons. This was the case of the cardinals of Rhodes, Ferrara, and Naples, who attended the Congregation only once. Moreover, many of the Curia cardinals who resided in Rome were members of several congregations, and so could not always attend every meeting. If, as happened in some cases, they were also prefects of another congregation, it could occur that they might never participate in the sessions of the Index. Such, for example, was the case of Cardinal Miecislas Ledóchowski (1822–1902), prefect of the Holy Congregation for the Propagation of the Faith.[6] In the years mentioned, only seventeen cardinals took part in the meetings (including the three just mentioned, who participated in one meeting only), with attendance varying between five and ten cardinals at each meeting.

Attendance at the Preparatory Congregation was not much more regular. A total of forty-five consultors are mentioned in the annual Pontifical volumes (*Annuario Pontificio*) for the period 1893–1900, but only twenty-eight of these actually participated. The number of consultors present, including the master of the Sacred Palace, varied between seven and fourteen on each occasion.

We have used principally two sources to follow developments in the Preparatory Congregation. First are the printed folios of convocation for the General Congregation, indicating the place, day, and time of the meeting (habitually they took place in the Apostolic Palace of the Vatican, Wednesdays, at 9:30 in the morning), the works that were to be examined, and the names of the consultors who had prepared the respective recommendations and those who would expound them at the meeting. The recommendations of the consultors were included as well. The second page contained an extract of the meeting of the Preparatory Congregation: its date, where the meeting had taken place, who had attended, and, in schematic form, what were the recommendations of the consultors for each of the books examined. The summary concluded with an expression of submission to the decision of the cardinals and of the pope.

The second source comprises the summaries of the Congregations, both particular and general, that the secretary of the Index wrote in the Diary of the Congregation. These generally provide the same data as the informative sheets but sometimes with greater detail.

The habitual practice of the Congregation of the Index was that, after the General Congregation, the secretary was received in audience by the pope, for him to confirm the decisions taken. The secretary would explain the cases studied and the decisions taken and the pope would then order the publication of the decree that converted such decisions into Church law. After being informed of the decisions, the pope customarily gave his approval. But this was not a simple bureaucratic action, because the pope, with some frequency, ordered other works added to the decree, or otherwise bypassed the regular Index procedure. As we will see in one of the cases that interest us, the pope intervened personally, more than once, to stop the publication of a decree of prohibition and to give other instructions.

The decree was then printed in large format and posted in the appropriate places in Rome, including the Vatican and the Palace of the Chancery. All decrees had the same structure. At the top of the page was the pope's escutcheon, flanked on either side by the figures of Saint Peter and Saint Paul. The text began with a formulaic paragraph, always the same, stipulating that the cardinal members of the Congregation had met on a specific date; nothing ever changed in this opening paragraph except for the date. Next came the list of books that the decree prohibited. Only the author and publication data of the book were given, and nothing more. This is extraordinarily important, because nothing was ever said about the reason for prohibiting the book, which is why some cases have been unexplained until the present. When the author was a Catholic and had accepted the Congregation's decision, the following sentence was added: "The author, in a praiseworthy manner, has submitted [to the decree] and has repudiated his book." At the end of the document came the date the decree was ordered and its publication date in Rome, together with the signatures of the cardinal prefect and the secretary of the Congregation.

Such a procedure meant that the decisions of the Congregation of the Index were based on already existing doctrine. The Congregation could not, of its own accord, decide whether a doctrine was acceptable or not: it could only apply already existing doctrine to specific books. Clearly, when the consultors examined books, they used their own arguments, but the decisions themselves had to be based on already existing decisions of popes, Church councils, or of the Congregation of the Holy Office: although theological arguments might be adduced, these did not have the value of public doctrinal authority, because they were never published and were only known to those who participated in the activities of the Congregation. Still they are a valuable source of information. The reports preserved in the archive reveal the arguments used in each case.

In the specific case of evolutionism, the Congregation of the Index continually found itself in an area where no doctrinal judgment had been clearly defined by the relevant authorities, and it therefore had to base its decisions on the arguments that surfaced within the Congregation itself. Perhaps this explains why the decisions of the Congregation of the Index, although on several occasions contrary to evolutionism, were always put forth moderately. As we will see, such moderation also is explained by the considerate way in which Catholic authors were treated, as well as the orders or religious congregations to which certain authors belonged: on some occasions the public condemnation of the book was replaced by a brief retraction by the author or by some phrasing that fell short of a retraction.

The Archives of the Congregation
for the Doctrine of the Faith

The archives of the present-day Congregation for the Doctrine of the Faith, opened to scholars in 1998, includes the archives of the two congregations mentioned: the Holy Office and the Index.[7] Each of the two archives has different sections: for example, in the Index archive, there is a series of volumes called Protocolli, which contains all the material produced during the examination of a book from the letter of the accuser to the final decree prohibiting the book, including the reports of the consultors, the summaries of the meetings of the consultors and cardinals, and the report prepared by the secretary for his audience with the pope. There are other volumes called Diarii, which are the diaries or calendars in which the different events and decisions are noted, with the relevant dates.

Even though this material is clear, ordered, and informative, that does not mean that everything that happened was written down. For example, of the two standard meetings where the books were discussed (that of the consultors and that of the cardinals), only a brief summary was recorded. At times it is stated in such summaries that the debate was long and heated, and, although the final result is always indicated, we still want to know in detail what was said and who said it. In some cases it is possible to fill in these gaps; for example, when some consultor or cardinal published his thoughts about evolutionism separately, one can infer the tenor of his remarks from the public record.

Owing to various historical circumstances, the complete archive has not been preserved. The greatest losses are owing to fires and to Napoleon's transfer of the Vatican archives to Paris. That mission was entrusted to the French army, but when—years later—the French authorities decided to return the archive to its legitimate owner, the Vatican did not have the means to carry out the mission, and its representative in Paris, using the authority given him, disposed of a great portion of the archive, which was either destroyed or sold as paper (one such lot is preserved in Dublin).

While these losses are irreparable, they still do not affect our story. Although evolutionism has ancient roots, its modern scientific formulation dates to the nineteenth century, especially after 1859, when Charles Darwin published *The Origin of Species.* Most of evolutionism's problems with the Roman authorities date to the second half of the nineteenth century. Many of the original documents have been preserved, having escaped the particular scourges mentioned.

None of Darwin's works were placed on the *Index.* His grandfather Erasmus Darwin's didactic poem *Zoonomia,* which contained a certain formulation of evolution along with doctrines held to be materialist, was indeed listed.[8] This fact should suggest that there was no systematic investigation of publications. Whether a book was examined by the Congregation or not depended, in great part, on someone formally denouncing it.

The second rule decreed by the Council of Trent prohibited all books on religious matters written by heretics. When Pope Leo XIII introduced his revision of the Index's rules in 1897, the same criterion remained in force: "Books written by non-Catholics that treat religion *ex profeso* are prohibited, unless it is ascertained that there is nothing in them contrary to the Catholic faith."[9] It is unsurprising, therefore, that some of the evolutionist authors who led the confrontation between evolution and religion were not placed on the *Index:* Darwin was not listed and neither were Thomas H. Huxley, Herbert Spencer, or Ernst Haeckel.

The Holy See granted special importance to books written by Catholic authors and in Catholic countries, because they were more likely to disturb the life of the Church. When Catholic theology books mentioned interventions by the Holy See, they were referring to books written by Catholics. For this reason, we here consider the cases of six Catholic authors whose publications were the object of Vatican scrutiny. These authors were mentioned in theology textbooks for several decades after 1859, and they continue to be cited whenever the history of the relationship between evolutionism and the Catholic Church is discussed.

Evolutionism and Christianity: Six Cases

The six cases are quite different and display the range of the Vatican's reactions to evolutionism in the final decades of the nineteenth century. Of the six authors, two came from Italy, two from England, one from France, and one from the United States. They held different ranks in the Church: two were bishops, two were members of religious orders (a Dominican and a priest of the Congregation of Holy Cross), one a diocesan priest, and the last, a layman.[10]

L'ÉVOLUTION RESTREINTE AUX ESPÈCES ORGANIQUES

PAR

LE P. M. D. LEROY DES FRÈRES-PRÊCHEURS

———

Emi e Rmi Padri

I.

Oggetto dell' esame

1. Un certo signore Ch. Chalmel, francese, ha denunziato alla Sacra Congregazione dell' Indice un' opera del dotto Domenicano Padre Leroy, muovendogli contro l'accusa di professare le dottrine seguenti:

A) Che nel racconto della Genesi nulla v'ha di ortodosso, se non la creazione divina e l'azione della Provvidenza. *Dans le récit de la Genèse il n'y a d'orthodoxe que la création de l'univers par Dieu et l'action de sa Providence.*

B) Che il come della creazione è lasciato alle dispute degli uomini. *Le comment de la création est abandonné aux discussions des hommes.*

C) Che il racconto di Mosè è un tessuto di metafore, un vecchio canto patriarcale. *Que le récit de Moïse est un vieux chant patriarcal, tissu de méthaphores.*

Index Report on Dalmace Leroy's *L'évolution restreinte aux espèces organiques,* fol. 128. ACDF, Index, Protocolli, 1894–96, fol. 128, p. 1. By permission of the Archives of the Congregation for the Doctrine of the Faith.

The first case is that of Raffaello Caverni (1837–1900), an Italian priest from Florence interested in science and its history. Toward the end of his life he wrote a monumental history of the experimental method in Italy. In an 1877 book, he proposed the reconciliation of evolution with Catholic doctrine. This book was immediately denounced to the Congregation of the Index by his own archbishop, was condemned by the Congregation in 1878, and was listed in all the later editions of the *Index*. Nevertheless, the case has been practically ignored because the decrees of the Index never explained the reasons for the censure, and evolution is not mentioned in the title of Caverni's book. But the archive contains an ample report written by a consultor of the Congregation, one of the most important Catholic theologians of the period who later became a cardinal.

The second figure is Dalmace Leroy (1828–1905), a French Dominican. The second edition of *The Evolution of Organic Species,* his book favorable to the reconciliation of evolution and Catholicism was published in 1891. Four years later, a Parisian newspaper published a letter written by Leroy from Rome, in which he retracted his position, explaining that his hypothesis had been judged untenable, having been examined in Rome by the relevant authority. This case has been mentioned frequently but on the basis of virtually no information. The archive permits us to reconstruct the case in its full complexity and to describe exactly what happened. We discovered that the Congregation of the Index decided to condemn Leroy's book, but the corresponding decree was never published out of consideration for both Leroy and the Dominicans. Instead, it asked Leroy for a public retraction, and he complied with the letter to the press. The archival documentation is abundant, containing six reports on Leroy's book, some of them very extensive.

The accompanying figure depicts the first page of a report on the work of Dalmace Leroy, drawn up by Teofilo Domenichelli, concluding that no further action should be taken against the book. This copy was used by the prefect of the Index, Cardinal Serafino Vannutelli, during the second General Congregation in which Leroy's work was discussed. The annotation on top is by the secretary of the Index, Marcolino Cicognani, indicating that on the last page of the report he had included a brief summary of the decisions of the first General Congregation. The sideways annotation on the left is an adumbration by Cardinal Vannutelli of the final decision to condemn the book, but without making the decree public, and to ask Leroy to make a public retraction. It appears to have been penned during the discussion. On the right is a reference, also in Vannutelli's hand, to a comment on Raffaello Caverni's sentence made by the consultor Tripepi in his report on Leroy's book.[11] The archive also contains Leroy's correspondence with the Congregation of the Index after 1895, including his proposal for a revised edition of his book and the responses he received.

The protagonist of the third case is John Zahm (1851–1921), an American priest of the Congregation of Holy Cross, professor of physics at the order's University of Notre Dame. Zahm published a number of volumes on science and religion, one of which, published in 1896, was condemned by the Congregation of the Index in 1898. But, again, the corresponding decree was never published. In this instance, there was no retraction, but we know that the Holy See opposed the diffusion of the book because Zahm himself said so in a private letter to his Italian publisher, which was reproduced in the press. Here is another complex case, with much maneuvering in the Vatican both against Zahm and on his behalf. Zahm's case is intimately related to the issue of "Americanism," a movement in Catholicism that involved some outstanding American clerics. Much was already known of this case based on abundant correspondence of the principals, but the archive provides new data that resolve some problematic issues.

The Italian bishop Geremia Bonomelli (1831–1914), an important and controversial public figure in the last part of the nineteenth century and beginning of the twentieth, is the protagonist of the fourth case. A book in which Bonomelli promoted a solution to the conflict between the new Italian state and the papacy was listed in the *Index of Prohibited Books;* the solution he proposed was quite close to the present arrangement, but it was highly controversial then. In an appendix to one of his books, Bonomelli praised Zahm's views on evolution. Then, without any official act by the Holy See, he found out from a friend—a cardinal—that his stance was viewed negatively in the Vatican; he then published a letter of retraction on his own initiative. There were no further repercussions.

Another bishop, the Benedictine John Hedley (1837–1915), bishop of Newport, Wales, is the subject of the fifth case. He praised several of Zahm's books, including the one on evolution, in an article and was himself favorable to evolutionism, although he was more reserved when it was applied to the origin of the human body. The Holy See took no action. Rather, Hedley become entangled in a polemic with *La Civiltà Cattolica* and published a letter that many interpreted as a retraction, although it really wasn't. An analysis of this case permits us to identify the origin of misconceptions about it that have persisted till the present and to clarify important aspects of some of the other cases.

St. George Mivart (1827–1900), protagonist of the sixth case, was an important English biologist who accepted evolution and published, in 1871, a book in which he held that biological evolution was compatible with Christian doctrine. Although the volume aroused some opposition, it was never condemned by the Vatican. Years later, however, the *Index* listed three of his articles on hell that bore no relation to evolution. Toward the end of his public life, Mivart wrote several articles that were quite critical of Catholic doctrine and the authority of the Church. Cardinal Vaughan, Mivart's bishop, after an exchange of letters containing three formal, explicit warnings, forbade him to receive the sacraments. Mivart died shortly thereafter. His confrontation with Church authorities is frequently attributed to evolution. In this case, an ample Holy Office dossier invites a reconsideration of documents known previously and permits clarifications of confused interpretations of Mivart's case.

The Catholic Reception of Evolution

The atmosphere surrounding these six cases was shaped by the tension between science and Christianity prevailing during the second half of the nineteenth century. The enormous advances of the natural sciences, together with archaeological discoveries that permitted greater understanding of ancient cultures, were perceived as a threat to Christianity and the privileged position it had held in European culture for many centuries. Theological positions were attacked in the name of science, although in many instances points of conflict were not owing to science per se but rather to doctrines (such as agnosticism and materialism) that appealed to scientific findings for support. Publications hostile to Christianity in general and the Catholic Church in particular multiplied precipitously. John William Draper's *History of the Conflict between Religion and Science* (1874) and Andrew Dickson White's more ambitious work, *A History of the Warfare of Science with Theology in Christendom* (1896), held that there was a permanent and inevitable conflict between science and theology. Both books enjoyed great success, and not only in the United States, where they were originally published.

In France, Louis Jacolliot argued that Christianity lacked any historical basis and was no more than a variant of Indian myths. From the first page of any one of his books, he presented his ideas as a consequence of scientific rigor that undermines all such baseless myths: "A new world is emerging. Science, with its rigorous methods, has dealt a mortal blow to religious poetry and historical legends, and the day is approaching when only sensate, rational, and human phenomena will be believed."[12]

Evolution was widely seen, by friend and foe alike, as threatening the status quo.[13] The Vatican thus responded to a whole series of books assaulting Catholicism and Christianity that were prejudiced and filled with exaggerations. The literal interpretation of the Bible was attacked everywhere. In such an environment, the enthusiasm that evolutionism sparked, particularly with the publication of Darwin's *Origin of Species* in 1859, was frequently mixed with attacks on religion and, in the name of evolution, an equally vigorous defense of agnosticism, atheism, materialism, and free thought. Furthermore, these arguments promoting evolution often overlooked the theory's many lacunae. All this helps to understand why Catholics frequently assumed a stance hostile to evolutionism, which they could picture as an instrument used by materialists to attack religion. The ideological baggage that evolution had acquired made it an obvious target for believers.

In the epoch of Mivart, Caverni, Leroy, Zahm, Bonomelli, and Hedley, Catholic theologians applied a severe critique to evolutionism, especially as it treated the origin of Adam's body. Nevertheless, only a few asserted that the direct divine creation of Adam's body was Catholic doctrine. One of these was Matthias Joseph Scheeben (1835–88), one of the most important Catholic theologians of his time. Scheeben was a kind of theological mystic who acquired great fame after his death and whose books are published to this day. On human evolution, he wrote: "It is heresy to pretend that man, insofar as concerns his body, 'is descended from monkeys' as a consequence of a progressive change registered in forms, including the supposition that in the complete evolution of man's form, God has simultaneously created a soul."[14]

Advancing a somewhat softer line but one still strongly critical of evolutionism was the important Jesuit theologian Camillo Mazzella (1833–1900). He was professor of theology in the United States (first at Georgetown University, then at College of the Sacred Heart in Woodstock, Maryland) from 1867 until 1878, a period in which he published several books, including a treatise on God the Creator that went through a number of editions into the twentieth century. In 1878 Mazzella was appointed professor at the Gregorian University in Rome and was named a cardinal in 1886. He was cardinal prefect of the Congregation of the Index from 1889 until 1893, when he was named prefect of the Congregation of Studies.

Mazzella plays a key role in our story because, after his service as prefect of the Congregation of the Index, he was still serving as a cardinal member when Leroy's book was examined in 1894–95. He not only participated in the two meetings of cardinals that examined the book but was also the most qualified theologian among the cardinals present. One of the arguments of the experts used in those meetings was strongly rooted in Mazzella's book on God the creator.

In this book, Mazzella criticized evolutionism and, in response to those who claimed that Adam's body could have evolved, he invoked the positions of two important theologians, one "classical," the other modern. The classical theologian was his Jesuit predecessor, Francisco Suárez (1548–1617), who held that the direct production of Adam's body by God was a Catholic doctrine. According to a third Jesuit, Giovanni Perrone (1794–1876), the proposition was likewise doctrinal. Mazzella added that, even though the statement might not officially be a Church doctrine, no Catholic could deny it, inasmuch as it constituted a "rash" doctrine lacking a solid basis. Mazzella defended his thesis on the basis of both revelation (Scripture as well as tradition) and arguments based on reason, which included extensive critiques of evolution.[15]

The Eclipse of Darwinism

Charles Darwin had proposed in the *Origin of Species* that the primary mechanism of evolutionary change was natural selection. The logic of the hypothesis was simple. In spite of the tendency for organisms to increase at a geometrical rate, populations of plants and animals in a given habitat tend to remain stable from year to year. Malthusian checks keep the numbers down. Out of the ensuing struggle for existence, "favourable variations would tend to be preserved and unfavourable ones to be destroyed. The result of this," Darwin continues, "would be the formation of a new species."[16] That is, small and relatively insignificant variations of character would confer an adaptive advantage to certain individuals in a given population, so that those with favorable variations would tend to survive in greater numbers than those without. Over many generations such adaptive changes would produce new varieties, subspecies, and, eventually, species.

Even such a stalwart defender of Darwin as Thomas H. Huxley was not willing to sign off on natural selection, as persuasive as Darwin's argument was, until it could be proved under laboratory conditions, something that was not in fact possible until the rediscovery of Mendel's laws of genetics in 1900 opened a way to the development of statistical tests of selection.

Because there was no agreement on the mechanisms of evolution among naturalists who were convinced that evolution was a fact and because Darwin himself, in successive editions of *Origin* (in particular the fifth and sixth), retreated on his claim that natural selection was the sole mechanism of evolution, the debate produced a plethora of competing mechanisms, including the neo-Lamarckism of American disciples of Louis Agassiz and orthogenesis or straight-line evolution.[17]

This situation prevailed in biology throughout the 1890s when all but one of our cases transpired, a period that Julian Huxley later characterized as the "eclipse of Darwinism" because of the uncertainty surrounding natural selection. According to Huxley, the reaction against natural selection set in during the 1890s as Darwinism became more and more theoretical, with no apparent support from experimentation—"merely case books of real or supposed adaptations."[18] In 1894 William Bateson published his *Materials for the Study of Variation*, which stressed evidence for discontinuous, large variations, rather than the continuous small ones that lay at the core of Darwin's theory. Bateson's approach was fortified by the mutation theory popularized by Hugo de Vries, one of the rediscoverers of Mendel's laws, after 1900. Bateson was then to conclude, in a famous address at the British Association for the Advancement of Science in 1914, that all evolutionary change is owing to mutation. "Selection and adaptation were relegated to an unconsidered background," and the death knell of Darwinism was sounded at the highest levels of British biology.[19] Of course, the patient had plenty of life remaining.

Three outcomes of the eclipse of Darwinism are important to our story. First, the abstract quality of evolutionary discourse in the 1890s made it an obvious target for the kind of Catholic apologetics that we observe in the six cases before us. Second, the absence of consensus on how competing hypotheses might be tested created a climate propitious for Catholics to advance their own hypotheses, such as Dalmace Leroy's "mitigated evolution" or St. George Mivart's "specific genesis." As a scientist himself, Mivart was immediately attacked, but he was not the only biologist to be so treated. Third, the lack of agreement among scientists on evolutionary mechanisms gave rise to one of the great shibboleths of religiously based anti-evolutionism, both among Catholics and Protestants: the disagreement of scientists was interpreted as proof of an inherent weakness of the general theory of evolution.

The Council of Cologne

In 1860 a provincial council held in Cologne had debated evolutionism. Its conclusions must be introduced in any discussion of the Catholic Church's stance with respect to evolution in the nineteenth century, because it is the most explicit statement on the subject made by any official Catholic authority.[20]

The council in question was not even a gathering of all German bishops but rather a provincial council that only affected the ecclesiastical province of Cologne. The archbishop of Cologne, Cardinal Ioannes von Geissel, sought the Holy See's authorization to hold the council on June 6, 1859. Pope Pius IX authorized it on July 30. Geissel convoked the council on February 25, 1860, and it was in session between April 29 and May 17. Following the norms regulating provincial councils, bishops of three other dioceses that made up the ecclesiastical province of Cologne—those of Trier, Münster, and Paderborn—participated along with the cardinal archbishop of Cologne, plus three more bishops, four auxiliary bishops, twelve canons, three representatives of Catholic faculties and universities, and some seminary rectors and superiors of religious orders.[21]

The Council of Cologne met in five general sessions. On June 15, 1860, Geissel sent the acts and decrees and of the council to the Holy See for official review. Pius IX replied on July 19 to say that he would transmit all the material to the Congregation of the Council, which was the competent authority in the Vatican. Cardinal Prospero Caterini, prefect of the Congregation of the Council replied a year and a half later, on December 19, 1861, praising the work of the council and forwarding a few notes.[22] Pius IX also replied, on April 7, 1862, adding his words of praise. Finally, the acts and decrees of the council were promulgated by the cardinal archbishop of Cologne on July 23, 1862.[23]

The Holy See's approval did not affect the outcome of the council's pronouncements. It was still a provincial council whose writ ran only in the dioceses of the ecclesiastical province of Cologne, and its decrees, though recognized by the Holy See, were promulgated by the authority of the archbishop of Cologne. In the document of promulgation of the acts and decrees of the council, addressed to the Catholics of the ecclesiastical province of Cologne, Cardinal Geissel wrote:

> Therefore, we have submitted to the supreme authority of the Holy See, so that it might recognize [*recognoscenda*] all the decrees which were approved in the Council by us and our venerable brother bishops of this province by unanimous vote, and which now through these letters we present these same decrees scrutinized and recognized [*revisa et recognita*] by the same Holy See, and we promulgate them by our metropolitan authority [*auctoritate nostra metropolitica promulgamus:* the authority proper to the archbishop, head of the ecclesiastical province].[24]

The Holy See's approval of provincial councils was known by the technical term, *recognitio* (recognition). Franz Xaver Wernz, who appears in the Mivart case as a consultor of the Congregation of the Index, wrote an extensive textbook of canon law that went through several reeditions and updates over several decades. On the question of provincial councils, he says that according to the rules in force it is not necessary that the pope confirm them and that, in any case, the pope does not generally give any *specific* assent. He adds that the metropolitan archbishop, before promulgating the acts and decrees, should transmit them to the Holy See, where they are normally reviewed (recognized) by the Congregation of the Council. This review, even when it includes corrections,

> does not cause the decrees of the Council to be transformed thereby into *pontifical* decrees extending to the *universal* Church, or that each and all of the matters mentioned in the *acts* and established in the decrees are themselves thereby approved or seen as *valid*, supposing that they may [also include] false or invalid measures; still, one must not underrate [the review] as if it *added no authority* to the acts and decrees of the provincial Council. In effect, it is an authentic *witness* of the higher authority that the Council was *legally convoked and held*, and that nothing *worthy of censure* was found in the correction of the decrees. Whence such decrees, although *still decrees of the Council*, are nevertheless easier to execute and better defended in the face of impugnations.[25]

This process describes exactly what occurred with the Council of Cologne, as we have just seen. Cardinal Geissel sent the acts and documents to be *recognized*. The pope responded, confirming that he had transmitted the documents to the Congregation of the Council to be *recognized* by it and, after informing the pope, to respond to Geissel. The prefect of the Congregation of the Council sent a letter of *recognition* of the acts and decrees. The pope then dispatched a letter in which he praised the council in a generic way. And the cardinal of Cologne promulgated the decrees on his own authority, noting that they had been sent to Rome for their *recognition* and that they had been *reviewed* and *recognized* by the Holy See (the italicized expressions appear in the original documents). It is clear that the process followed is exactly that described by the canon lawyer Wernz.

The Council of Cologne deals with evolution in part I of its decrees ("On Catholic Doctrine"), title IV ("On Mankind"), chapter XIV ("On the Origin of the Human Species and the Nature of Man"). The first paragraph states: "The first parents were created [*conditi*] directly by God. Therefore, we declare as contrary to Sacred Scripture and to the faith the opinion of those who are not ashamed to assert that man, insofar as his body is concerned, came to be by a spontaneous change [*spontanea immutatione*] from imperfect nature to the most perfect and, in a continuous process, finally [became] human."[26] After asserting that all humankind descends from Adam, the chapter continues with extensive considerations of the human soul, confirming its spiritual quality and its immortality.

The strong tones of the conciliar document can give the impression of a dogmatic declaration of faith. But the council lacked such authority, not even when armed with Rome's "recognition." Without doubt, rejection of evolution was quite general among theologians, but there existed no such general consensus to the effect that the direct creation of the body of the first man was an article of faith: many preferred to consider the divine creation of Adam's body, without intermediaries, as a "certain" or "common" doctrine, holding the contrary thesis to be a "rash" opinion that should be avoided, even though it did not reach the level of heresy. The position of the Council of Cologne reflects the opinion of the German theologian Matthias Joseph Scheeben and his Italian counterpart Giovanni Perrone, but, as we have seen, Camillo Mazzella admitted that perhaps it was not a matter of faith, adding that, in any case, no Catholic could deny it, since to accept the opposite was, at the very least, a rash or uncertain doctrine.

Those who supported the compatibility of evolution and Christianity found an easy solution. The Council of Cologne denied that the body of Adam arose from lower beings by means of a *spontaneous* transformation. This meant that the council did not condemn an evolutionary origin outright, but only opposed those who asserted that this evolutionary process had taken place *without the assistance of divine action*. On the contrary, there would be no problem in accepting evolution so long as one recognized simultaneously the necessity of divine participation for the process to take place, in such a way that the secondary causes might join with continuous divine action in giving being and activity to all organisms. All Catholics, moreover, agreed that a special divine act was required for the infusion of the soul: only the origin of the body was in play.

Theology Textbooks

After the actions of the Holy See against Leroy and Zahm had run their course, as well as the presumed action against Bonomelli, for forty years thereafter generations of seminarians, priests, and professors the world over studied Catholic theology textbooks in which it was explained that the Holy See had acted against evolutionism in these cases of Leroy, Zahm, Bonomelli, and Hedley, based almost exclusively on information gleaned from *La Civiltà Cattolica*, a journal directed by a group of Jesuits in Rome, whose contents were scrutinized in the Vatican. Such references were always brief, because not much information was provided. The textbooks customarily proposed the simple thesis that God formed the body of Adam directly and immediately, without any evolutionary process. The principal proof adduced was a very literal interpretation of the narrative of the creation of man found in Genesis. This interpretation of Genesis was confirmed by that offered by the majority of the Church fathers. They also stressed the lack of proofs in favor of evolution. And, in a section devoted to decisions of the Church, the cases of Leroy, Zahm, Bonomelli, and Hedley were cited to show that the position of the Holy See was contrary to evolution. The majority of the authors characterized the divine creation of Adam's body as a "certain" or "common" doctrine, and the contrary position was considered a "rash" opinion that, even though it did not constitute heresy, ought to be avoided.

These references to Leroy, Zahm, Bonomelli, and Hedley usually contained errors of fact. The principal one was to attribute the Holy See's intervention in these cases to the Holy Office, and to conclude that the Holy Office, which was the principal doctrinal organism of the Church, opposed evolution. The source of this confusion was a series of commentaries by the Jesuit Salvatore Brandi in *La Civiltà Cattolica*, which were taken as true owing to the peculiar prestige of *La Civiltà*. Some authors simply referred to earlier textbooks, so that these imprecisions were transmitted uncritically from generation to generation. The authors of these textbooks were unable to obtain better data, because such data could only be found in the inaccessible archives of the Vatican.

A typical case is that of the Jesuit Christian Pesch who, in the 1908 edition of a textbook that had many editions, wrote:

> The opinion rejected here had already been repudiated several times by the Roman authorities. In 1891 Leroy published a book in which he defended the opinion of Doctor Mivart. But he went to Rome in 1895 "to receive a warning" [*ad audiendum verbum*], was ordered to retract his views, and he did so. Some years later Zahm wrote a book in which again he defended Doctor Mivart's opinion as probable. But in 1899 he too was ordered by the Congregation of the Holy Office to withhold his book from sale. Therefore, it is clear that the Congregation of the Holy Office opposes this opinion.[27]

Another author of widely read theology textbooks, Adolphe Tanquerey, in a work published in 1913, devoted a section to Mivart, Leroy, and Zahm. According to Tanquerey, Mivart's views were not heretical because the Church, up to that time, had reached no conclusion about it. But, Tanquerey continues, the traditional meaning of the Bible should be maintained if there is no prudent reason to abandon it. He adds that the Roman congregations had on various occasions criticized the evolutionary origin of the human body, and in a footnote he says that Leroy and Zahm, both of whom held this opinion as probable, were ordered to withdraw their books from sale. As the source of this information he refers the reader to Christian Pesch's textbook. This information, while not wrong, is imprecise.[28]

Our examination of other textbooks widely used in seminaries and universities revealed a similar pattern. We found that the sources usually lead back to *La Civiltà Cattolica*, the information is scant and not very precise, and there are at times patent errors of fact. For example, the Jesuit Blas Beraza stated that the ordinary teaching (Magisterium) of the Church on this issue is clear, because the Holy Office ordered Zahm to withdraw his book, while Bonomelli and Leroy had retracted; but his only sources are *La Civiltà* and older textbooks by Pesch and Hurter.[29] In 1940 the Jesuit Charles Boyer, professor of the Gregorian University of Rome and author of philosophy and theology textbooks, warned that over the past twenty years a way to the acceptance of evolution had been opened. When he alluded to the history of the problem, he included the inevitable reference to Leroy, Zahm, and Mivart: "According to what was written in *La Civiltà Cattolica* in 1902 (vol. 6, p. 77), on the occasion of a letter from Monsignor Hedley, bishop of Newport, we learned that the orders, to which fathers Leroy and Zahm submitted in a manner worthy of praise, came from the Holy Office."[30]

In 1953 Karl Rahner was still attributing to the Holy Office a decision that never existed.[31] The same occurs in the 1959 edition of Pietro Parente's well-known textbook.[32] Brandi's mistake, whereby he gives the actions of the Holy See greater significance and authority than they really had, has been repeated to the present time. For example, in the third edition (1996) of a book by the theologian Juan Luis Ruiz de la Peña it is stated that "in February 1895, P. Leroy was required by the Holy Office to correct his stance. The identical requirement was asked of Dr. Zahm four years later, just as in the case of two bishops, the Italian Bonomelli and the American Hedley. Still, the Holy See did not hold it necessary to publicize the matter, which was aired in a purely private setting, without the intervention of doctrinal teachings."[33]

But the Holy Office had not ordered Leroy or Zahm to do anything, and in the cases of Bonomelli and Hedley (who, besides, was English, not American, the bishop of Newport, Wales) there was no official intervention at all by the Holy See. Ruiz de la Peña provides two sources: a 1967 article by Z. Alszeghy in the journal *Concilium*, which repeats the habitual errors about the Holy Office and then refers the reader to *La Civiltà Cattolica*,[34] and another article by R. Juste, which has better data,[35] because he cites Ernest Messenger's 1931 book. There, in a brief but dense section, Messenger provides the most accurate summary of the issues available until the opening of the archives.[36] Nevertheless, a good part of Messenger's account is based on data and commentary published by *La Civiltà Cattolica*.

'La Civiltà Cattolica' and Evolution

Even though *La Civiltà Cattolica* was not the only source of information on the six cases (some private letters were also known), it was the principal one. In less than eight months, the Jesuit journal published three retractions: those of Bonomelli (November 5, 1898), Leroy (January 7, 1899), and Zahm (July 1, 1899). Even though only Zahm mentioned the Holy See, and despite the absence of published official documents or public statements, the reportage gave the impression that the Vatican had intervened in the three cases.

In 1902 the *Dublin Review* published a letter by John Hedley, bishop of Newport, that cast doubt on the veracity of *La Civiltà Cattolica*'s information, while asserting that the Holy See had never intervened in any case concerning evolution. Salvatore Brandi responded in *La Civiltà Cattolica* on March 24, 1902, in a brief article titled "Evolution and Dogma: The Erroneous Information of an Englishman."[37] Although Brandi exhibited some respect for the veteran bishop, he clearly stated that Hedley was wrong. The information provided by *La Civiltà*, Brandi said, was correct, and Hedley could confirm this himself because, as a bishop, if he officially queried the Vatican, he would surely be supplied (though perhaps confidentially) with information at present unknown to him.

Brandi added a significant observation. He said that *La Civiltà* had not cited any concrete decision of the Vatican because "the Holy See, for the best of reasons, had not thought it opportune to condemn *in a public act* a theory which, in any case, was continually losing credit among true scientists."[38] The "theory" to which Brandi alluded was not evolution in general (although it also played a role in the dispute), but rather a highly specific aspect of evolutionism, namely the evolution of the human body from a lower animal—the central axis of these debates. Brandi clearly gave the impression that the Holy See had intervened against evolution in the cases of Leroy and Zahm, although it had not done so publicly. But he did not specify what the "good reasons" that had led the Holy See not to condemn evolution publicly were, although he did provide a clue, inasmuch as he alluded to the prudent norms that regulated the operation of the various congregations and to their habit of acting benevolently in cases involving works written by Catholics of some fame.

La Civiltà Cattolica had published articles that were very critical of evolutionism, and in the decade of the 1890s there arose a certain sense of alarm because the theory kept attracting more and more followers, including among Catholics. As we will see in greater detail when considering our case histories, between 1897 and 1902 *La Civiltà* published a highly critical review of Zahm's book by F. Salis Seewis; an article by Salvatore Brandi attacking Zahm's book, in which Brandi reproduced Leroy's retraction; Bonomelli's retraction; Zahm's presumed retraction; Hedley's reply; and several articles more—for example, one titled "The Dissolution of Evolution,"[39] which argued that evolution was a failed theory according to the judgment of science itself.

The Authority of 'La Civiltà Cattolica'

No one considered *La Civiltà Cattolica* the official organ the Vatican, because it clearly was not. Nevertheless, it enjoyed a singular level of authority, owing to certain features that dated back to the origin of the journal in the mid-nineteenth century. In that epoch, different revolutionary movements affected various European countries, including the Papal States, obliging Pius IX to leave Rome for the Kingdom of Naples, where he enjoyed the protection of King Frederick II. There, Pius IX gave his unconditional support to the founding by the Jesuits of a general cultural journal of Catholic orientation, aimed at a broad readership. The original idea was that of the Jesuit Carlo Maria Curci, and the journal was named *La Civiltà Cattolica*. The first number was published in Naples on April 6, 1850. Later it moved to Rome, and then was published in Florence from 1871 to 1887, the year in which it returned definitively to Rome. It is the oldest journal currently published in Italy. It has always appeared on the first and third Saturday of each month, except for the three months following the occupation of Rome by the new Italian state on September 20, 1870, when the journal moved to Florence.[40]

Conceived and published with the complete support of the papacy, *La Civiltà Cattolica* has had a singularly idiosyncratic relationship with the Holy See. While it was still in proofs, each number of the journal was sent to the Vatican Secretariat of State for its approval. This official scrutiny of the contents was to ensure they were in accordance with the teachings of the Church on faith and morality and to see that nothing was published that could create difficulties in the Vatican's relations with the Italian state, given that political topics were also discussed in its pages. Each Monday prior to the first and third Saturday of each month, the editor is received in the Secretariat of State and is advised of the Vatican's observations (indeed, until Pius XII, the pope himself received the editor).

Therefore, although the journal is not an official organ of the Vatican, it has a special relationship with the Holy See that gives it a certain authority, however hard to define. This explains the self-assuredness with which the articles just mentioned were written. The journal is run by a tightly knit team of Jesuits, sometimes referred to as a Jesuit "college." For many years the articles were not signed (we can identify the authors we cite because many years later the journal itself published some indexes that provide authors' names). It is evident that the journal's editorial board was hardly trembling in fear when it responded to Bishop Hedley with the title "Erroneous Information of an Englishman," nor when it harshly attacked the work of Zahm, a man with great prestige among American Catholics.

But *La Civiltà Cattolica* could well make mistakes, even in matters pertaining to the Vatican. Although the journal's board may perhaps have had privileged information, and the Holy See's review helped to eliminate gross errors, the editors' scope was quite limited and not privy to the intimate workings of the Vatican, whose internal activities are subject to a reserve that, in some cases, is very strict. It is no surprise that some of the articles mentioned might not be completely free of mistakes.

For example, there are imprecisions in Brandi's response to Bishop Hedley. According to Brandi, a Catholic should act reasonably and not accept as scientific a theory that, as "Zahm himself says, is not demonstrated nor is there expectation it may be." To back up this claim, Brandi provides a direct quotation from Zahm's book: "We have already learned that, as a matter of fact, no positive evidence has been adduced in support of the simian origin of man, and that there is little, if any, reason to believe that such evidence will be forthcoming." The citation is correct, but it is taken out of context. In his book, Zahm goes on to ask: "may we not, nevertheless, believe, as a matter of theory, that there has been such a link, and that, corporeally, man is genetically descended from some unknown species of ape or monkey? Analogy and scientific consistency, we are told, require us to admit that man's bodily frame has been subject to the same law of Evolution, if an Evolution there has been, as has obtained for the inferior animals. There is nothing in biological science that would necessarily exempt man's corporeal structure from the action of this law. Is there, then, anything in Dogma or sound metaphysics, which would make it impossible for us, *salva fide*, to hold a view which has found such favor with the great majority of contemporary evolutionists?" Zahm mentions several Catholics who have supported the compatibility between evolutionism and Christianity (the English scientist Mivart, the Spanish cardinal Zeferino González, and the French Dominican Leroy), and he says that, even though improbable, it is possible that with the passage of time the origin of the body of Adam by evolution may be accepted, in the same way that those who come after us might accept as scientific truth that which we now only hold to be an unproved hypothesis.[41] It is easy to see that, in this case, Zahm said something quite different from that which Brandi makes him say.

Other imprecisions have to do with the activities of the Vatican. In the same article, Brandi responds to Bishop Hedley, who had complained that *La Civiltà* had not specified which Vatican authority had intervened in the cases under consideration. Brandi wrote:

> Zahm's work met the same fate that had befallen, four years before, another work with the same argument, written by Father Leroy. He had also defended the theory of the origin of man's body out of that of a brute animal; his work was also denounced to the Holy Office and he too, in order to avoid public censure, made a public declaration, "unauthorizing, retracting and condemning the work in question" and making clear his desire "to withdraw the copies of the book from circulation, insofar as that were possible." In both instances, the "competent authority" that examined and judged the works and whose orders Leroy and Zahm obeyed in a praiseworthy way, was the Supreme Tribunal of the Holy See.[42]

But these words of Brandi, designed to clarify the situation, only confused it more, because they embody serious mistakes that, from then on, were transmitted from generation to generation.

In fact, neither Leroy's book nor that of Zahm was denounced to the Holy Office but rather to the Congregation of the Index. There is a notable difference between them. The Holy Office, the highest among all the congregations of the Vatican, was presided over directly by the pope, concerned itself with all kinds of doctrinal problems, and gave orders to other congregations: for example, it could order the Congregation of the Index to include concrete works on the *Index of Prohibited Books*. By contrast, the jurisdiction of the Congregation of the Index was much more limited. It only examined published works and decided whether they were to be put on the *Index*. In the decrees of the Index the motives for the prohibition were not even mentioned. Sometimes the reason simply was that it was deemed to be inopportune to circulate the book in particular circumstances. Therefore, a decree of the Index generally carried much less weight than did a doctrinal declaration of the Holy Office.

But there is more. Brandi said that both Leroy and Zahm made public declarations unauthorizing their books and stopping their further diffusion. This is true in Leroy's case but false in that of Zahm. Zahm's letter to his Italian translator was a private document. *La Civiltà Cattolica* converted it into a public retraction that never happened (as we will see, Zahm wanted to avoid a public retraction at all cost, and he did so).

Finally, Brandi said that the "competent authority" that examined the books and whose orders Leroy and Zahm obeyed was that of the "Supreme Tribunal of the Holy See," a confusing identification, to say the least. A benign interpretation would be to think that he used the expression "Supreme Tribunal of the Holy See" to mean simply the Holy See. But that interpretation cannot be sustained, because Bishop Hedley had publicly raised doubts about which concrete congregation of the Holy See had acted. In this context, Brandi's "Supreme Tribunal of the Holy See" was the Holy Office, or Congregation of the Inquisition. We have seen that Brandi had mistakenly said that the works were denounced to the Holy Office, and now he increased the confusion by having us believe that the actions were indeed taken by the Holy Office and that it too made the final decision, which we will see was not so. Brandi's confusion can be partly explained by the fact that the members of the Holy Office were held to a strict rule of secrecy. On the basis of information from his Vatican friends, Brandi must have guessed that the actions were those of the Holy Office, but the archival documents show that was wrong—a mistake transmitted to many generations of Catholic priests and professors across the world.

To sum up, *La Civiltà Cattolica* published, between November 1898 and July 1899, Bonomelli's and Leroy's letters of retraction and the letter intended by Zahm to halt the distribution of his book. None of these letters was a Vatican document, and only in one of them is the Holy See even mentioned. *La Civiltà*'s information was basically correct but included significant imprecisions, especially when it attributed to Zahm a nonexistent public retraction, and to the Holy Office some activities that in reality were those of the Congregation of the Index. Despite these shortcomings, *La Civiltà Cattolica* served for many years as the principal source of what was written about these issues in the standard theology textbooks.

More Confusion

Whoever might think that the opening of the archives of the Holy Office and the Index has cleared up all of the confusion would be wrong. The author of an article published in 2001 on the activities of the Congregation of the Index in the cases of Leroy, Zahm, Bonomelli, and Hedley argues that evolutionism and the possibility of reconciling it with Christianity had been subjects of free debate among Catholics throughout the nineteenth century when suddenly, toward the end of the century, the Roman authorities hardened their position, thus producing the various retractions of theologians favorable to evolution.[43] The reality is quite different. The cases we have been discussing all took place in the decade of the 1890s but quite a bit before the Roman authorities had intervened against evolutionism. Our first case, that of Raffaello Caverni, ended with the condemnation of his book in 1878. The book was placed on the *Index of Prohibited Books*, which did not happen in the cases heard toward the end of the century.

Besides, in the case of Leroy, which is the one we present here in greatest detail, there were harsh polemics between Leroy and the Jesuits of the journal

Études of Paris, whose stance on evolutionism was just as hard as that of the Roman Jesuits. Indeed, one of them had to defend himself from the suspicion that it was he who had provoked the condemnation of Leroy's book.

Another study, also published in 2001, using documents of the Archives of the Congregation of the Index as if they were from the Holy Office, wrongly asserts that Leroy's book was placed on the *Index of Prohibited Books*, and it contains other imprecisions too.[44] These examples demonstrate the need to present the new archival data objectively. Only in this way is it possible to get beyond erroneous ideas that have held sway for more than a century and to prevent the emergence of new myths.

An Ineffective Decree

Raffaello Caverni

One of the biggest surprises emerging from the Archive of the Index is that the Vatican condemned a book favorable to evolution, but no one ever knew it. For more than half a century, theology texts mentioned the cases of Leroy, Zahm, Bonomelli, Hedley, and Mivart, in none of which did the Vatican pronounce publicly. But they never mentioned Raffaello Caverni, the only instance in which the support of evolutionism by a Catholic merited a public condemnation. Even *La Civiltà Cattolica*, which at the height of the evolution controversy—roughly between 1897 and 1902—never let a relevant article or even letter go unmentioned, failed to note Caverni's book even once, although it had been put on the *Index of Prohibited Books* by a decree issued by the Holy See in 1878. How can this omission be explained? Did *La Civiltà*, so attentive to this topic, even know about Caverni's book or its listing?

Inside the Index

To explain this odd situation we must refer to the peculiar operations of the Congregation of the Index. We are not referring to the secrecy that members practiced with respect to their work; that was also the rule in other congregations, owing to the logical reserve merited by matters that affected persons' reputations. What is relevant here is the way in which decisions were made public. The decision to place a book on the list of prohibited books was taken by the member cardinals of the Congregation and then submitted for the pope's approval, but it did not take effect until a decree was publicly announced. Moreover, the decree never explained the reasons for the listing.

In the case of Caverni, the decree stated that the meeting of the cardinals had taken place on Tuesday, July 1, 1878, in the Vatican. It was signed by Cardinal Antonino de Luca, prefect of the Congregation, and by the Dominican friar Girolamo Pio Saccheri, secretary. The decree was dated in Rome, July 10, 1878, and was published on July 31. Seven books by six authors are listed. The first is Caverni's and the entry reads as follows:

> Caverni Raffaello. *De' nuovi studi della Filosofia. Discorsi a un giovane studente.* Firenze, 1877. Auctor laudabiliter sed subiecit et opus reprobavit.[1]

> Caverni Raffaello. *New Studies of Philosophy. Lectures to a Young Student.* Florence, 1877. The author has submitted and has repudiated his own work, in a manner worthy of praise.

Note that the decree says nothing about why the book was prohibited, and that its title does not mention evolution. Only the members of the Congregation of the Index knew the reasons that led to the ban. No one else would be able to confirm that the reason for the condemnation was the promotion of evolutionism, but the members of the Congregation never commented on their work: they were not allowed to.

Those outside the Congregation might well have thought there were various reasons for prohibiting the book. It is true that Caverni favored evolutionism, but he also criticized several aspects of Church life, such as the education of seminarians in Italy and the scholastic method applied in these programs, his criticisms having been leveled against the Jesuits in particular. One might well have thought that those critiques were what provoked the prohibition of the book. And this is not our hypothesis: according to a study of Caverni published in 2001, "The reasons behind the prohibition of Caverni's book cannot be found in the theory of evolution as he presented it, or at least not only in it. In that book Caverni leveled a harsh critique at various aspects of the ecclesiastical world, its culture in particular."[2]

This approach probably originated in some articles on Caverni published between 1910 and 1920 by Giovanni Giovanozzi (1860–1928). According to Giovanozzi, Caverni's book was not listed in the *Index* because of his defense of evolution, but rather due to his caustic and harsh attacks on specific institutions, methods, and persons in the ecclesiastical world. This idea turns up again in the article on Caverni in the *Dizionario biografico degli italiani*.[3] Such a conclusion has scant basis in fact. Both Caverni's critique of the ecclesiastical culture and of the Jesuits no doubt played a role in the denunciation of the book. Nevertheless, documents in the Archive of the Index show that evolutionism was the primary motive of the prohibition.

A Singular Village Priest

Raffaello Caverni was born on March 12, 1837, in San Quirico di Montelupo, near Florence. At the age of thirteen he left to study in Florence. He was ordained as a priest on June 2, 1860. After a ten-year stint as professor of physics

Raffaello Caverni. Reproduced from Sara Pagnini,
*Profillo di Raffaello Caverni (1837–1900). Con appen-
dice documentaria* (Florence: Pagnini e Martinelli,
2001), p. 5.

and mathematics in the seminary of Firenzuola, he was from 1871 on parish
priest of the village of Quarate, whose proximity to Florence permitted him to attend simultaneously to his ecclesiastical duties as well his scholarly pursuits and
publications.[4]

Caverni published a number of books related to the sciences. His principal effort was the six-volume *History of the Experimental Method in Italy*,[5] a work of
colossal proportions (some 4,000 pages) that won first prize in a contest spon-
sored by the Royal Institute of Science of Venice. Caverni's *History* was not
merely descriptive. It contains personal interpretations, including some critiques
of Galileo, that led quite a few specialists to reject the work. Caverni was a friend
of Antonio Favaro, who was editor of the famous national edition of the complete works of Galileo. Favaro was in part responsible for the award of the prize
to Caverni, but subsequently they became estranged.[6] Because of the polemical
atmosphere surrounding its reception, Eugenio Garin has argued that Caverni's
History has been unjustly forgotten.[7]

There is general agreement that Caverni was an independent thinker who was
not very fond of social convention, which explains both his turning down oppor-

tunities to occupy positions in a university or scientific society and his criticisms of various aspects of Church life.

'Lectures to a Young Student'

Since 1868 Caverni had been interested in evolution and the possibility of reconciling it with Catholic doctrine and the biblical narrative of Creation. The publication of Darwin's work in Italy had provoked heated debates, in Florence as elsewhere.[8] In 1875 and 1876 Caverni wrote a series of articles in the *Rivista Universale* on the philosophy of the natural sciences that, in 1877, he published as a book, with a different title: *New Studies of Philosophy: Lectures to a Young Student.*[9] The central theme of the book was that it was possible to harmonize evolution and Christianity.

To achieve this reconciliation, Caverni emphasized the need to uphold divine Creation and the active intervention of God, who, by his providence, guides natural processes so that they achieve their foreseen end. Human beings, moreover, were left out of the process of evolution. In this way Caverni sought to avoid the principal problems that evolution presented for Catholics.

The first difficulty with which Caverni had to contend was the literal interpretation, predominant in this period, of the biblical narratives of Creation. Toward this end he distinguished between two aspects of Scriptures, one divine and the other human. According to Caverni, the divine aspect concerned the truths of faith and is infallible, whereas the human aspect reflected notions acquired through study, which, like all that is known through human reason, might be either true or false. Caverni cited in his favor the ideas that Galileo had proposed for the interpretation of Scripture, insisting that there was no intent in the Bible to teach us scientific facts, but rather to show us the road to heaven.[10]

Caverni held that believers need not fear science, which ought to have complete freedom to investigate the origin of living species. But he cautioned that the natural sciences have certain boundaries that they should not cross: although science concerns the material world, it can tell us nothing about the spiritual one.[11] In this way he excluded human beings from the process of evolution. Caverni opposed an evolutionary theory that denied purpose, and promoted a theistic, finalist evolution that preserved basic religious notions about Creation and God's action in the world.

'Civiltà Cattolica''s Review

Caverni's book drew harsh fire from *La Civiltà Cattolica*, which in those years happened to be published in Florence, the center of Caverni's intellectual activities and the place where his book was published. Before the book was condemned by the Congregation of the Index, *La Civiltà* published a long, two-part review, in two successive numbers of the journal. The author was a Jesuit, Francesco Salis Seewis.[12] At the beginning of his review, he praised both Caverni's intention and his knowledge, and he acknowledged that Caverni had exempted the human species from the evolutionary process. But this was just a preamble to a harsh theological and philosophical critique of Caverni's ideas.

The first part of the review focuses on the criteria that Caverni advanced for interpreting the Bible and harmonizing it with science. Caverni said that we should distinguish between the human and divine aspects of Scripture, asserting that the details of the Genesis narration of Creation are owing to the former and thus are not to be considered as a matter of faith or Catholic doctrine. The fact that there are coincidences between biblical texts and others just as ancient shows that such narratives are human inventions. Salis Seewis responded that such coincidences reveal traces of a pristine revelation, prior to both texts, and adds that Caverni's ideas would have disastrous effects if accepted, particularly with regard to the infallibility of revealed texts and the beliefs based on them. The point reflects a burning issue of the 1870s. Salis Seewis takes the more traditional stance, according to which every sentence of the Bible, interpreted in a strictly literal sense, must be true because each is the word of God, whatever stylistic or literary considerations one might add. Were that not the case, he continued, the word of God would be reduced to a function of human taste. The ultimate consequence would be a devaluation of the divinely inspired nature of the entire Bible: Caverni's approach, which Salis Seewis characterized as "unheard of," could only have a "sacrilegious" outcome.

The second part of the review tackles the scientific and philosophical aspects of the book. In Salis Seewis's florid language, Darwinism is a bush sprouted from a bad seed sown in the wild, which has grown poorly and yielded only poisonous fruit, at least up to now. Caverni has tried to improve its chances by attempting a transplant into philosophy. Moreover, in becoming a partisan of Darwinism, Caverni has ignored the many criticisms launched against it. According to Salis Seewis, "Darwinism is a seed of unbelief, the result of considering nature without God and of the tendency to exclude God from science. All the laws that Darwin dreamed up have the result of making divine action superfluous. Darwin scarcely concealed the equivalence of his theory and the atheistic, materialist principles that have been professed afterward by his followers and successors without any qualification."[13]

Salis Seewis's critique can be reduced to two principal issues. The first is the atheistic, materialist character of evolution, which explains nature without reference to God. The second is the materialism implicit in the inclusion of human beings in the evolutionary scheme. Salis Seewis admits that Caverni intends to avoid these two consequences by asserting that God is author and governor of natural laws and by excluding the evolutionary origin of man. But he argues that Caverni's arguments are insufficient and contradictory.

La Civiltà Cattolica's response to Caverni's book was thus immediate and clear in its opposition. But there was more. In his review, Salis Seewis stated that in the scant space afforded by a book review he had been unable to treat the problems that Darwinism posed in sufficient depth. Thus, in the same issue of the journal the reader might consult a detailed study of the hypotheses of Darwinism and the evidence for them, and find an adequate discussion of what was scientific about "that tissue of ridiculous suppositions, intolerable analogies, and patent errors."[14]

A Treatise in Installments

The first installment of the treatise mentioned by Salis Seewis appeared in the same number of *La Civiltà Cattolica*, just before the second part of his review of Caverni. Its author was the Jesuit Pietro Caterini. He wrote thirty-seven installments in all, in successive numbers of the journal throughout 1878, 1879, and 1880.[15] In 1884 these articles were collected and published as *On the Origin of Man According to Transformism*, a total of 383 pages.[16]

Caterini laid out the problems of evolutionism in an insistently critical mode, with no concessions offered to the enemy. This was an all-out war, in which the objective was to discredit Darwinism on scientific grounds, with the last part of the volume reserved for the contradiction between Darwinism and Christianity. In his book, Caterini cites Caverni in three passages. In the first, Caterini recalls that, according to the Darwinists, animals have souls and minds just like men do, whereas Caverni, in order to rescue the higher essence of mankind, denies that animals have knowledge, properly speaking. On this point Caterini gives the nod to the Darwinians *contra* Caverni: in accordance with traditional Aristotelian philosophy, animals feel and "know"; however, sense knowledge is limited to the singular, concrete, and material, whereas intellectual knowledge surpasses this level.[17] In the second passage, Caterini approves of Caverni's reflections on the difference between the vocal organs of animals and those of man, and of their respective functions, so that one cannot conclude that human language has evolved from animals.[18] On the third occasion, Caterini laments that Caverni lavishes praise on Darwin's book, to which Caterini denies the scientific merits that Caverni attributes to it.[19] These three references first appeared in articles published after the condemnation of Caverni's book. Besides these concrete allusions, there runs throughout Caterini's exposition a harsh critique of the reconciliation of evolution and Christianity as proposed by Caverni.

In his last three installments, Caterini tries to show that such a reconciliation is impossible. Section XXXIX is an attempt to prove the thesis of the origin of the human body by the direct action of God: "It is revealed doctrine that the first parents of humankind were produced by the immediate operation of God, not only insofar as the soul is concerned, but the body also."[20]

To confirm this thesis, Caterini proposes a literal interpretation of the Genesis narrative, cites additional scriptural texts, asserts that this thesis is accepted unanimously by the Church fathers, adds that scholars and theologians are also in agreement, and refers especially to the theology treatises recently published by Hettinger, Perrone, Palmieri, Mazzella, Hurter, Dupasquier, and Berti. At the end of his lengthy study, he states that the theses of transformism are absolutely contradictory to the teachings of Catholicism. Therefore, those who say there is no opposition between it and evolution are deceived. The doctrines of transformism, Caterini concludes, are filled with theological errors, and in addition, if examined from the perspective of science and reason, they are exposed as a set of philosophical absurdities, confused concepts, gratuitous assertions, exaggerated data, and false conclusions. In sum—and these are the closing lines of this long series of articles—transformism "considered *theologically* is a gross and manifest error contrary to faith. Examined *philosophically* it is a plainly absurd variant of *materialism*. Assessed *scientifically*, it is a fantastic dream, a strange *a priori* system, opposed to observations and facts of nature."[21] This is the stage on which the denunciation and condemnation of Caverni's book took place.

The Archbishop's Denunciation

The first earliest document on the Caverni case in the Archive of the Index is a very short letter (one page with wide margins) from the archbishop of Florence, Eugenio Cecconi, dated November 9, 1877, addressed to Girolamo Pio Saccheri, the Dominican friar who had been secretary of the Congregation of the Index since 1873. It is not the original denunciation but a letter in which the archbishop responds to a letter the secretary had written him two days before: "I have the honor to transmit to your most reverend father a copy of the two reports relating to Caverni's book, that you, also in the name of the Cardinal Prefect of the Holy Congregation of the Index, requested of me in your letter of November 7. I enclose a copy of the book, which is the one used by the first consultor, because his report refers to certain passages of this book that he himself marked in red."[22]

That it was the archbishop himself who had denounced Caverni's book to the Congregation of the Index is not in doubt, because the consultor Tommaso Zigliara says so explicitly at the beginning of the report. In his denunciation, the archbishop must have referred to two reports on the book commissioned in Florence; the secretary of the Index requested them, and the archbishop sent him copies, along with the letter just cited.

The copy of the two reports, preserved in the Archive of the Index, is written in the same hand, unsigned and undated. Therefore, it contains no indication as to the identity of the reports' authors.[23] The first report was commissioned by the archbishop (so it states) and its copy is nine pages long. There we read that Caverni's book fails in both theology and philosophy, because it promotes a grave error, specifically that the divine inspiration of Scripture is relevant only to faith and morality, and therefore the rest of the biblical text can contain errors of fact. Moreover, the author is prideful and lacks a properly ecclesiastical attitude. Caverni accepts Darwinism, a very dangerous system, even though he limits its action to species lower than man. The first part of Caverni's book examines the evidence for Darwinism, and the second seeks to show that the evolutionary process does not include mankind. But, the report goes on to say, it is strange that a priest defends ideas that have lost prestige even among unbelievers. On this issue, the report refers to reviews of two anti-Darwinian books in *La Civiltà Cattolica*.[24] The report laments that Caverni does not summarize the arguments contrary to Darwinism. The author of the report recognizes that, if man is excluded from evolution, there is no rule of the Church that condemns evolutionism as heretical, but still he deems it contrary to the faith because it is opposed to Scripture, the Church fathers, and tradition. Moreover, even though Caverni excludes human beings, it would be more consistent with the principles he expounds to admit the evolutionary origin of man. The report also alludes to problems related to Caverni's reduction of animals to machines. He ends by asserting the need to choose between a Darwinism that denies the spirit and the rejection of Darwinism: having stated the choice starkly, the report concludes by rejecting the validity of Darwinism.

The second report appears to have been written by a monk, because he wishes the bishop well for his diocese, phrasing that a diocesan priest would not have used. The copy is six pages long. Three of them are a critique of the erroneous criteria of scriptural interpretation proposed by Caverni; two more address Darwinism. The report states that Caverni's argument stops before including man because doing so would conflict with the faith, according to which man was formed directly by God, his body included. He adds that evolutionism has been opposed by the most notable figures of science. The notion of animals being reduced to machines is an odd one. He goes on to say that it was easy to criticize Darwin's materialism without referring, as Caverni does, to the doctrine of ontologism. "The volleys launched against the Jesuits, so popular nowadays," he says, "are at the very least uncivilized." He laments that Caverni criticizes in public the education given in Italian seminaries and concludes that Caverni is lacking precision in theology, is too tenacious with his philosophical preferences, and betrays a surfeit of acid in his zeal.

The denunciation followed the route prescribed by the Congregation of the Index. The next step was to have Caverni's book examined by a consultor of the Congregation, designated by the cardinal prefect or by the secretary. In this instance the consultor was the Dominican Zigliara, one of the most important churchmen of the second half of the nineteenth century—hence, the special interest of his report.

A High-Profile Consultor

The Dominican Tommaso Maria Zigliara (1833–93) was one of the leaders of the "Neo-Thomist" movement promoted by Pope Leo XIII in order to revivify Catholic thought. The idea was to orient the study of philosophy and theology around the thought of Saint Thomas Aquinas. Neo-Thomism received its initial stimulus with the publication of Leo XIII's encyclical of 1879, *Aeterni Patris.*

Zigliara both finished his theological studies and was ordained a priest in 1856, in Perugia. He was ordained by the bishop of Perugia, Gioacchino Pecci, the future Pope Leo XIII. Later on, Zigliara was professor of philosophy and theology in the Dominican house of Santa Maria sopra Minerva, in Rome, and he also served as rector of the College of Saint Thomas, another Dominican establishment in Rome (which later became the Angelicum University, from Saint Thomas's cognomen "Doctor Angelicus"). In 1879 Leo XIII named him cardinal. He also became president of the Pontifical Academy of Saint Thomas (also founded in 1879), and cardinal prefect of the Congregation of Studies He was widely esteemed both for his intellectual prowess and his character.

Among Zigliara's many publications, perhaps the one that was best known was his *Summa philosophica,* first published in 1876, which went through nineteen editions. It was written in Latin, a language still highly cultivated in the ecclesiastical world, and for many years it was used as a textbook in both European and American seminaries.

In six pages devoted to evolution in the second volume of his *Summa,* Zigliara renders a very clear judgment: transformism is a materialist doctrine that only differs in its details from ancient forms of materialism that sought to explain everything through chance interactions of atoms. Lamarck had explained evolution by the relationship of organisms with their external circumstances, which stimulate new needs in the organism, which, in turn, provoke the appearance of new organs. Darwin's answer is natural selection, and mankind is nothing more than the result of this process. But theories of this type had already been refuted in antiquity by Aristotle. Zigliara introduces various arguments to show that the theory of evolution is metaphysically absurd because it is based on false principles, is an arbitrary and even contradictory hypothesis, and is also absurd from a physiological perspective.[25]

The particular evolutionism that Zigliara criticizes in his *Summa philosophica* is "spontaneous evolution," which is achieved by virtue of the forces of nature only and which excludes the possibility that God, as primary cause, worked through natural, secondary causes, including evolution. Zigliara had in fact contemplated such a possibility in another of his works, an introduction to theology for classroom use. He rejects it, however, on the basis of a literal interpretation of the Genesis story, according to which God created different species from the start. He adds that the human soul cannot develop out of matter, nor can life itself. To the objection that the scholastics held that animal life can be explained by the potentiality of matter, Zigliara replies that the scholastics did not refer to any active potentiality or capacity but rather to a passive kind of explanation: life could arise from matter only if some causal agent acted upon it, not through forces proper to matter itself.[26]

Zigliara's Report

Zigliara's report on Caverni's book (nineteen pages long, dated May 25, 1878) was predictably negative, as was his conclusion: he proposes, first, to list it on the *Index of Prohibited Books* and, second, to request that the archbishop of Florence see to it that Caverni not only accepts the condemnation but also refrains from publishing a second book, on the origin of man, that he announces in his preface.[27]

Zigliara's clear and methodical report is divided into three parts: about Darwinism, on Darwinism in relation to Genesis, and on the origin of man. Zigliara says that Caverni's intention is good but not his results. Caverni laments the poor condition of philosophy, owing to its estrangement from the natural sciences, and proposes, by way of remedy, to center philosophy in the study of the origin of man. So Caverni begins by introducing Darwinism. But, Zigliara comments, everyone knows that Darwinism is not a new system but rather an improvement of Lamarck's system (not to mention the work of the ancient materialists, such as Democritus, Leucippus, and Lucretius).[28] Lamarck held that the first living cell was formed out of inorganic material, and from there the different forms of life originated, through transformations that responded to environmental circumstances: to survive, new organs and habits were developed, and useless ones lost. Lamarck and Darwin started from the same principle: after primitive cells formed through chemical reactions, successive development produced the diverse types of living forms. But, according to Zigliara, who cites the French naturalist Cuvier, both points have been rejected by famous scientists.

Caverni thinks that Darwin's system appears as probably true to those who examine it without prejudice, and that it is supported above all by embryology, inasmuch as the embryo of every mammal passes through the same phases as the organism has during the course of its evolution. In this way, nature achieves in a few months what evolution accomplished over many centuries. Although Zigliara does not use the terms, he is of course referring to the biogenetic law of Ernst Haeckel and Fritz Müller according to which "ontogeny recapitulates phylogeny." It is precisely at this point, says Zigliara, "where all the poison of Darwinism is concentrated," namely "materialist pantheism in its embryogenic form." Why so harsh a judgment?

Perhaps it would have been better for Caverni had his book not been scrutinized by so famous an expert. It is conceivable that an ordinary theologian would have had fewer pretensions, but the prestigious Zigliara had to push the argument further. Citing other authors and pursuant to a fairly audacious logic, he asserts that Darwinism can essentially be reduced to Hegelianism, because from the primitive cell it jumps to a potentially universal cell, that is, Hegel's Absolute, which undergoes differentiation in a process of becoming, in the mode of Heraclitus. Zigliara reproaches Caverni for not making this logic explicit. Nevertheless, Zigliara himself provides a possible solution. Caverni asserts that a process of evolution that develops blindly by virtue of purely natural forces is the equivalent of a kind of pantheism (everything shares a divine nature), but he asserts that primitive organisms possess the capacity to develop gradually because God had granted them that capacity. If God creates matter and infuses it with such a capacity for development, and also guides its evolution through his provident action, the difficulty disappears, and evolution need have nothing to do with either materialism or pantheism: rather, it responds to a divine plan and is possible thanks to God's action. Zigliara acknowledges this but still reproaches Caverni for imprudence because he concedes to the Darwinist the core of this thesis on the primitive cell, and once this is conceded, everything is reduced to the evolution of pure matter. Isn't this, Zigliara asks, a kind of pantheism, even if sui generis? The first part of his report ends a bit inconclusively because, at least at first glance, Caverni has resolved the difficulties raised by Zigliara, as he himself states, yet Zigliara continues to warn that Caverni's position is dangerous.

In the second part of his report, Zigliara examines Darwinism in relation to the Creation narratives in the book of Genesis. He criticizes the canons that Caverni proposes for biblical interpretation, with his distinction between texts that are human and fallible and those which are divine and infallible, with questions of natural science (when raised in Genesis) in the fallible category. Zigliara's critique is harsh, because he views this distinction as beset with several serious difficulties: what Caverni categorizes as a problem resolvable through science—Creation for example—could not also be resolvable by faith; the Church could say nothing about notions of rational science, and if it did pronounce, it would not be infallible. Such would be the case of the Church's concepts of the immortality of the soul, its union with the body, the creation of souls, and so forth. Who, exactly, would set the boundary between the objects of science and those of faith? Finally, the first Vatican Council (1870) would have erred when it asserted that there are certain revealed truths that are also accessible to natural reason.

These objections were serious. Biblical interpretation had become an important issue in the nineteenth century, provoking interventions by the popes that, at the end of the process, led the Catholic Church to entertain a more nuanced interpretation of sacred texts, one that took literary genres into account. Zigliara characterizes Caverni's ideas as "an awful theory" that limits the truths revealed in Genesis to the original creation of the world and the preservation through divine action of all the beings created, as well as similar phenomena, and completely leaves aside any reference to any particular mode by which God had formed the world. Zigliara was especially irritated by some of Caverni's careless allusions to Moses, which constituted nothing less than "an impious reproach aimed at God himself." Zigliara had no lack of reproaches himself, stating that a child could respond to Caverni's ideas, whose "horribly false and scandalous" approach to scriptural interpretation is at the core of his book. Zigliara rejects Caverni's view that the Church fathers were influenced by the ideas of their epoch, even pagan traditions. "These assertions of Caverni," Zigliara concludes, "condemn him, and free me from the task of refuting the incredible temerity of his exegesis."

As for the third section of the report, on the origin of man, Zigliara emphasizes that Caverni is only a halfhearted Darwinist: the anatomical arguments that appear to him convincing, that establish the genealogical relationship among diverse types of animals, become simple analogies when he comes to the case of man. Zigliara opines, probably correctly, that this is not very convincing. Why deny the continuity between the two cases? Caverni takes pains to show that thought is not an emanation of the brain. But, for Zigliara, Caverni's approach is overly reliant on "ontologism." Ontologism was a philosophical approach developed around that time by Vincenzo Gioberti and Antonio Rosmini, according to which we have a general or indeterminate intuition of God, which does not arise from sense knowledge. But ontologism was itself an object of criticism within the Church, suggesting that the remedy was worse than the sickness: in an attempt to avoid falling into materialism, Caverni ends up with ontologism. Zigliara's point is that for a Catholic priest to rely on ontologism for proof of the spiritual character of human beings is "impertinent and ignorant."

One remaining point was the reduction of the animal to a machine, whose mechanisms—Caverni tells us—are for now unknown. Zigliara says that this error is even worse than the previous one. Caverni thinks that if animals have passions, a soul, and mind, then there is no distinction between them and human beings. This leads him to propose that animals are automatons, like machines. But that suggestion, Zigliara admonishes, is tantamount to materialism, however much Caverni might protest, because even the Bible itself attributes a living soul to animals. Moreover, this doctrine is contrary to the most evident facts, because animals possess memory, imagination, and "estimative,"[29] and if all this can be reduced to matter, then why not do the same for human animality?

Caverni says he is not going to debate Darwin on the origin of the human body, but he gives us to understand that the difference between man and animal might originate simply in a "flash of brilliant light," of divine origin, thanks to which the brain became human; perhaps a new brain was formed naturally, or perhaps the brain of an animal became human upon receiving this light. In either case, even though it appears that Caverni does not want to discuss the origin of the body of the first man, he also indicates that evolution accounts for it. Zigliara reproaches Caverni for losing himself in a conceptual maze when explaining, with scant luck and little clarity, the difference between man and beast: "however much he tries to escape [from the maze], he falls further into materialism."

Prohibition Proposed

Zigliara sums up all of his arguments in three points, which constitute the substance of Caverni's book:

a. Caverni endorses the Darwinian position on primitive cells and their evolution and transformation into the species of brute animals, by natural selection. And Darwinian evolution . . . is nothing more than the material part of total evolutionism, which is the same as Hegelian pantheism.

b. Caverni's rules for biblical exegesis are absurd, omitting any divine inspiration, and therefore infallibility, from anything that can be considered the object of natural science. A corollary is that Darwinism or any other physiological, geological, etc., system is all admissible, even though manifestly opposed to the Bible. And when, in spite of all, Caverni intends to reduce Moses' Genesis to Darwin's, he himself admits that he is in disagreement with the common interpretation of the exegetes and the Church Fathers.

c. When afterward he intends to exclude man from Darwinian embryogenesis, he does so in vain, having already granted Darwin's premises. He says that Darwinism can only be refuted with ontologism; he is obliged to deny animal life to brutes; and he admits that the human soul might well be a divine irradiation upon the brain of animals. In sum, Caverni's doctrines, in opinion, lead straight to materialism, not because that was his intention, because he tries to refute materialism, but because of the real nature of his doctrine.[30]

Evidently, the practical consequences of this judgment could not be positive:

These [three points] are the substance of Caverni's book. More points could be added, but they would be secondary by comparison. There is no need to list them because the first [three] seem to me sufficient to conclude that the book, in spite of its good parts, merits inclusion in the *Index of Prohibited Books*. It is written in an imaginative and poetic style, which makes it even more dangerous. Moreover it strikes me as most difficult, if not impossible, to consider correcting it, as can be appreciated from the analysis I have just made. I do not know whether if it would be appropriate to suggest that the archbishop of Florence approach the author, not only to obtain his submission, but also to persuade him not to publish, without prior examination, the other work on *The Origin of Man* that he promises in his preface and I do not know whether it has been sent to the printer yet. As I have said, Caverni is smart and, versed as he truly is in physiology, he could, if he were to take the proper direction, make many contributions to science.

In submitting this humble opinion of mine to the wise judgment of your most reverent excellencies, I kiss the holy purple and remain

Your Humble, Devoted and Faithful Servant

Fr. Tommaso M. Zigliara, Dominican

Consultor

Rome, May 25, 1878[31]

Caverni had run into a tough opponent, who did not hesitate to follow his arguments to their ultimate conclusions. The philosophical part of the report might seem quite debatable, but the critique of Caverni's exegetical canons was important in this period, and the section on the origin of man was one of the book's weak points.

As the norms of the Congregation established, the report was printed and debated in two successive meetings. In the Preparatory Congregation the consultors were present along with the secretary of the Congregation. The results of this gathering were then debated by the cardinal members of the Congregation at a General Congregation.

The Preparatory Congregation, June 27, 1878

A printed sheet preserved in the archive advises that the Preparatory Congregation in which Caverni's book was discussed was held on Thursday, June 27, 1878, at the house of the secretary (40 Via Sudario, in Rome). Attending were the secretary and thirteen consultors, whose names are listed.[32] It is germane to our argument to note that among the consultors present were Zigliara, author of the report on Caverni's book, and Tripepi, whom we will meet again years later in another case related to evolution. Both Zigliara and Tripepi were made cardinals.

As was customary, there was a list specifying the order in which the books were to be discussed, and the same order was maintained later on in the General Congregation. The books were referred to by their order in the list. In this instance six books and one pamphlet were discussed; Caverni's book was number 5.

The printed summary of the Preparatory Congregation recorded, as was customary, only the results of the scrutiny. All agreed to propose the prohibition of book number 1. On number 2, all but two agreed with the conclusion of the consultor. Eight thought number 3 should be prohibited, while the rest were content to simply ignore it (*spernendum*), and therefore take no measure. The discussion of each book ended with a vote, whose result was reflected in writing for the benefit of the succeeding actions. On number 5, Caverni's book, "They said it should be prohibited, adding: the author should be heard before the Decree is published." This was a unanimous decision. The desire to hear the author was customary when the author was, as in this case, a priest. The decision was communicated to the author so that he might accept it, in which case the published version of the decree would include the additional clause "the author has submitted and has repudiated the work in a praiseworthy manner," which was a good sign, because it made clear the good intentions of the author and his intent to obey the disposition ordered by the authorities of the Church. The results of the Preparatory Congregation were next examined by the Congregation of cardinals or General Congregation, held a few days later.

The General Congregation,
July 1, 1878

The General Congregation took place in the Apostolic Palace of the Vatican, on Monday, July 1, 1878.[33] This meeting represented a higher level of authority. The cardinals listened to the reports prepared by the "relators" and consulted the results of the Preparatory Congregation, and then they alone discussed each book and took a final vote to establish a decision to be transmitted by the secretary of the Congregation to the pope.

It was normal practice for the secretary to write a synthesis of the discussions and the final vote, indicating the principal arguments made. This document, in which the books are listed in the same order followed in the Congregations, was used by the secretary when he briefed the pope, so that the pope could make an informed decision. He almost always approved the decision of the cardinals.

We know the cardinals' decision in this case, because the manuscript report of the General Congregation of July 1, 1878, is preserved in the archives.[34] There we find the names of the nine cardinals who attended the meeting. Although at times there were long and heated debates (as appears in reports of other such meetings), the final report usually contained only a brief synthesis of what went

on in the meeting. In any case, the part relating to Caverni's book is quite extensive, filling more than a page handwritten in smallish script. This is the text:

> *De' nuovi studi della filosofia. Discorsi di Raffaello Caverni a un giovane studente.* This
> work merits serious and special attention. In it, Darwinism is expounded and
> partly approved, [stating] that it has many points of contact with religious doc-
> trine, especially with Genesis and other books of the Bible. Until now the Holy See
> has rendered no decision on the system mentioned. Therefore, if Caverni's work is
> condemned, as it should be, Darwinism would be indirectly condemned. Surely
> there would be cries against this decision; the example of Galileo would be held
> up; it will be said that this Holy Congregation is not competent to emit judgments
> on physiological and ontological doctrines or theories of change. But we should
> not focus on this probable clamor. With his system, Darwin destroys the bases of
> revelation and openly teaches pantheism and an abject materialism. Thus, an in-
> direct condemnation of Darwin is not only useful, but even necessary, together
> with that of Caverni, his defender and propagator among Italian youth.
>
> No less reproachable are Caverni's canons of scriptural exegesis, as wisely
> noted by the consultor Father Zigliara in his report, inasmuch as he seeks to limit
> the divine inspiration of the Holy Scripture to revealed dogmas and to morality,
> leaving unaffected by infallibility everything that sacred writers might teach with
> respect to the natural sciences. This exegetical system has recently been re-
> proached by this Holy Congregation in its examination of, and observations on,
> the work of Canon Wies.[35]
>
> Thus the decision of both the consultors and of the Most Eminent Cardinals
> was unanimous in their conclusion that the work be proscribed, along with
> two notices concerning the representation that should be made to the author
> in order to obtain his submission and to induce him to desist from his plan to
> publish his work on the origin of man.

Thus we know that that the decisions of both Congregations were unanimous. It also appears that Zigliara's authority played an important role, inasmuch as in the final summary he is mentioned by name and his reasoning is accepted, including his seemingly most drastic conclusion, identifying Darwinism
as the equivalent of materialistic pantheism. It is also clear that the issue of
scriptural interpretation loomed large. It is asserted that Darwinism clashes with
the foundations of Christian revelation and this, together with its consequences
(it leads to materialist pantheism), is what led to the condemnation of Caverni's
book.

Moreover, this manuscript report assumes unusual importance owing to the
significance that it attributes the decision of the Congregation. We have stressed
that until this moment (July 1, 1878), the Holy See had made no pronouncements on Darwinism. And it also asserts that the prohibition of Caverni's book

would amount to an "indirect condemnation" of Darwinism. What does that mean?

According to this report, the Congregation of the Index wanted to condemn Darwinism. But its competence was ordinarily limited to examining publications and determining whether they should be listed on the *Index of Prohibited Books.* Thus it speaks of an "indirect condemnation": if a book defending Darwinism is prohibited, the book is condemned directly, but the ideas it contains are condemned indirectly. Nevertheless, bearing in mind that the decree of condemnation was all that was made public, and that the reasons for the condemnation were not specified, it's hard to see how one can speak of an indirect condemnation of Darwinism. The only decision made public was the prohibition of the book. One might equally conclude that the book was condemned for its critique of the ecclesiastical world, or for the criteria it proposed for scriptural interpretation, which greatly influenced the condemnation.

The Papal Audience and the
Promulgation of the Decree

Now only the last step remained. The decision of the cardinals did not acquire the status of an official decision until the pope approved it and it was promulgated publicly. The pope's approval took place in the course of an audience in which the pope received the secretary of the Congregation of the Index, who transmitted the decision of the cardinals to the pope. Probably he would read aloud the summaries he had prepared in writing before the audience. In the present case, he would have also read the summary on Caverni's book along with those of the deliberations on the other books considered.

Ordinarily the pope approved what the cardinals had decided. The audience used to take place shortly after the meeting of the General Congregation. In this case, Leo XIII, who had only been pope for a few months, received the secretary of the Index on July 10, 1878, and he approved the decree.[36] Caverni's book was placed on the *Index of Prohibited Books,* after an approach was made to Caverni to secure his retraction and to dissuade him from publishing the new book on the origin of man.

Caverni Reacts

Several months after his book was published, Caverni began to hear negative rumors. The book was denounced to the Index on November 9, 1877. Although the denunciation was made discreetly, word got around. We know that rumors reached Montelupo, Caverni's birthplace, because on November 29, 1877, Caverni wrote a long letter to the parish priest there. He sent him a copy of the book so that he could study it himself. He goes on to say that there were murmurs abroad, and he attributed both the denunciation and the murmurs to "some fanatical and very ignorant priests." He adds:

> I am sending you this letter along with my book, which I direct to you, first, and after to any of my countrymen who might want to read it or are able to read it. The source of all the fuss involves a few phrases where I intended, perhaps too openly, to relieve the secular clergy and especially parish priests from the yoke that the friars and particularly the Jesuits have wanted to place on their necks, and I say that it would be better to read the Bible and the Church Fathers than *La Civiltà Cattolica*. The priests who have felt attacked have roared like wild animals irritated by an open wound and, looking for revenge, have rummaged around in my book. . . .
>
> You should know that in the current controversies between science and faith, I have sought to intrude a bit as a peacemaker, and with regard to Darwin I have said that once it is demonstrated that man cannot come from the monkey, it matters little for religion to accept the Darwinian system of transformation of species for the other animal species. This is the substance of all my heresies. . . .
>
> The ignorant fanatics who have spread the scandal in your town are overjoyed because they said my opinion will be condemned. . . . I think it is true that the book is being examined in Rome, but for now I do not know that it has been condemned. In any case, if it can be shown to me that I am mistaken and I confess my error, I do not believe I merit the insults of these people.[37]

As we have seen, the condemnation of the book was not decided until July 10, 1878. Two days later, on July 12, Caverni was summoned by Monsignor Amerigo Barsi, general vicar of the diocese of Florence, who had always treated him amicably. Barsi informed him of the condemnation and also requested that he not publish his new book on the origin of man. He no doubt also indicated to him the advantages of declaring his submission to the decision of the Congregation of the Index. On July 13, Caverni wrote to the archbishop of Florence, then in Rome, to say that he planned to submit, even though he was surprised by the mention of the other book, most of which was still in his head. The archbishop expressed his satisfaction in his own name, in that of the Congregation, and of the pope himself. On August 31 Caverni wrote him again, thanking him for having sent him the original letter of the secretary of the Congregation of the Index about the condemnation of his book and expressing his good wishes.[38]

Once Caverni made known his acceptance of the condemnation, the decree containing the prohibition of his book, and of five others, was published on July 30, 1878.

Three years later, Caverni published his book on the origin of man.[39] He seems to have had no problems with it, which is logical because he held that the available scientific data were uncertain and did not permit any hypothesis on the age of mankind. Caverni declared that believers could follow scientific debates on this issue without fear, because science had no basis for contradicting that which God had wanted to reveal. Caverni would wait even longer in writing and publishing his great history of the experimental method in Italy, whereby he won a worthy place in the historiography of science. He never skimped in his ecclesiastical duties.

The Meaning of the "Indirect Condemnation"

To evaluate the effect of the condemnation of Caverni's book, we must return to the "indirect condemnation" of Darwinism that the Congregation of the Index thought it was emitting.

We have seen that the secretary of the Congregation, in the summary he prepared for his audience with the pope, noted that the prohibition of Caverni's book constituted an "indirect condemnation" of Darwinism. But that condemnation was so indirect that no one not directly involved in the matter would even have suspected an intent to condemn Darwinism. The book's title, as long as it was, still made no mention of evolution or Darwinism. The decree of the Congregation of the Index provided no explanation, simply making public the prohibition of the book without explaining its motives. Doubtless, the prohibition induced Caverni not to disseminate his ideas—the book was not printed again —but that was the same consequence as befell any prohibited book and says nothing about Darwinism. Caverni himself attributed the condemnation to the ignorant fanaticism of those who felt attacked by him, and that opinion has lasted until the present. Few mention Caverni when speaking of the actions of the Church against evolutionism, and that is logical if one bears in mind the peculiar circumstances of the case.

Years later, in the decade of the 1890s, there were various actions of the Holy See with respect to evolution that reached the general public. *La Civiltà Cattolica* gave them the maximum possible public exposure, but not even then was the Caverni case mentioned. Was no one aware of it, including *La Civiltà*? We know that is not the case. Soon after its publication in 1877, Caverni's book provoked a very critical two-part review by the Jesuit Salis Seewis. In 1897 this same author published in the same journal another review, this one no less critical of a book by John Zahm, who, like Caverni, tried to reconcile evolution and Christianity. Still, neither in this case nor in the other articles then published against evolutionism did *La Civiltà* refer to the condemnation of Caverni's book, which could have been useful for its argument.

What is the reason for this important and probably voluntary omission? We can only guess. The simplest explanation is that the only public notice of the book's condemnation was the decree, in which the motives for the condemnation were not specified. Such motives might not be related, or only partly related, to evolutionism. Not even the Jesuits of *La Civiltà Cattolica* could have said any more.

The "indirect condemnation" of Darwinism was ineffective. The title of Caverni's book did not mention evolution, and the decree that placed it on the list of prohibited books did not specify the reasons for the condemnation. When actions of the Holy See adverse to evolutionism have been discussed, the case of Caverni is almost always omitted. Only since the Holy Office archives were opened has it been possible to dust off the documents and see what they reveal.

Retraction in Paris

Dalmace Leroy

The Leroy affair is very well documented in the Vatican archives and can be reconstructed in fine detail. Early on, it looked as though the denunciation of Leroy's book, *The Evolution of Organic Species,* would lead nowhere. But then the book was examined in greater detail, and four written reports, some very long, were produced. The final decision was to condemn the book but, out of consideration for the author—a Dominican—and for his order, the prohibition was never made public. Rather, Leroy was asked to write a letter of retraction, which he did in 1895.

The first report proposed that no action be taken against Leroy's book. At this point in time, the secretary of the Congregation simply assumed that no measure would be adopted. The second report proposed that the book not be prohibited but that the author be warned. The third proposed to prohibit the book or to warn the author, inviting him to retract, but he personally favored the second, more benign option, again out of consideration of the author and his order. Still to come was a fourth report, very critical of Leroy. Finally it was decided to condemn the book but without publishing the decree. Instead, Leroy was asked to retract publicly. After the retraction, Leroy tried to get permission to publish a revised version of the book, which triggered two additional written reports. It was not permitted. Although a long and complex case, it was little known until the opening of the archives of the Holy Office. The Holy Office published no document pertaining to this case.

Leroy's Career

Dalmace Leroy was born in Marseilles in January 1828,[1] with baptismal name François Marie. Upon being ordained a priest, he received the habit of the Order of Saint Dominic on August 28, 1851, feast day of Saint Augustine, around 11:30 A.M., and took the religious name Dalmace (in French; Dalmacius or Dal-

L'ÉVOLUTION RESTREINTE

AUX

ESPÈCES ORGANIQUES.

PAR

Le P. M. D. LEROY

DES FRÈRES-PRÊCHEURS

Mundum tradidit Deus
disputationi eorum.
(ECCL., cap. III, V. 2.)

DELHOMME & BRIGUET, ÉDITEURS

PARIS
13, rue de l'Abbaye, 13

LYON
3, Avenue de l'Archevêché, 3

1891

Title page of Dalmace Leroy's *L'évolution restreinte aux espèces organiques* (1891). Courtesy of the Biblioteca de la Universidad de Navarra, Pamplona, Spain.

matius in Latin). Exactly one year later, on August 28, 1852, he made his profession as a Dominican. In the Dominican Archive in Paris the relevant documents are preserved, drawn up in Latin. The profession is an entire page handwritten by Leroy and signed "Fr. Marie Dalmace Leroy, of the Order of Preachers"—thus he is usually referred to as M. D. Leroy.[2]

For the Dominicans these had not been easy years, especially in France, where the order had been suppressed as a result of the French Revolution at the end of the eighteenth century. In France the Dominicans almost disappeared. It was Henri Lacordaire (1802–61), a priest famous for his preaching in the Cathedral of Notre Dame de Paris, who undertook the task of restoring the Order of the Dominicans in France. Upon entering the order, he underwent his novitiate in Italy, returned to France in 1840, restarted his preaching in Paris, attracted many followers, and in the 1840s opened a novitiate and several religious houses, having been named provincial when the Dominican Province of France was restored on September 15, 1850.[3] Leroy took the Dominican habit several months after the restoration of the province. He was one of the eighty-two first to profess as Dominicans of this province between 1840 and 1854.[4]

Leroy was a tidy, methodical person. After he professed as a Dominican, he was named assistant teacher of novitiates and exercised his ministry in Paris. Later, he was elected prior of Flavigny, a post he held from 1864 till 1867. He then returned to Paris and for years worked as assistant to the provincials of the Dominicans and held other posts, such as chaplain, in accord with whatever the political circumstances permitted, inasmuch as there were further decrees against the religious orders, one of them in 1880. Leroy had chronic bronchitis, and he experienced a serious deterioration when he was around eighty years old. He was living in the outskirts of Paris at the time. He was taken to a hospital in Paris, where he died a much beloved figure, on May 19, 1905.

In an obituary printed in the *Annals* of the Dominicans it says that, whatever else were his duties, Leroy never abandoned his studies of philosophy and natural history, which had enthused him since his youth.[5] In 1887 he published a book entitled *The Evolution of Organic Species.*[6] The reviews it received induced him to prepare a new edition, corrected and expanded, in which he attempted to explain further certain controversial points. This new edition was published in 1891 under the title *Evolution Limited to Organic Species.*[7] The title was changed to draw attention to the exclusion of man from Leroy's revised vision of evolution. This 1891 edition is the one that was examined in Rome and is the focus of our discussion.

'Evolution Limited': A Prophetic Book?

In his book, Leroy proposes to demonstrate that evolution is compatible with Christianity in such a way that it might be confined to the realm of science and need not be converted into a materialist, atheist philosophy. In his introduction he points out that, unfortunately, a spirit of partisanship has exploited the theory of evolution to attack religion, but this is no reason to reject the theory. In prophetic words, he invokes Galileo: "I think the idea of evolution will run the same course as that of Galileo; it will initially alarm the orthodox, but once emotions have calmed, truth will be distinguished from the exaggerations of both sides, and a way will be opened. . . . Let us know how to give to Caesar what is Caesar's, and to invite Caesar that he, in turn, give to God what is God's" (p. 2).

In 1891 the concept of evolution was widely discussed. In France, however, the introduction of Darwinism into scientific circles occurred quite late. According to Yvette Conry it was not until the twentieth century that the conditions necessary for its introduction arose in France.[8] The distinction between evolutionism and Darwinism is important generally, but even more so in France, home of Lamarck (1744–1829), who proposed in his *Zoological Philosophy* of 1809, fifty years before Darwin's *Origin of Species*, an evolutionary theory based on the inheritance of acquired characteristics. Darwin himself—especially in the sixth edition of *Origin* (1872)—conceded a certain importance to Lamarckian inheritance, although he continued to hold that the principal factor of evolution was natural selection.

Evolutionism was viewed by many Christians as a scientifically dubious, even extravagant, theory, which moreover had been appropriated by materialists and atheists as a weapon against religion. Still, from the very beginning there were Christians who accepted evolution, having isolated it from its ideological baggage. Among such Christian evolutionists was Dalmace Leroy.

At the beginning of Leroy's 285-page book there are two letters. The first is from the Catholic geologist, Albert Auguste Cochon de Lapparent (1839–1908), applauding the publication of the book. The other, equally laudatory, was by Jacques-Marie-Louis Monsabré (1827–1907), a Dominican well-known for his sermons at Notre Dame de Paris. Both Monsabré and Lapparent vouched for the orthodoxy of Leroy's views.

In his first chapter ("Atheistic Evolutionism: A Refutation," pp. 4–29) Leroy proposes to refute the godless brand of evolution, according to which matter is eternal and has formed everything that exists in the universe through a process in which God does not intervene at all. The second chapter ("Limited Evolution and Religion," pp. 30–50), a reflection of the Catholic doctrine according to which there can be no conflict between scientific and revealed truth, proposes to test whether evolutionism contradicts Catholic doctrine. Leroy notes that his book (as its title indicates) is about "limited evolution," that is, evolution limited to organisms lower than human beings. Leroy also calls this stance "mitigated evolution." Human origins are excluded from his purview, although in reality what is meant is a provisional exclusion. Like everyone else, Leroy was especially interested in the origin of man, and even if he brackets the question throughout most of his book, he does devote its tenth and last chapter to the question ("Evolution and the Human Body," pp. 240–73). The intervening chapters (3 through 9) consider scientific and philosophical arguments for and against evolution.

Mitigated Evolution and Religion

After the tenth chapter Leroy supplies a "Summary" (pp. 274–83), where he states:

> Along with all spiritualists, I obviously have rejected deterministic, uncontrolled, godless evolution. I have proposed a limited evolution, and I think I can defend it. But before I embarked upon such a venture, I wanted to assure myself that the way was open from the perspective of religious orthodoxy. Thus, I submitted my thesis to the judgment of Scripture, Tradition, the Church, and Theology, to establish that none of these authorities condemns my opinion. All the authors, including those most hostile to transformism, recognize this unanimously. It is appropriate to emphasize in passing that the matter has been judged, and there is no need to return to this subject. Thus, the question is reduced to a problem of philosophy and natural history. (pp. 274–75)

Why was Leroy so sure of himself in commenting on an area that was about to cause problems for him in the Vatican? The answer is clear. At the beginning of his second chapter he asserts:

> I must declare at the outset that I agree with the ideas and principles of the partisans of the fixity of species on the most important points. Like them, I accept primordial creation *out of nothing* from cosmic matter through all-powerful, divine action, as well as the creation of physical forces together with the admirable laws that govern them. Like them, I attribute the introduction of life on Earth to the special intervention of a prime cause. Like them, I recognize and salute the incessant action of divine Providence in the universe, unfolding in accord with a grand plan that reveals an infallible intelligence—according to the energetic expression of Aristotle: "the work of an intelligence that does not err." Like them, in sum, I admit the existence of specific types susceptible of perpetuating themselves by generation; but, instead of attributing the origin of each one of the *five or six hundred thousand* known organic types to a special act of creation and of declar-

ing them immutable, I only recognize a relative fixity in them, and I consider them the result of evolution from previous types with the exception of the first ones. (p. 31)

In support of his position, Leroy cites the opinions of Fulcran Vigouroux, Albert Farges, and Joseph Brucker, three orthodox Catholic writers; he adds his own comments on the relevant biblical passages, and concludes:

That Scripture does not condemn limited evolution is the least I can infer from the passages that I have just cited. Nevertheless, I must say that the same authors hold that the interpretation [of Scripture] in the sense of special Creation and the fixity of types is more natural and in greater conformity with the inspired text. . . . The Bible does not teach anything concrete about the fixity of species, and tells us nothing about how the Creator made them. This is the least we can conclude from the witnesses and authorities we have just cited; remember this, and from now on it will not be necessary to return to this point. (pp. 35–36, 41)

Leroy next turns to tradition. According to Leroy, Catholic authors agree that the Church fathers, who lacked the scientific data necessary to resolve the problem, never developed a doctrinal position on it. He cites Farges, who conceded that Saint Augustine's opinion came close to evolution, and it had not been disputed by either Aquinas or Suárez. Brucker stated that the fathers insisted that plants and animals (except for man) were formed through the action of secondary causes and declared that there was no patristic tradition bearing on the issue. Leroy's conclusion is clear: "Thus, everyone agrees that there is no tradition, in the theological sense of the word, on this matter, and that the fathers would not be competent here" (p. 43).

The Origin of the Human Body

Just as Leroy observed, the origin of man was the problem that provoked the greatest interest: materialists thought that evolution proved their point, while spiritualists were wary of evolutionism for just this reason. Leroy states that both positions are wrong and sets out to establish, above all, that human beings have a spiritual and immortal soul: if this is established (and Leroy expounds some of the traditional proofs), the origin of the body matters little. But if that is *not* established, it little matters that the human body is created by the direct action of God.

Once Leroy makes the case for the immortal, spiritual soul (pp. 240–56), he proceeds to the origin of the human body and asks whether it could be considered a product of evolution. His answer is a clear negative:

My reply leaves no doubt: no, this is not admissible . . . in religious teaching, which here takes precedence over all the facts of science. There is an order of things that so proclaim. No, man's body does not derive from animals but is the result of a direct intervention by the Creator. We have said that religious teaching is based on four sources: the Holy Scriptures, the Church Fathers, the decisions of the Church, and theology; if I interrogate each one of them separately, I might not find the question completely decided; but taken together, they assume a luminosity that banishes all doubt. (p. 257)

With regard to Scripture, Leroy notes that there is an explicit indication: God formed the human body from the mud of the earth. But nothing is said about the manner of this formation: it could have been a direct and instantaneous operation of the Creator, or it could have been produced by secondary causes, by evolution. The Church fathers add no further clarification.

Neither is there any teaching of the universal Church regarding the origin of man. However, a provincial council in Germany (held in Cologne in 1860) did discuss the topic. There it was affirmed that God created the first parents of humanity directly.[9] Leroy resolves this difficulty in a few lines. The Council of Cologne condemns those who claim that the human body arose as the result of a *spontaneous and continuous change* starting from a lower animal, but clearly, says Leroy, that condemnation refers to the radical transformists who propose evolution without God and does not apply to those who propose an evolution linked to the divine plan and to the action of the Creator: in this case there is no *spontaneous change*, because change occurs under divine inspiration.

Leroy says that the reason that finally convinced him to oppose the evolutionary origin of the human body is found in the theology of Saint Thomas Aquinas. Here we find an argument that plays a fundamental role in Leroy's thought:

According to the Angelic Doctor [Thomas Aquinas], every corporeal substance is composed of two agents, one essentially active and the other purely passive, form and matter. In this association, it is form that supplies all the nature or *quiddity* of the substance. Saint Thomas says: "*Cum forma sit tota natura rei . . .*" [As the form is, so is the whole nature of the thing . . .]; matter, as absolutely neutral, is capable of taking any form. It is scarcely necessary to say that in the living compound called the human body, the soul is the form. In these conditions, it is evident that the human body could not arise from the transformation of an animal. For it to have effectively derived from animality, it would be necessary that the human soul, that is, the form which constitutes the whole nature, also originate in the transformation. But this is impossible because, in contrast to what happens in animals, the [human] soul can only come directly from God. Thus, Catholic theology . . . formally rejects the opinion that would have the human body come from animals. (pp. 259–60)

Scripture says that God formed man from the mud of the earth and breathed the spirit of life on his face, and thus the first living man was made. Here Leroy distinguishes between two operations: "formation," and "breath." The breath or insufflation is the "infusion" (to use the classical expression) of the soul, which comes directly from God. Therefore:

> It is only after the infusion of the soul, and because of the infusion itself, that man is constituted a living being. Before infusing the spirit, there was nothing human, not even the body, inasmuch as human flesh cannot exist without the soul, which is its substantial form. . . . Thus, the Bible—interpreted by theology—tells us that man's body cannot be derived from lower nature. (p. 261)

The Church fathers are unanimous in this respect, Leroy continues. They see in the formation of the human body a special intervention of the first cause— that is, of God—that is not found in the creation of animals. Theology tells us that this intervention is the infusion of the rational soul, by which the Creator has transformed mud into human flesh. The conclusion reached at the Council of Cologne has the same meaning. "The body of man is not derived from animal; it is the direct product of the Divine power through the infusion of the rational soul" (p. 262).

Leroy devotes a few pages (pp. 262–66) to demonstrate that science supports the conclusion of theology, emphasizing the great differences that exist between human anatomy and that of the animals most like us.

But now, only seven pages before the end of the book, Leroy introduces a new consideration that adds a twist to the problem:

> Now it may seem that everything has been said on this subject. Nevertheless, there is one more point to consider. The human body is composed of matter and form, and the soul, in substantial form, comes directly from God, of course. But where does the matter come from? It is also certain it comes from the mud of the earth—Scripture and tradition clearly say so. But does this mud receive the infusion of the human soul instantly, that is, without any preparation? And if it has undergone some preparation, as Genesis indicates, might this have been done through evolution? Such is the question that we can still ask. (pp. 266–67)

Leroy identifies two proponents: those who think that divine action was exercised directly on the mud of the earth, and those who opine that the *substrate* destined to receive the immortal soul was the work of God, but that it was prepared through secondary causes, that is, by means of evolution. Leroy says he is not presenting himself as a champion of this second hypothesis, but only seeks to analyze it to see at what point one could tolerate or reject it (p. 267). But the pages that follow (pp. 267–73) contain a defense, however one might view it, of this very hypothesis. Leroy concludes the chapter with an explicit insistence that the human body is not derived from animals but is the direct product of Divine power by the infusion of the rational soul (p. 273). At the end of his summary, in the final paragraph of the book, he asks:

> Have I succeeded in making the theory of evolution less suspect? I hope so, but I have few illusions. Although it may be shown that certain objections are insubstantial, there will always be some who persist in repeating them. I expect, therefore, that I will still hear it repeated that evolution, even the limited form, is in opposition to the Bible and the teachings of the Church, that it is not supported by any scientific fact; and that to seek to explain, insofar as possible, the formation of the world without miracles, is to propose Creation without God. For this there is only one remedy: time. It is too much to expect that the problem could be discussed freely at this time. Thirty years ago one could not have done it without risk. Each generation needs to become accustomed to new ideas before being able to do them justice. This is the case of the system of limited evolution. . . . I expect, nevertheless, that it will survive the test, and who can say if perhaps some day it will strike us as strange that it could have encountered such antipathy. (p. 283)

It was evident that Leroy was defending evolution and its compatibility with Catholic doctrine. On the one hand, he denied that the human body proceeded by evolution from lower animals, and at the same time he asserted that the substrate could have provided such an evolution and then be converted into an authentic human being through the divine infusion of the rational soul.

Inevitable Polemics

In the second edition of his book in 1891, Leroy intended to clarify the criticism aimed at the first edition of 1887. Among these polemics, two merit special consideration. The protagonists were two Jesuits, Bonniot and Brucker, both writing in the journal *Études religieuses,* published since 1856 by the Society of Jesus in Lyons and, later, in Paris.

Joseph de Bonniot (1830–1889), a French Jesuit, generally wrote on problems affecting the faith that arose from scientific issues.[10] Already in 1873 he had written an article against evolution for *Études.*[11] His ideas leave no room for doubt. He begins by stating that evolutionism enflames the spirit, something that does not happen with pure science, and then he offers an explanation for why this happens:

> Transformism, a nebulous, floating theory, seems to create difficulties for spiritualism and Christian faith, to favor atheism and impiety, that is to delight the bad instincts of the human heart. This is the principle cause of all the noise produced when the names of Darwin and Lamarck are heard. But passion tends to create il-

lusion: it impedes us from seeing what exists and makes us see what does not exist. If transformism were as rigorously demonstrated as the theorem of the square of the hypotenuse, it still would have no power against the great dogmas on which reason and morality rest. But in addition this theory has the misfortune to be in direct opposition to reality.[12]

Bonniot had also published two books on animals ("beasts"), to show the difference between man and animal.[13] Leroy praises him here, observing that Darwin wanted to establish that man is nothing more than a perfected animal and stretched the point, using a hypothesis to establish similarities to the four distinctive characters of humanity: reflection, language, religiosity, and morality. But then Leroy launches into a broad criticism of Bonniot for failing to recognize that animals have knowledge. According to Leroy, the reason for this error is his yearning to refute Darwinism: to counter the Darwinists, he denies that animals have any kind of mental faculty; but the best philosophers, like Saint Augustine, Saint Thomas Aquinas, and Suárez, without conceding anything to materialism, all admitted that animals have (nonintellective) knowledge (pp. 169–83). Bonniot could not respond to Leroy's critiques, because he died in 1889, before the second edition of Leroy's book was published.

In the second edition, Leroy criticized another Jesuit, Joseph Brucker (1845–1926).[14] Born in Wintzenheim (Alsace), Brucker was on the editorial board of *Études*. His association with the journal lasted until 1920, and he was editor from 1897 to 1900. He wrote mainly about biblical questions and opposed the reconciliation between evolution and Christianity.

In April and May of 1889, Brucker published two articles on evolution in *Études*. In the second of them he criticized Leroy. Speaking of the formation of the bodies of the first couple, he says: "There are today apologists who enjoy certain prestige in theology who dare to say and to write that the Church Fathers did not *dogmatize* on those issues *in which they were not competent*. This assertion is completely false as we shall see."[15]

And, in a footnote, he cites P. M.-D. Leroy, *L'évolution des espèces organiques* (1887), pp. 15–16, adding that "the author, a limited transformist, attributes this assertion, which he approves, to Father Valroger, who is less affirmative."[16] In *L'évolution restreinte* of 1891, Leroy answered, arguing that the texts cited by Brucker to prove the immediate formation of man's body by God, excluding all other factors, did not strike him as conclusive (pp. 257–59). Brucker reviewed this edition in *Études*.[17] Most of the review's nine pages seeks to rebut Leroy and to clarify matters referring to the Bible. Brucker states that, despite the corrections introduced by Leroy in his new edition, and even recognizing that at times his responses to the critics of transformism are correct, Leroy is still a transformist and transformism is wrong.

With respect to the origin of the human body, Brucker congratulates Leroy for the clarity with which he affirms that the human body does not come from animals by way of evolution. But two problems remain. In the first place, he would like to know what Leroy thinks about the formation of the first woman, because he makes no reference to that biblical text, which is so at odds with any transformist explanation. Still, Brucker seems to have read Leroy's book too quickly, because Leroy had in fact written: "No believer could doubt that woman's body had been formed from that of Adam" (pp. 156–57). Leroy does not examine the difficulty that this raises for evolution, but he accepts the literal interpretation of this passage.

The second objection is that Leroy's theory is not very clear. Brucker mentions the article by Bonniot (who, by then, had died) on the immutability of species, characterizes the article as "excellent," recalls that Leroy had criticized it, and adds that, in his opinion, Leroy had not responded to Bonniot. He says that, if divine intervention refers only to the infusion of the soul, then the soul is produced indirectly by evolution. A transformist might say that, when the animal has become, by evolution, a human organism, the corresponding substantial form— that is, the human soul—necessarily must have been given to him by God. Bonniot thus concludes that Leroy's ideas on this issue are unacceptable on all counts: "Above all, the *special* intervention of God in the formation of Adam's body, so clearly emphasized by the Bible and the Church Fathers, is reduced to nothing, inasmuch as *evolution,* that is, a purely natural cause, had achieved everything, except for the infusion of the soul."[18]

In the final part of his review, Brucker excuses himself for having become involved in polemics, and asks that readers not think he acts out of personal motive, but rather only out of zeal to defend Scripture and the traditional interpretation of the Church. Brucker ends his essay with some words that are highly evocative in the light of events that soon unfolded:

> I well conceive that, even without sharing the ideas of a work like this, one might think it good that it is published. The Church has always tolerated great freedom of discussion among its theologians, leaving itself the power to close down the debate when it judges that it has lasted long enough. This liberty, in great measure, is one of the conditions of the life and of the progress of theology and exegesis, just as happens with any other science. We will never apply pressure so that the anathema of the Index falls on a book that is a serious work, a conscientious search for the truth, even though it might seem to stray a bit from the straight path. But, for their part, those who publish new doctrines, which are more or less estranged from the most authoritative traditions, should not be surprised when obliged to pass through the control of a severe critique. This is the indispensable counterweight of the freedom extended to them. And if they appeal justly to their

orthodox reputations, they still may not ask that we treat them in detriment to our own orthodoxy, and above all in detriment to the faith of many readers who find their publications too liberal.[19]

Was Brucker's reference to the Index a call to attention? Was it pure coincidence, bearing in mind that Leroy's book would be denounced to the Index three years later? A small piece of evidence seems to show that the denunciation was not promoted by Brucker or his colleagues at *Études*. In point of fact, the denunciation that reached the Index in 1894 referred to the first edition of Leroy's book, not the second revised edition that Brucker reviewed. If Brucker or an associate of his had wanted to denounce the book, common sense would lead one to think he would have denounced the second, more complete edition.

The Polemic Broadens

In 1893 the Dominicans began to publish, in Paris, a journal titled *Revue Thomiste*. In the first number (January–June) there appeared an article favorable to evolution, which would continue in eight more installments, until 1896. The author was Ambroise Gardeil (1859–1931), one of the founders of the journal. He had joined the Dominican Order in 1879 and in 1904 founded an academic center at Le Saulchoir, Belgium, which soon became famous. Gardeil knew and appreciated Leroy's work, which he cited explicitly in the first of his articles.[20]

Another article in the first number of the *Revue Thomiste* was a review of a book written by a Jesuit about the old polemic on the nature of grace and freedom that had brought the Dominicans and Jesuits into conflict centuries before. The Jesuits of *Études* were offended by the tone of some of the criticisms, and one of them, E. Portalié, responded. Portalié reproached the Dominican journal, which in its inaugural number had bestowed a certificate of Thomist orthodoxy on Leroy with respect to evolution, even when applied to the human organism.[21] Leroy soon responded with a four-page letter to the editor of the *Revue Thomiste*, published in the July–August issue.[22] In this letter Leroy summarized the arguments that he had expounded in the last chapter of his book. By now, the polemical tone had reached a shrill pitch:

> It is pure calumny when [the editors of *Études*] say that I apply evolution to the human organism. Leaving aside bad intentions, I repeat, it is in itself an absolutely gratuitous calumny. . . . In sum, there is cause for me to be astonished that the serious and learned editors of *Études Religieuses* might confuse the *substrate* of my hypothesis with the *human organism*. Once again, leaving aside possible ill will, I see no explanation for this illusion other than intellectual color blindness.[23]

To assert that the Jesuits who had slandered him were intellectually color-blind must not have pleased the editors of *Études* in Paris, nor those of *La Civiltà Cattolica* in Rome.

Leroy now appeared as the champion of a somewhat suspicious cause that, moreover, was attracting more and more Catholics, as demonstrated at a series of International Scientific Congresses of Catholics, and furthermore, Leroy was publicly criticizing those who opposed his approach.

The Origin of Eve

Still another Jesuit, François Dierckx, authored a pamphlet critical of Leroy.[24] This critique is especially interesting, because it highlights an important difficulty for Catholic evolutionists—what the Bible says about the formation of Eve from Adam's rib. We have already seen that Brucker, in his review of Leroy's book, asked what Leroy thought about the origin of Eve, and that in his book, Leroy had asserted, in passing, that he accepted the immediate formation of the body of the first woman from Adam. A bit later Leroy was more explicit when he replied to Brucker through some correspondence published in February 1892 in *La Science Catholique*. Leroy again stated that God proceeded differently in the case of Adam and Eve, but still left unanswered why, if he did not accept a literal interpretation of the Bible with respect to Adam's body, would he accept it for Eve? If Adam's body was prepared through evolution, did there then exist no female body apt for the infusion of the soul, as in the case of man? Leroy confesses that he does not know how to respond, but adds that a similar difficulty occurs for his critics, because they too ought to admit that God acted differently in the two cases. In his pamphlet, Dierckx observes:

> His reply is ingenious, but sadly it only saves the appearances without resolving the core of the difficulty. What is distasteful in Leroy's theory is that, wishing to please all schools, he takes with one hand what he discards with the other, all arbitrarily, running the risk of not pleasing anyone. What spoils it is that, in order not to break violently with Catholic doctrines, he limits himself to applying evolution to Adam's body, while admitting for Eve the so *feared miracle* of immediate creation by God. It is disappointing that he interprets the biblical text about Eve literally, while he strives to seek a metaphorical meaning for the formation of Adam. . . . *What should be demonstrated is that the Creator wanted to form Eve with his own hands and to leave to the care of secondary causes the formation of the body of Adam.* That is what Leroy should have proved, and he does not prove it.[25]

Moreover, the difficulties did not arise only from the formation of Eve. Dierckx also alludes to monogenism, stating that every Catholic should admit the existence of a first father, common to all mankind, because were the contrary true, the unity of humankind would be lost and, along with it, the solidarity of all people in the sin of Adam. But, if transformism is true, then collateral branches of men would exist. What purpose did they have? What happened to them? Would they be capable some day of giving rise to species as intelligent as or more intelligent than present-day man? Surely this is not what Leroy advocates.[26]

Dierckx stressed that Leroy's ideas suggest problems that he does not resolve. But here too is an echo of the ideas of the prestigious Spanish cardinal Zeferino González, who, although critical of transformism, declared that he did not condemn from the theological perspective those who used it to explain the origin of the human body so long as the Church respected or tolerated this opinion.[27] The defenders of traditional orthodoxy claimed to respect their opponents so long as the Church did not pronounce on the subject. But in 1894 Leroy's book was denounced to the Congregation of the Index. The intent of the action was to force the Roman authorities to make up their minds.

Leroy Denounced to the Congregation
of the Index

At the beginning of the summer of 1894, the Congregation of the Index received a short accusatory letter against the first edition of Leroy's book.[28] The author of the letter, who signed as "Ch. Chalmel, Officier d'Académie" was, in all probability, unknown in Vatican circles. Somewhat later when the examination of the question began, a report referred to the author of the denunciation as "a certain Mr. Ch. Chalmel, a Frenchman."[29]

Chalmel, in a highly respectful tone, submitted two questions for the consideration of the Congregation. The formal language with which the letter began concealed an intent to accuse, as was quickly revealed in the letter's first question:

> Monsignor: May I beg to submit to the Holy Congregation two questions of extreme importance for the faith?
>
> A Dominican scientist, Father Leroy, a friend of Father Monsabré, who shares his opinion, has published a book (*Evolution of the Organic Species*). In this work, which upholds the opinion of Darwinn [*sic*], he claims that in the Genesis narrative there is nothing dogmatic "except for the creation of the universe by God and the action of his providence"; that interpretation of Creation be left to the "discussions of men"; that the narrative of Moses is "an old patriarchal poem, a tissue of metaphors"; and that, in consequence, science cannot take the literal sense of Genesis seriously. Has this new interpretation of Genesis been approved by the Congregation of the Index, and should the literal interpretation that the Church has always given be abandoned?

The second question refers to the Galileo case, but it was never considered. In the original letter it is crossed out with two lines of red pencil. The same pencil drew attention to Leroy's book. We think that these annotations were made by the cardinal prefect.

The Denunciation Proceeds

In 1894 only three persons worked in the Congregation of the Index with consistent regularity: the cardinal prefect, the secretary of the Congregation, and an archivist.[30] Most likely the letter was read first by the secretary, who passed it along to the cardinal prefect to decide what specific measures would be taken.

The secretary of the Congregation was the Dominican Marcolino Cicognani (1835–99), recently named to the post on March 22, 1894. Cicognani was also procurator general of the Order of Preachers (Dominicans), the religious order to which the secretary of the Congregation had always belonged. Following traditional practice, the master general of the Dominicans, Andreas Frühwirth, had presented to the Congregation on March 19 a list of candidates for the office of secretary, indicating that, in his judgment, the best suited was Cicognani, adding that if Cicognani were elected he should also continue as procurator general of the order for one more year.[31] The Congregation accepted the proposal and, on March 22, the secretary of state, Cardinal Rampolla, communicated to the cardinal prefect of the Congregation of the Index the appointment of Cicognani as secretary.[32]

The cardinal prefect, Serafino Vannutelli (1834–1915), had also occupied his post for only a few months. He had been a professor in the Roman Seminary and had held several posts in the nunciatures of the Holy See in Mexico City and Munich, had been apostolic delegate in various countries of Central and South America, and nuncio in Belgium and the Austro-Hungarian Empire. Created cardinal in 1887, he was named prefect of the Congregation of the Index on December 6, 1893, succeeding Cardinal Camillo Mazzella, who became prefect of the Congregation of Studies. In all certainty, Chalmel's accusation ended up on the desk of Cardinal Vannutelli who, besides red-penciling it, wrote Chalmel's name and address on the last page. It was his responsibility to set the course of action.

It is not possible to determine exactly when the letter reached the Congregation, for its receipt was not calendared immediately, as would have been the normal procedure. Perhaps in those first months of his tenure Cicognani had not yet organized his work routine, because he held two important posts simultaneously. Following the custom of most of his predecessors, Cicognani kept a diary in which he personally noted the various tasks related to his work: appointments, Preparatory and General Congregations, audiences. The first mention of Leroy appears on September 13, when he recorded the results of the first Preparatory Congregation in which the book was discussed. Toward the end of the year he decided to correct his omission, and on December 16 he recorded the denunciation, using his characteristic abbreviations: "The work of F. Leroy, *L'évolution* etc., was denounced on June 20 of this year by Monsieur Ch. Chalmel, Official of the Academy, etc., which was discussed by the Congregations of September 13 and 19, and in the next."[33]

In reality, the denunciation must have reached Rome quickly. After a rapid examination of the accusation, Cardinal Vannutelli discarded the second point, on Galileo, but decided to examine Leroy's work. The customary procedure was to entrust the examination to one of the consultors of the Congregation. On occasion, the cardinal prefect himself chose the consultor, although the more usual practice was for the secretary to make the designation, taking into account all the relevant factors (circumstances of the case, competence and interests of the various consultors, etc.). In this instance the task was assigned to Teofilo Domenichelli, an Observant Franciscan. He was one of the newest consultors of the Congregation of the Index, having been appointed on June 20, 1894, the same day on which Chalmel had written his letter of accusation.[34]

Domenichelli's Favorable Report

Domenichelli must have received the assignment at the beginning of the summer. He probably did not consider it especially urgent, taking into account the traditional summertime inactivity in Rome. But his report was finished in late August and is dated August 30, 1894.

It is worth noting that the report does not refer to the first edition of Leroy's book (*L'évolution des espèces organiques*, 1887), the one that had been denounced by Chalmel, but rather to the new edition *L'évolution restreinte aux espèces organiques* (1891), quite expanded and also with a few points corrected. Leroy and his work were not unknown in Rome. The controversies provoked by his works in France most likely had had some echo in the Roman Curia. We know, moreover, that the secretary of the Congregation himself had received a copy of the work, probably sent him by Leroy. That is what Cicognani wrote later on in an advance of the report he prepared for an audience with the pope, in September of the same year.[35] Even though he does not so indicate explicitly, he seems here to refer to the second edition. In any case, at the time they went to commission the examination of the work, both the prefect and the secretary were aware of the second edition, because that was the one given to Domenichelli, along with Chalmel's concrete accusations.[36]

Domenichelli's report is quite complete, occupying twenty-seven printed pages, divided into six parts and twenty-two sections, numbered in the margins.[37] The first part of the report (§§1–3) is titled "Objective of the Examination." After summarizing the contents of denunciation, he states that, by a decision of the prefect, the study would not be limited to the questions raised by Chalmel but would range over the entire book, in its second edition. This does not mean, as Barry Brundell seems to suggest, a sudden change of mind or an exception to the rules of the Congregation.[38] Unlike the Holy Office, the Congregation of the Index did not have the task of examining or condemning a specific doctrine, only that of assessing the danger that certain published works might hold for Catholic readers. Thus even though a denunciation might refer to only one doctrinal point in a given book, the whole work was examined. In this case it seemed logical to scrutinize the second edition, as Domenichelli wrote, "so that the verdict might be more complete and fair" (p. 2).

Next (§§4–7), Domenichelli reviews the contents of the book. After introducing Leroy as a learned and pious Dominican with profound knowledge of theology and the sciences, he says that Leroy in great measure accepts evolutionary ideas, and that he expects that evolution will run the same course as had the ideas of Copernicus or Galileo, that is, once it has been shorn of the exaggerations of both its partisans and critics, and when emotions have calmed, it will become widely accepted .

In the book's first chapter (still following Domenichelli's report) Leroy refutes "atheist evolutionists who, with the intention of explaining the world without God, want the limitless variety of nature, from inorganic matter to man, to derive from uncreated and eternal matter, by means of natural, necessary laws, with no need for an ordering mind" (p. 3). In the second chapter Leroy seeks to demonstrate that evolutionism is not opposed to faith: it has not been condemned by the Church, which till now has emitted no judgment on this subject; it is not contrary to Scripture or tradition. Domenichelli reproduces from Leroy's book the following paragraph, the one behind the first accusation of the denunciation:

THE FIRST PAGES of Genesis present the oldest compilation of sacred teachings. THE STYLE OF THESE PAGES is not that of a textbook of cosmography, geology, or zoology; IT IS THE POETIC STYLE of a primitive poem, of a patriarchal psalm. Thus it would not be reasonable to apply to these METAPHORS THE LITERAL EXEGESIS that should be applied to the prosaic text of a modern book on physics. . . . Thus, Scripture teaches us that the plant and animal species are the work of God, but it does not EXPLAIN HOW GOD MADE THE TOTALITY OF THOSE BEINGS (pp. 44–45 of Leroy's book). (p. 6)

In the third part of his report (§8, "Examination of the Points Denounced"), Domenichelli asserts that "the greater part of these censures [of the denunciation] are the result of having misunderstood the author, or of having given to his expressions a much broader meaning than, in reality, they have" (p. 7). For example, Leroy does not say that Genesis *is an old patriarchal poem* and must be read metaphorically. He applies these expressions only to the *style* of the first pages of Genesis, and only with regard to the *way* in which God created material things. Understood in this limited way, these expressions would be condemnable only if, as the denunciation seems to suppose, the Church had always given a literal meaning to these pages. But Domenichelli says that is not so and proceeds to demonstrate the point in the next part (§§9–15): "The six days of creation in the Genesis narrative may freely be understood in an allegorical sense" (p. 8).

The report was drawn up shortly after Pope Leo XIII published the encyclical *Providentissimus Deus* (dated November 18, 1893) on the interpretation of Scripture, a subject that provoked many problems in that period. Domenichelli cites it and takes it into account in his analysis. With respect to interpreting the Genesis Creation narrative, Domenichelli alludes first to the theological authority of Saint Thomas Aquinas and Saint Bonaventure (pp. 8–10). Following Aquinas (*Super Sent.*, lib. 2, d. 12, q. 1, a. 2, co.), he credits only the fact of Creation as essential to faith. The way by which God created touches faith only accidentally, and thus it has been interpreted differently by the Church fathers. Saint Bonaventure and other scholastic writers agreed.[39] Citing a text of Saint Augustine (*De Genesis ad litteram*, lib. IV, 52), to which he adds a long list of other fathers, Domenichelli seeks to show that, in fact, the Church fathers accepted the possibility of diverse interpretations of the Creation narrative. Domenichelli concludes: "To state that in the first chapters of Genesis Scripture uses a figurative language *in a human mode, in order to condescend* (as Dante would say) *to our own understanding* is completely correct" (p. 10).

Numerous theologians of various epochs, according to Domenichelli, agree with this opinion, so that one can conclude that the question of whether the biblical narrative of Creation should be interpreted literally or allegorically is open to the free discussion of theologians. It cannot, therefore, be censured. Domenichelli recalls the warnings of Saint Augustine and other fathers not to be facile in presenting a literal interpretation of the first pages of Genesis "in order not to expose oneself to ridicule in the presence of unbelievers" (p. 13).[40] Nowadays, he adds, with our present knowledge, such a literal interpretation is simply absurd. Attempts at harmonization are demonstrably false and can only harm the Church and believers.

On this basis, Domenichelli rejects Chalmel's accusations, which he considers lacking in foundation (p. 14), and goes on to the next part, where he tackles the question of "Evolution from the Theological Perspective" (§§16–17). First, he says that Leroy's book, in spite of its brilliant arguments, "has not succeeded in inducing me to embrace these [evolutionists'] opinions which, at least in part, I consider hypothetical and lacking a solid footing" (p. 15). But recall that the Congregation of the Index does not judge books according to the criteria of science but rather by considering only the teachings of the Church. He mentions that many Catholic thinkers and theologians have accepted evolution as compatible with the Catholic faith, so long as they impose some limits, especially in the case of the direct creation by God of each human soul. Domenichelli cites, among others, John H. Newman and Maurice d'Hulst, and says that, with the nuances that Leroy covers in his book, the theory of evolution is completely immune from danger of negating the truth of God, Creator and ordainer (p. 17). In this context, he adds a clear call for prudence:

> The number and prestige of theologians who see in this opinion [evolutionism] no offense to the Catholic faith are not of little weight, and they caution us to proceed with great reserve. Because this is a relatively recent theory, which therefore is neither well developed nor well known (Lamarck wrote in 1809, but the first reasonably serious debate was that in 1830 between Cuvier and Saint-Hilaire, and first forceful exposition with the publication of Darwin's book in 1859), my report should be used only if you decide either not to prohibit the book or to postpone the judgment. If a decision to prohibit it were to prevail, than another report should be made, fuller and more profound than mine, which is very short and should not be used to condemn the work. (p. 16)

Domenichelli alludes to a typical objection against evolution, that is, that "lower species give rise to the higher ones; the lesser produces the great, which is absurd." Domenichelli says that this is a philosophical, not a theological, objection, which he will therefore not address. He adds that the Church fathers and scholastic doctors had accepted the spontaneous generation of certain animals from inorganic matter, which Pasteur had refuted. And if such "absolute heterogenesis" could not be censured, how much less might one censure "the much more limited heterogenesis of the evolutionists" (pp. 18–19).

The most controversial point, however, is the evolution of man, which Domenichelli presents with a certain breadth and attempts to judge from the theological point of view in the last section of his report titled "Evolution and Man" (§§18–21, pp. 19–25), that is, "the last and harshest chapter of Leroy's book."

In this chapter, Domenichelli notes, Leroy demonstrates the existence, spiritual nature, and immortality of the human soul, and therefore its origin by direct creation by God. He also affirms the divine origin of the human body, inasmuch as it is not possible to split human nature in two. However, Scripture says that man was made from the mud of the earth, but it does not determine how: whether it was by an immediate act of God, or whether God let natural causes act. The Council of Cologne of 1860, approved by Rome, denies only that there was a "spontaneous" transformation, without God's participation. And because the body is human only when informed by the soul, one must agree that the body is also produced by God: the human body is likewise formed through the infusion of the soul. Thus man's body does not come from "animality"; it is always the direct product of divine action that infuses the rational soul. For Domenichelli interpretation is sufficient to refute what the Council of Cologne sought to condemn, that is, the formation of man's body by "spontaneous transformations."

After summarizing what Leroy said, Domenichelli begins to pronounce his verdict. In the first place, it seems clear that evolutionism does not need to deny the existence of the spiritual, human soul, created by God. Here he introduces an interesting reflection: if natural generation whereby some human beings engender others does not impede affirming the existence of the spiritual soul, why would the difficulty be increased in the case of evolution, whereby some species are transformed into others? It is certain, Domenichelli adds, that there are many materialists among the evolutionists, but materialism existed before evolutionism, and moreover there are not a few evolutionists who are believers. Domenichelli concludes that in this area there is nothing to fear: arguments for the spiritual nature of the human soul, splendidly summarized by Leroy, are not weakened by evolutionism (p. 22).

It appeared more difficult for Domenichelli to evaluate Leroy's idea that evolution could have prepared the matter destined to become the human body through infusion of the soul. It is possible to interpret in this sense the phrase from Genesis where it is said that God formed man from the mud of the earth, but one cannot doubt that the Church fathers, doctors, and theologians with near unanimity understood this phrase literally, even though there were some contrary opinions. Domenichelli states that, faced with this uncertainty, it is safer and more pious to hold to the common view and that here Leroy's stance "reaches a limit beyond which speculation is transformed into condemnable temerity." But has Leroy overstepped the bounds in such a way as to merit a theological censure? Domenichelli says he would not dare to make such an assertion, although here he didn't feel as sure as in his previous conclusions. He recalls that many distinguished theologians and other Catholics have accepted evolution. Leroy's book had circulated freely since its first edition of 1887 without meeting any obstacles and without the Church having said anything. Other books on this subject written by Catholics, which are less accurate than Leroy's, circulate without being censured.

Domenichelli continues his concluding remarks asserting that, if a special measure were taken against Leroy, then Leroy, those who had refereed his work, and his order (the Dominicans) would all be harmed. Moreover, the wrong would be increased by striking the least offensive book, while leaving unpunished more dangerous and exaggerated volumes. If the theory seems too dangerous, Domenichelli suggests, then it would make greater sense to identify the errors and condemn them with the appropriate measure, as the Church had done on other occasions, without condemning specific books, an impossibility owing to their great number. Moreover, among so many diabolical attacks on the Church, evolution is attractive as a useful arm against Catholics; but if it is tolerated, spirits will be calmed and the issue could be judged with greater civility and impartiality

The conclusion of the report Domenichelli submitted to the cardinals was favorable to Leroy and proposed that his book not be condemned (pp. 26–27). The Latin word used by Domenichelli, which we translate as "not be condemned," is *dimittatur.* The Latin verb *dimitto* means "say goodbye, dispatch, abandon, pardon," and the conclusion *dimittatur* was applied when it was proposed to take no measure.

The First Preparatory Congregation, September 13, 1894

Domenichelli's report was signed Thursday August 30, 1894, and most likely sent to the Congregation of the Index the same day. A Preparatory Congregation was convoked Thursday, September 13, 1894. Among the four books examined on this occasion, Leroy's was third.

The Preparatory Congregation took place in the residence of the secretary Marcolino Cicognani, which in that period was the General House of the Dominicans, since Cicognani was still procurator general of the order. It was located in the Piazza di Spagna, on Via San Sebastiano, number 10. At the meeting, besides Secretary Cicognani, were the master of the Sacred Palace, Raffaele Pierotti, and seven consultors: Luigi Tripepi, Guglielmo D'Ambrogi, Pio Arcangeli, Gioacchino Maria Corrado, Enrico Buonpensiere, Alessio Boccasso, and Teofilo Domenichelli. Two of them (Tripepi and Buonpensiere) would play key roles in the Leroy case later on. Three of the consultors had been named recently: Buonpensiere, Boccasso, and Domenichelli had received the nomination of the consultors on June 20, 1894, so they were participating in their first such meting. Tripepi, on the other hand, was a veteran and had participated in the Particular Congregation where sixteen years before, in 1878, Raffaello Caverni's book had been examined and condemned. Without doubt, Tripepi recalled the previous discussion of evolutionism and Tommaso Zigliara's report; in fact, in a report that Tripepi would write on Leroy's book, he mentions that report by Zigliara.

The conclusion of the Preparatory Congregation was not at all what Domenichelli had proposed. In his Diary of the Congregation, Secretary Cicognani wrote, "On the 3rd [book] all voted unanimously: Wait, and let someone else write a report in which the question will be reconsidered."[41] The consultors had agreed that Domenichelli's report was not sufficient. Tripepi and Buonpensiere later on revealed themselves to be clearly opposed to evolutionism. Tripepi, fifty-eight years old in 1894, had just been named secretary of the Congregation of Rites. But his career still had not reached its high point: two years later he was named substitute of the Secretariat of State and, in 1901, cardinal. His long standing as consultor and his role in the Roman Curia lent weight to his opinion.

The First General Congregation,
September 19, 1894

The General Congregation was convened in the Vatican Apostolic Palace on the following Wednesday, September 19, 1894, at 9:30 in the morning. The Diary of the Congregation lists as present, besides Cardinal Prefect Serafino Vannutelli, only four other cardinals: Parocchi, Granniello, Verga, and Mazzella. This was an unusually low number, but not unheard of. Also participating, as was the custom, were the secretary, Cicognani, and the master of the Sacred Palace, the Dominican Raffaele Pierotti.[42]

It is helpful to review briefly the careers of the cardinals who were about to pass judgment on Leroy's work and submit their decision to the pope. Lucido M. Parocchi (1833–1903), although only sixty-one, was one of the longest-standing cardinals. From 1884 he was the pope's vicar general for the city of Rome and had headed two congregations of lesser importance (Apostolic Visit and Bishops' Residence). He was also a member of eleven other congregations. Two years later, in 1896, he would become secretary of the Holy Office. He was one of the most learned and influential cardinals of the Roman Curia.

Giuseppe Maria Granniello (1834–96), a Neapolitan, belonged to the Regular Clerics of Saint Paul (Barnabites) and was one of the most recently named cardinals, designated in the consistory of June 12, 1893. Until then he had been secretary of the Sacred Congregation of Bishops, but had also been a consultor of the Congregation of the Index. His health was poor, and he died shortly after.

Isidoro Verga (1832–99) had been prefect of the Sacred Congregation of Bishops and Regulars since 1888, besides being a member of other congregations.

Camillo Mazzella (1833–1900), a Jesuit, was prefect of the Congregation of Studies and a member of other congregations. He had been prefect of the Index from 1889 until 1893, which, together with his prestige as a theologian, gave him special authority in the Congregation. He was openly opposed to evolutionism, as is obvious from his influential theology textbook on *God the Creator.*

The cardinals present had received the reports on the works they were about to adjudicate, including the report favorable to Leroy written by Domenichelli. But they also received information from the Preparatory Congregation, in which more study of the question was recommended. They agreed to the need for further study, but not before a full debate of the questions implied, to the point where Secretary Cicognani wrote in his Diary that a decision had been reached "after diverse argumentation on the subject and a serious debate." Concretely, two additional reports on Leroy's book were solicited from the consultors Ernesto Fontana and Luigi Tripepi. The cardinals also designated three questions to which the consultors should respond: the exegetical criteria that the author (Leroy) considers should be followed in the interpretation of Genesis; the system of evolution of organic species as proposed by the author; and his doctrine with respect to the formation of the first man.[43]

It is not easy to reconstruct the course of the discussion precisely, but Mazzella was the most authoritative theologian, and we can conjecture that his position, completely opposed to evolutionism, carried great weight. In any case it is clear enough, following Cicognani's summary, that not all those present were openly opposed either to Leroy's book or to the theory of evolution, inasmuch as a "serious debate" had taken place, and, in spite of Mazzella's influence, further study was called for. We can state this with confidence, at least insofar as concerns Cicognani, because in a personal memorandum (*pro-memoria*) he collected the arguments favorable to Leroy that Domenichelli had presented in his report and made them his own.[44]

On October 3, 1894, Cicognani noted in the Diary of the Congregation that he transmitted assignments to the two consultors: "The Secretary gave an order by letter to the Most Reverend Fathers Fontana and Tripepi, to examine the book of Father Leroy *L'évolution . . .*, according to the Decree of the Congregation of September 19 of this year."[45] So there commenced a second examination of Leroy's book. But before describing it, we must now consider a document that is surprising, because it appears to indicate that the secretary of the Congregation had predicted a completely different outcome of the case.

Cicognani's Memorandum

Because the decisions of the General Congregation were transmitted personally to the pope by the secretary of the Index, Cicognani customarily prepared a memorandum in which he summarized the conclusions of the meeting. On this occasion, the audience took place the same day, September 19. What is surprising is that Cicognani's memorandum did not at all agree with the decision of the General Congregation.

Cicognani's memorandum occupies four pages, written in his characteristic way, that is, using only the right-hand side of the page.[46] The left-hand side he reserved for annotations. It begins with the words: "Extraordinary Audience, September 19, 1894," followed by "On the four books examined—for the approval of the decree." The summary of what was said about the books is presented in four numbered paragraphs. That on Leroy is the longest, occupying more than half of the memorandum. But the principal surprise is that Cicognani's summary is completely favorable to Leroy's book and does not criticize even a single point.

After providing details about the book, the accusation, and Domenichelli's report, Cicognani mentions the various approvals the book had received and says that copies had been received by Cardinal Zigliara and himself: "This work by Father Leroy was approved by the Order's examiners, Father Monsabré, Father Villaud, and Father Beaudouin—[in the margin] It also bears an explicit approval by Lapparent, professor of geology at the Catholic Institute of Paris—and it did not pass unnoticed either by Cardinal Zigliara or myself, because we each received a copy."

This last point is interesting. Cardinal Tommaso Zigliara, who had died the year before, was a Dominican, as were Leroy and Cicognani, and he had been prefect of the Congregation of Studies, a member of the Congregation of the Index, and one of the most important philosophers of the Neo-Thomist revival spurred by Leo XIII. As we have seen, he played a direct role in Caverni's case. Knowing that Zigliara had received the book and had not considered it necessary to take any action was an important point in Leroy's favor. Even in the event that Zigliara had not read the book, its subject matter and orientation would have been evident at first sight. And yet Zigliara took no measure.

Next, Cicognani defends the book, recalling that the consultor had declared the points raised in the denunciation baseless. He then explains the kind of evolutionism that Leroy proposes:

Leroy professes the doctrine of Evolutionism in a most appropriate form, always admitting the creative and providential action of God. He completely excludes evolution by natural forces, as the materialists and atheists would have it, refuting them successfully, because they exclude and deny God. His system is that God has created everything, that he governs with his Providence, and in the successive development of organisms he makes use of created forces, or secondary causes, always guided by his Providence.

In the case of man, God alone is the creator of the soul and body, so that evolution had only "prepared the dust of the earth" for the divine creation of humanity:

Doesn't the human body derive from animality? But inasmuch as the human body is not human, according to Saint Thomas, until it is informed by the human soul, *non est caro humana, quæ non est informata ab anima rationali*, Leroy's basic question of evolutionism is this: "whether the material part of the human body has undergone transformation before the infusion of the human soul." And he resolves it with the words *formavit de humo* [he formed from dust], *formavit homi-*

nen de humo terrae, formatis igitur de humo cunctis animantibus terrae [he formed man from dust of the earth, for he formed from dust all the living beings of the earth] and permits, along with the creative force, a natural evolution which prepares the earth or mud destined to be a human body by the infusion of the soul. As the mud reaches that point nearest the human body, God only lays on his hand and infuses the rational soul: *et inspiravit in faciem eius spiraculum vitae* [and he inspired the spirit of life in his face].

Cicognani's conclusion is unambiguous; briefly and clearly he proposes that no measure be taken against Leroy's book: "This opinion was not found contrary either to theology or faith; and of this book it was said: *dimittatur* [that the case be dismissed]."

Clearly this memorandum, in spite of the date it bears, does not represent the conclusions of the General Congregation that Cicognani was to present to the pope, because there it had been decided: "Dilata et scribant alii duo" (that the conclusion wait, and two more persons write reports). Not even the conclusions of the Particular Congregation, where it was also proposed to assign another report, are represented. The memorandum is based only on the arguments and proposal of Domenichelli's report. It seems clear that Cicognani must have written this memorandum before the two Congregations. Examining the notes relative to the three other books scrutinized, one can see that Cicognani only presents the main ideas put forward by the consultors in their reports, sometimes with the same words.

Why did Cicognani write and then file this memorandum, in spite of the fact that the Particular and General Congregations had changed the decision there proposed? It was still early in Cicognani's tenure as secretary of the Congregation of the Index: he had been named scarcely six months before. And he continued to serve as procurator of the Order of the Dominicans. It was only the second time he had organized the meetings of the Index, and he must have considered it appropriate to study the various cases before the Preparatory Congregation. The same memorandum could serve afterward for the General Congregation and then the audience. Perhaps he did not think it necessary to prepare another. The decisions taken were simple: only in Leroy's case was a decision different from that proposed by the consultor in each case taken. In the other three cases, Congregations confirmed the proposal. Cicognani must have thought, therefore, that it was not necessary to write a new memorandum. It would be sufficient to take the brief note with the three points concerning which the General Congregation had sought a more inclusive study. The memorandum was still useful in the event the pope should ask for more information on the contents of the book.

What conclusions, then, did Cicognani present to the pope at the September

19 audience? Did he use any of his favorable memorandum, or did he limit himself to the conclusions of the General Congregation? The only document referring to the papal audience is a note Cicognani entered in his Diary. After noting the meeting of the General Congregation, he writes: "The same day: The Secretary was admitted in audience with His Holiness and, when the actions of the Congregation just held were related, His Holiness approved the decree and ordered it published; the following works were added: . . ."[47]

The only certain conclusion is that Cicognani presented the decision of the Congregation of the cardinals to the pope. He could not have acted otherwise. We cannot know whether he used the contents of his memorandum in the papal audience.

But we can extract one more interesting conclusion. In the case of Leroy, Cicognani was not limited to copying or summarizing some points of Domenichelli's report. Cicognani had received a copy of Leroy's book, probably a short time after its publication in 1891. The memorandum perhaps reflects as well, at least in part, his own reading of Leroy's work. It permits us to conjecture that Cicognani's stance was not, at least in principle, contrary to the reconciliation of the theory of evolution with Christian faith. It is possible that this was one of the causes of the "serious discussion" in the General Congregation.

A Fresh Examination:
Fontana's Report

The General Congregation had requested that Cicognani commission two new reports from the consultors Fontana and Tripepi, which he did on October 3, 1894. Three weeks later the first of these was ready: Ernesto Fontana signed his on October 24.[48] It was a short report of less than five pages, in which the author responds briefly to the three questions formulated by the Congregation. The reason for Fontana's brevity may simply be that he had been named to a new position a short while before. In June of that year he had been named bishop of Crema, a small city near Milan. Although he continued as a consultor of the Index for several years more, his new occupation would leave him less time for this task.

Fontana's report was less kind to Leroy than Domenichelli's had been. Yet it too did not favor condemnation. The position he adopts is in a certain sense prudent: while he sees no reason to condemn Leroy's thesis, it strikes him as dangerous, on account of which the author should be warned. It is interesting to examine his reasons in detail, for, as we will see, his attitude, although with different nuances in each case, is that which Congregation would adopt with respect to evolutionism.

After a brief presentation of his report, Fontana sets out the dilemma it raises. On the one hand, he does not find anything directly reproachable in Leroy's book: "I find nothing to censure against faith and good custom: I am fully in accord with the two referees, Fathers Villard and Beaudouin [the two censors who had given permission to publish the book in Paris]" (p. 1).

Nevertheless, Fontana adds, it is distasteful to him to see such a zealous defense of a hypothesis that, even if not contrary to faith, is uncertain and is opposed by many naturalists.[49] And it is contrary to the way in which, till now, believers have commonly understood the words of the Bible with respect to Creation. The three questions reflect the same dilemma: Leroy's position is risky and even dangerous although it does not contradict the faith.

The first question referred to the exegetical criteria proposed by Leroy for the interpretation of Genesis, here summarized by Fontana: "The author says explicitly and absolutely that, when the biblical text is susceptible to diverse interpretations and neither dogmatic definitions nor the consensus of the Church Fathers determine one of them, scientists are free to apply to it the meaning that science gives it" (p. 2). To be sure, according to Fontana this criterion "cannot be reproached or condemned; but I think it dangerous and somewhat impertinent." Where is the danger? Fontana says that Leroy applies this principle rashly, inasmuch as "the meaning that science gives it," that is, the progressive evolution of organic species, "is not scientific theory, but rather a pure scientific hypothesis."

The response to the second question is more positive. He had been asked for an account of "the system of evolution of organic species, as it is proposed by the author." Fontana recognizes that, when the human species is excluded, as Leroy expressly does, this system "is neither absurd nor condemnable." Fontana argues from classical philosophy, according to which the forms of animals and plants are not subsistent (inherent in matter). It can therefore be accepted that they might undergo modifications owing to the environment in which they live:

> It is admissible that forms that are not subsistent, inherent in matter, and, even more, immersed in matter and unable to exist and act if not in and with matter, might suffer those affections that matter suffers owing to climate, nutrition, and the circumstances in which the individual organism lives. Supposing, as the author supposes, that the creative act may have placed in the individuals of the primitive types aptitudes and potentials as embryonic, to be developed, evolution is not impossible. (p. 3)

Could evolution, beyond introducing accidental variations, also modify species? Leroy thinks so. Fontana considers that the majority of naturalists deny it, but there is no certain answer, because there are no clear criteria for defining what a species is, and how it is distinguished from a race, genus, or variety.

Evidently, the most difficult was the third, the formation of the first man. Fontana finds no clear error in the position of Leroy, who holds that the human body has not been formed from inorganic matter, but from organic (that is, living) matter, in which God infused the spiritual soul. Leroy excludes any transformism in what touches the human soul. Fontana, nevertheless, finds an objection: Leroy does not explicitly state that the form formerly possessed by this organic matter, a purely sensitive soul, disappears totally when the spiritual soul is infused. This could support "Rosmini's fatal error that when the idea of being becomes manifest to the soul of man, it changes from sensitive to intellectual." Nor, secondly, would the unity of the substantial form of the human body be clear.

Fontana next adds a third objection, which, although it seems similar to the preceding ones, is quite distinct:

> The human body would have its origin in the body of a beast, raised, all that one might like, to the highest grade of perfection, but still a *beast*. The author does not ever use, here, the word beast or animal; he always uses the term *organized matter*; but matter most perfectly organized (precisely as is required here, by the exigency of the whole system), is it not a beast? Now, this hypothesis is repugnant and clashes with Christian and human feeling; in what refers to myself, I cannot tolerate it. Darwin and his followers tolerate it, because in some way it is similar to theirs. (p. 4)

It is not clear what the objection here really is. What is clear is the rejection of evolution, which still presents an animal organism, that is, a *beast*, as the origin of the human body.

Fontana also alludes, though very briefly, to other difficulties, such as calling the cogitative faculty of animals *intelligence;* to speak of their *consciousness;* and to call the body of man a *human compound*, instead of reserving this expression for the combination of the human body and soul. But he considers that he is not competent to examine these points. It is worth pointing out that what Fontana identifies as difficulties seem to have come from a not very attentive reading of Leroy's book or from having taken several of his expressions out of context. For example, the term "intelligent" is attributed by Leroy to animals in a literal citation taken from Quatrefages, in a passage that Leroy goes on to refute.[50]

He concludes that some measure short of condemnation should be taken against Leroy: "I dare not propose that the book of the learned Leroy be placed on the *Index of Prohibited Books*, but I declare my wish that the author be seriously warned for the intemperance and impertinence of his ideas, which will please those evolutionists who are also atheists and materialists, but cannot be acceptable to true Catholics" (p. 5).

Fontana was the second consultor to recommend that Leroy's book not be prohibited and who did not regard it as opposed to the Catholic faith.

Leroy's Book Examined by Tripepi

The examination of the third consultor, Luigi Tripepi, was a slower process. He did not finish his report until December 8, and it was much more extensive, filling fifty-four printed pages.[51] It was also more severe than the two previous reports. Although he did not exclude the possibility, he too did not propose to condemn the book in any decisive way, but his argument tended to demonstrate the complete incompatibility of evolution and Catholic doctrine. The most important question for him is the formation of the body of the first man. In his opinion, this point alone would suffice to determine the decision on Leroy's work (p. 2).

The Origin of the First Man

Tripepi presents the most heated point, which is none other than the debate between the partisans of a certain level of evolution that leads to the formation of the body of the first man and those who uphold his direct and immediate formation by God. Leroy's stance in favor of the first alternative is very clear, and Tripepi proposes to refute it, on both scientific and theological grounds.

From the scientific point of view, Tripepi's arguments are brief, general, and fall short of the proposed objective. He states that evolutionists frequently fall into contradictions and that all of their arguments are reduced to hypotheses, without their ever proving either the fact or the logical necessity of evolution. Moreover, he adds, the arguments are old ones, contrary to experience, based on arbitrary principles, and with mechanisms that are completely insufficient and often ridiculous. According to Tripepi, evolutionism "is nowadays abandoned and refuted as false and absurd by the same rationalists and unbelievers who previously supported it" (p. 9).

In his intent to refute evolutionism from a rational perspective, Tripepi does not display much brilliance. He adopts a doctrinal stance, expounding the opinion of theologians, closely (sometimes literally) following Cardinal Mazzella's textbook *De Deo creante* (On God the Creator).[52] He states that the action whereby God formed the body of the first man is distinct from the first creation of matter; therefore, it cannot be said that God created man because he created the matter from which man's body is derived. This divine action is also distinct from the concourse that God provides, as first cause, to the actions of secondary causes. In sum, this is a matter of an *immediate* act of God, as a *unique efficient cause*, not the same as creation out of nothing, but something that is proper only to God himself (p. 14).

This was, Tripepi recalls, the opinion of all Catholic theologians until very recently. But Tripepi says that now there exists a diversity of opinion: "It is true that nowadays some Catholics, whether in writing or at some Congress of Catholic Science or in some Catholic Institute in France or elsewhere, have expressed diverse opinions" (p. 14). Who are these Catholics? He can name a few right off the bat: Fabre d'Envieu,[53] Gmeiner,[54] Zahm, and Mivart. Allusions to the Institut Catholique de Paris, and to the recent Scientific Congresses of Paris and Brussels (the latter mentioned explicitly in the same breath as Zahm) are also clear. Yet these cases, for Tripepi, are not significant:

> These cannot at all diminish the complete, solemn, uninterrupted, and universal agreement (at least, *until just recently*) of theologians on this issue. They can be considered erudite, eloquent, ingenious men; but certainly they are not great or profound theologians, at least on this subject. Their names alone tell us that their philosophical findings cannot carry too much weight with those who in Rome have pursued serious ecclesiastical studies on the Church Fathers and the great philosophers and theologians who flourished over the many centuries of the Church; and still less can they boast of authority above the elevated wisdom of the most eminent Judges of the Roman Congregations. (p. 14)

This passage, with its rhetorical tone, seems directed at persuading the cardinals that cases like this should not modify what has been common opinion in theology just because some theological parvenus say so. Furthermore, Tripepi adds, some of these same persons have articulated dangerous opinions on the Bible, which have been repudiated by the pope's recent encyclical,[55] and one of Mivart's books has even been placed on the *Index* because he takes evolution to the point of asserting that it even exists in hell, so that the punishments of the condemned gradually diminish (p. 15). We discuss Zahm and Mivart in separate chapters, but here only observe that Mivart's articles on "Happiness in Hell" were placed on the *Index of Prohibited Books* for his ideas on the nature of hell, not for his views on evolution.

Once he has established the agreement of theologians on this doctrine, Tripepi asks himself whether it is possible to uphold the contrary opinion. To answer this question, he returns to Cardinal Mazzella's book, reproducing long passages almost literally and with only a few short comments interspersed (pp. 15–16).[56] This text has a double objective: to show, first, that the doctrine examined is taught by all theologians and can therefore be considered as a doctrine *of Catholic faith;* and, second, that although it might not be a matter of faith, it is not permissible to negate it, since any proposition contrary to the unanimous consensus of the fathers and doctors of the Church cannot be *certain.*

With respect to the first point, Mazzella observes, referring to Vatican Council I, that for a truth to be considered *of the Catholic faith* it is not necessary that it be solemnly declared so. It is sufficient that it be proposed by ordinary and universal Magisterium as revealed by God. But Tripepi goes even further. He insinuates that agreement among theologians achieves, in practice, the same value as the ordinary and universal Magisterium of the Church hierarchy. In reality, the consensus of theologians is not a cause of Magisterium but a result of it. When Vatican I approved the constitution *Dei Filius* in 1870 and referred to this subject, it made no mention of the consensus of theologians.

Moreover, Tripepi continues, even in the event that the immediate formation of the body of the first man were not a doctrine of faith, it would not therefore follow that it was permissible to deny it. Besides properly heretical propositions, there are others that are *nearly heretical, erroneous,* or *imprudent,* and for a Catholic it is not permissible to sustain them. Tripepi follows Mazzella on this score, supplying only a paraphrase suitable to the context: "Whence we see to what point some people who say they are *Catholic writers, Catholic scientists, Catholic attendees at enlightened science meetings* are incautious and how some are self-deceived. They allow themselves be carried away by their studies, and first establish their theories, without consulting the teachings of revelation, and later strive to accommodate revelation to their theories" (p. 17).[57]

Following Mazzella, Tripepi interprets the two texts of Genesis on the creation of man in favor of his stance, and adds that when Scripture says that after God insufflates the breath of life "man was made a living being," it is affirming implicitly that the "dust of the earth" had no life before: "Therefore, before it had no life; that is, it could not come by evolution from any animal" (p. 18).

Once again paraphrasing Mazzella's text, Tripepi collects other scriptural references that present God's special action in the creation of man. And, the same as Mazzella, he argues that the creation of Eve through Adam's rib, which should be understood literally, implies a special action by God:

> Thus a special and immediate act of God in the formation of Eve is clear in Scripture, as F. Leroy also recognizes in a recent work. But this necessarily casts new light on the creation of Adam by a special and immediate act of God, and not through a natural evolution *by means of* which God would have formed man ... as the moderate transformists would like, along with those who, in this aspect, agree with Leroy. (p. 20)[58]

Tripepi next turns to the opinion of Church fathers, insistently arguing that "the judgment of all the Fathers to exclude the imaginary development of natural forces is energetic, constant, complete, and unanimous; when they speak of the formation of the body of man before it is animate, they all speak of a special and immediate act of God, distinct from the first creation of matter and of natural forces" (p. 21). Here too, the texts that Tripepi presents in favor of this thesis are selected from the corresponding part of Mazzella's textbook.[59] Tripepi goes beyond it, however, asserting that there is a consensus of papal councils and documents, although he provides no concrete reference:

I must add, because it is undeniable, that whenever, not one, but all the Councils and documents of the Roman Pontiffs have had occasion to refer to the origin of the bodies of our first ancestors, prepared to receive the soul, always (and not just—as Leroy would have it—the Council of Cologne in 1860, but *all*, absolutely all of them, whether Ecumenical or particular), have referred to the immediate act of God. *Always and all* the Councils and documents of the Popes. And they never manifest not even the suspicion of the evolutions that nowadays are speculated. In this there is no doubt nor exception of any kind. (p. 24)

For Tripepi, the conclusion is clear. He cannot understand how Leroy could say that Scripture openly favors the system of evolution. Nor how this case could be compared with that of Galileo, by those who hoped that evolution would finally triumph. For Tripepi there is a great difference between the two cases, because in that of Galileo, although there was an error in the way he presented his theory, it was not in itself punishable: "He found support not only in scientific arguments of fact and reason, but also in many well-interpreted scriptural texts, in ancient Fathers and Doctors, Pontiffs and theologians, even some before Galileo" (p. 26). This assertion is surprising, inasmuch as the prohibition of Copernican books in 1616 was in fact motivated by their presumed opposition to Scripture: in that case, the decree of the Congregation of the Index says so explicitly.

Before concluding his examination, Tripepi interrupts the theological argument to recall that several years before there had already been a similar prohibition: "I must not omit that my reflections up this point were confirmed in the prohibition that the Congregation of the Index ordered of a book by the priest Caverni, *On New Studies of Philosophy.* On this book, the late Cardinal Zigliara wrote a learned report, in 1878" (p. 27). Caverni, like Leroy, was a moderate evolutionist, as we have already seen. Tripepi recalls a central theme in Zigliara's old report: the theory of evolution is absurd, and even its moderate version "falls into the trap of pantheism-materialism in its embryogenic form, and the snares of Hegelianism" (p. 27).

Tripepi includes the testimony of several modern authors close to Leroy who criticize evolutionism and Mivart's approach to it: the Jesuit Devivier, author of a widely diffused apologetics course; a "learned scientist of the *Dublin Review,*" whose name he does not indicate; Mazzella; the Jesuits Domenico Palmieri and Joseph Knabenbauer; and Cardinal Zeferino González, who died November 29, 1894, a few days before Tripepi finished writing his report. Nevertheless, Tripepi recognizes that, although some Catholics oppose the evolution of the human body, they do not go so far as to say that it is contrary to revealed doctrine, so long, at least, as the Church does not render a determined judgment on it: "They think that the theory of Mivart and Leroy may clash with Catholic doctrine, but that this is still not demonstrated, and that tradition perhaps cannot resolve the question, inasmuch as in the epoch of the Church Fathers [evolution] had not been proposed as it is nowadays" (p. 30).

This is the case of the Jesuits Dierckx and Corluy, and of Cardinal González: although they refuted evolutionism, they did not want to condemn it so long as the Church had given no clear definition with respect to it. But Tripepi is not comfortable with such a stance:

> They all speak that way because they believe that the Church lets pass or tolerates books that uphold such opinions. It seems to me, with all that has been said, that the doctrine generally professed by the Church is clear. It is also certain that some books, at the time they were denounced, had already been censured, like that of Caverni, which promoted a moderate evolutionism with respect to the origin of man's body. . . . And I conclude that, even though these matters have not yet been defined dogmatically, transformism, if it is limited only to the origin of the human body, prepared and destined to receive the soul, does not seem reconcilable with Catholic doctrine. (p. 31)

The Value of Transformism

The analysis of the second question (the value of transformism) is shorter. From the start, Tripepi does not hide his negative attitude toward evolution, which, as he says, is a theory that is being abandoned even by the oldest and most famous followers of Darwin and is considered a mere hypothesis, lacking any solid foundation (p. 31). An examination from a rational perspective demonstrates, according to Tripepi, the falsity of the evolutionist system:

> If one considers it only as an aspect of human natural science, leaving aside the teachings of revealed doctrine, this system can be considered and shown to be false. Furthermore there are many contradictions among its followers; its arguments can be considered gratuitous hypotheses and prove nothing either in regard to *facts* or to *logical* reasoning. The system is contrary to great and firm philosophical principles and to facts which show that specific types [species] are absolutely irreducible, and varieties of a species should not be confused with transformations of species. (p. 31)

Tripepi bases his assertion on a series of authors, many of them philosophers: Mazzella, Brucker, Rossignoli, Ballerini, Vigouroux, Bonniot, De Mandato, Agassiz, and Lavaud de Lastrade. The last-named philosopher, Tripepi says,

> demonstrates at length that the doctrine which teaches that animal and plant species can change from one to another through a slow transformation, viewed scientifically and independently of its consequences, is a false doctrine that must be rejected, because it contradicts: 1. experience and observed facts; 2. history; 3. paleontology; 4. common sense; 5. the most important naturalists. (p. 35)

Tripepi next expands his list of naturalists, including even some supporters of evolution, who have acknowledged the problematic nature of evolutionary theory: Lyell, Haeckel, Huxley, Buchner, Moleschott, Darwin, Agassiz, Thomson [Lord Kelvin], Wright, Elam, Tyndall, Bischoff, Aeby, Frédault, Cuvier, and Quatrefages. The promiscuous mixing of scientific authorities—dead and alive, pro- or anti-Darwin, Protestants and Catholics—was typical of apologetic literature in the 1890s, when Darwinism was still in "eclipse" and there was common confusion over the possible mechanisms of evolutionary change.

Finally, Tripepi cautions that the acceptance of spontaneous generation by scholastic philosophers should not be considered as supporting evolutionism, and he reproduces, though not literally, a text of Mazzella in which, with some quotations of Saint Thomas and Suárez, it is shown that the scholastics thought that higher animals can only be generated by individuals of the same species (pp. 36–37).

Turning to theological arguments, Tripepi asserts, first, that the theory of evolution is not in agreement with the natural sense of Scripture. In the Creation narrative, even though the creation of each species is not specified, a direct act of God creates animals and plants. And, in any case, it says that God created the plants and animals "according to their kind." For Tripepi this is incompatible with the hypothesis of the transformation of species, and he adds that the Church fathers and doctors—Saint Augustine, Saint Thomas, and Suárez—understood just that (pp. 37–41).[60]

Two other arguments raised by Tripepi against evolution seem not to have been taken from Mazzella's textbook. First, no being tends naturally toward his own destruction; but the transformation of species means exactly that, inasmuch as a change of essence destroys the preceding being. And, second, transformation of species would mean that living beings would have forces or powers beyond their own nature (pp. 41–42).

Still, Tripepi admits there is a difference between man and the other organic species. With respect to organisms inferior to man, Tripepi recognizes, following the Jesuit Devivier, that the expressions used in Scripture, such as "that the earth might germinate," "that the waters might produce," "that the earth might produce," leave open the possibility that secondary causes might have contributed to the formation of organic species. But "the primordial determination of species by means of the act of God the Creator seems to indicate, rather, a law whereby each species is, from that moment, fixed and immutable"[61] Other Jesuits, like Ignace Carbonelle and François Dierckx, also agree that revelation does not exclude the hypothesis of the transformation of organic species inferior to man and that only science can resolve the question. Thus, Tripepi does admit that in the case of beings lower than man there is more room to accept evolution; here, he cannot find definitive arguments (pp. 43–44).

Exegetical Criteria

Finally, Tripepi confronts the first question posed in the Congregation: what exegetical criteria should be followed, according to Leroy, in the interpretation of Genesis? He sees no difficulty in accepting, with Leroy, that the style of the first pages of Genesis is not that of a book of science and that the six days of Creation can be understood in an allegorical sense: this doctrine had already been sustained by some Church fathers. But Leroy exceeds those criteria and, following the Oratorian biblical scholar Hyacinthe de Valroger (1814–76), considers that Genesis is poetic in style and thus should not be interpreted literally. For Tripepi, Genesis is a historical book, and so it is false to say that it explains nothing about how God has created the diverse species, and he insists that the Creation narrative speaks with total clarity of the plant and animal species (p. 46). Neither can the truth of the Genesis narrative be disqualified by alluding to its metaphorical language, as Leroy does. Every language, including that of science, needs to use metaphors: "But it is completely different to say that there are metaphors in the first pages of Genesis, and to say that these pages are metaphors, *ces métaphores*, so that the literal meaning would be irrational. Not this: these are two immensely different things" (p. 47).

Tripepi emphasizes that the literal meaning can be maintained in the first pages of Genesis. He cites the encyclical on the interpretation of the Bible, recently issued by Leo XIII, where it is recalled that the hermeneutic rule is that the literal meaning of Scripture should not be abandoned, unless it leads to a patently absurd conclusion, or when other scriptural passages or the traditional interpretation indicates an exclusively allegorical meaning, and he adds: "Neither of these two cases applies here; thus it cannot be said that it is irrational to take the first pages of Genesis literally. If Creation is accepted, it cannot be said it is irrational to admit the particular mode given by the literal sense. . . . God could create. Could he not have created and acted in the way indicated by the literal sense? Who might dare to say so?" (p. 47).

For Tripepi the literal sense should be followed, the conditions for allowing an allegorical one not having been met. As for the problem of days or epochs, both interpretations are acceptable, but not as concerns animal and plant species. Tripepi concludes: "On this point, up to now the Church Fathers and theologians are morally agreed to exclude evolutionary theory and take the words of Scripture in a literal sense, for that is precisely the most natural, obvious and proper sense, as the late Cardinal Zigliara also demonstrated, in May 1878, in his learned report on a book by Caverni" (pp. 48–49).

Tripepi's formulation of this hermeneutical rule is much stronger than that of Leo XIII himself, who, in his recently published encyclical *Providentissimus Deus*, had written:

But he [the exegete] must not on that account consider that it is forbidden, when just cause exists, to push inquiry and exposition beyond what the fathers have done; provided he carefully observes the rule so wisely laid down by Saint Augustine—not to depart from the literal and obvious sense, except only where reason makes it untenable or necessity requires (Saint Augustine, *De Gen. ad litt.* I, viii, c. 7, 13); a rule to which it is the more necessary to adhere strictly in these times, when the thirst for novelty and unrestrained freedom of thought make the danger of error most real and proximate. Neither should those passages be neglected which the Fathers have understood in an allegorical or figurative sense, more especially when such interpretation is justified by the literal, and when it rests on the authority of many.[62]

Moreover, in the same encyclical, glossing Saint Augustine and Saint Thomas, the relationship between scriptural interpretation and the physical or natural sciences is considered:

There can never, indeed, be any real discrepancy between the theologian and the physicist, as long as each confines himself within his own lines, and both are careful, as Saint Augustine warns us, "not to make rash assertions, or to assert what is not known as known" (Saint Augustine, *In Gen. op. Imperf.* ix, 30). If dissension should arise between them, here is the rule also laid down by Saint Augustine, for the theologian: "Whatever they can really demonstrate to be true of physical nature, we must show to be capable of reconciliation with our Scriptures; and whatever they assert in their treatises which is contrary to these Scriptures of ours, that is, to Catholic faith, we must either prove it as well as we can to be entirely false, or at all events we must, without the smallest hesitation, believe it to be so" (Saint Augustine, *De Gen. ad litt.* i, 21, 41). To understand how just is the rule here formulated we must remember, first, that the sacred writers, or, to speak more accurately, the Holy Ghost "Who spoke by them, did not intend to teach men these things (that is to say, the essential nature of the things of the visible universe), things in no way profitable unto salvation"(Saint Augustine, ibid., ii, 9, 20). Hence they did not seek to penetrate the secrets of nature, but rather described and dealt with things in more or less figurative language, or in terms which were commonly used at the time, and which in many instances are in daily use at this day, even by the most eminent men of science. Ordinary speech primarily and properly describes what comes under the senses; and somewhat in the same way the sacred writers—as the Angelic Doctor also reminds us—"went by what sensibly appeared" (*Summa Theologica*, I, q. 70 a.1 ad 3), or put down what God, speaking to men, signified, in the way men could understand and were accustomed to.[63]

Tripepi's Conclusions

Before concluding his report, Tripepi acknowledges that Leroy's book contains much that is good: not only its good intention, which is not in doubt, but also his critiques of radical, materialist, or atheist evolutionism, as well as his arguments in favor of Creation, and of the spirituality and immortality of the human soul. He recalls the elegies addressed to him by Lapparent and Monsabré inserted at the front of his book (although he observes the elegies were of the first edition, in which there was practically no discussion of man's body, and he thinks they are not relevant to this second edition). He adds that these positive values, rather than redeeming the book, could make it more dangerous by favoring the diffusion of the errors it might contain (pp. 49–60).

What measure should be taken? For now, Tripepi says, the task is to judge the book and not the doctrine of evolution itself, although that could happen in the future. Two reasons lead him to think that it is necessary to take some measure. The first is the confusion and even scandal that is caused among so many of the faithful who read these doctrines so contrary to the natural sense of Scripture. Proofs of this, he says, are the recent protests against Antonio Fogazzaro's attempt to reconcile Catholic doctrine with evolution (for example, critical articles in *Civiltà Cattolica* and *Scuola Cattolica*). Moreover:

> The second reason arises from the need to restrain a certain, most deplorable freedom of thought and teaching that has insinuated itself among some Catholics, who call themselves scientists, in several countries, especially France, in Congresses called Catholic, and in *Institutes* or Universities for Catholics and even for clerics. Just recently we have seen some of these scientists, including professors and even rectors of Catholic universities, continue to reduce Scripture to a kind of tissue of errors . . . in passages that might have some relationship with the natural sciences. (p. 51)

There are clear allusions here to the Institut Catholique in Paris and to the International Catholic Scientific Congresses, held in Paris in 1888 and 1891 and in Brussels in 1894, and above all to the "biblical question," which in 1893 had reached a high point, precisely in the environs of the Institut Catholique in Paris. The rector of this institute, Monsignor d'Hulst, had published an article in *Le Correspondant* on January 25, 1893, titled "The Biblical Question," in which he tried to counter the criticisms of the way Alfred Loisy and others taught the biblical sciences at the institute.[64] The maneuver was unsuccessful, and Loisy was dismissed from his chair on November 15, 1893. Three days later, *Providentissimus Deus* was published. In this encyclical, one of the theses supported by d'Hulst, called "limited inerrancy," was explicitly rejected,[65] even though attempts to place d'Hulst's works on the *Index* failed.[66]

Therefore, Tripepi's concerns were not limited only to the problem of the formation of the body of the first man. He was also reacting to the fear that the Catholic world, obsessed with modern science, might progressively yield the truths of the faith. For this reason, he adds: "There are Catholics who value and fear so-called modern science *too much*, and to whom it seems that because of it they can almost put aside our beliefs. They concede too much to the word of man, and try too hard to associate it with the word of God, robbing it, at times, of practically all its meaning" (p. 52).

In such a situation, Tripepi concludes, it is necessary to take some measure against Leroy's book. The cardinals must decide whether to prohibit the book, or only to warn the author through his superiors and order him to withdraw the edition from sale and to retract the censured theories. This second option would have the advantage of showing consideration for the good intentions and intellectual and moral qualities of Leroy, who would not see his book explicitly condemned, while other books like it, that have not been denounced, circulate freely (p. 54).

The Second Preparatory Congregation, January 17, 1895

Tripepi completed his report on December 8, 1894. The next Preparatory Congregation was set for Thursday, January 17, 1895, again at the residence of the secretary, Cicognani, in the General House of the Dominicans, on Via San Sebastiano.

This time the attendance of consultors of the Congregation was especially high: fifteen persons appeared, including Cicognani and Pierotti, master of the Sacred Palace. The number of books to be examined was also high—works by six authors, including the complete works of Emile Zola, pursuant to a decision of the preceding General Congregation. Leroy's book was fourth on the list. Reports were to be heard from Ernesto Fontana and Luigi Tripepi; however Fontana was unable to attend, probably because of appointments in his diocese. His report may have been presented by Cicognani, the usual solution when a consultant could not participate in the Congregation.

Six of the consultors present (Tripepi, Corrado, Arcangeli, Buonpensiere, Boccaso, and Domenichelli) had participated in the previous Preparatory Congregation, and were already well acquainted with the case, as were Cicognani and Pierotti. Guglielmo D'Ambrogi, a consultor who attended the last time, was absent.[67] The new participants were Giuseppe Pennachi, the Dominican Pio Tommaso Masetti, the Capuchin Giacinto da Belmonte, Alfonse Eschbach, Josyf Sembratowicz, the Jesuit Franz Xaver Wernz, and Tito Cucchi.

Some biographical data helps to situate these consultors. Tito Cucchi was a

secular priest, named recently, at the same time as Buonpensiere and Domenichelli. Buonpensiere, who was to play an important role in the case, was one of
the youngest consultors, born in 1853, rector of the Dominican Studium in
Santa Maria sopra Minerva (one of the traditional seats of the Holy Office, where
Galileo's public retraction had taken place). Sembratowicz had been archbishop
of Leopoli (Lvov, Ukraine) and since 1882 had occupied the titular see of Teo-
dosiopoli; he was also a consultor of Propaganda Fide on matters related to his
own Eastern rite. His attendance at Congregations of the Index was not very regular: out of a total of thirteen meetings held between 1894 and 1899, he participated in only three. The Dominican Tommaso Masetti participated in this Congregation as a reporter for one of the accused books, an anti-Jesuit tract, but did
not participate in the subsequent Congregations, at least in the period covered
here, although he continued as a consultor. Eschbach, at the time rector of the
French Seminary, had been proposed by Vannutelli as a possible substitute for
Fontana as author of the second report on Leroy. Another consultor with poor
attendance at Congregations in this period was Da Belmonte, while Pennacchi
and Wernz participated regularly.

As was the custom, the result of the debate over Leroy's book was recorded by
the secretary, Cicognani: "As for number IV, all judged: the doctrine, as it is found
in the book, should be proscribed. Let the author be invited, via the Master General [of the Dominicans], to retract it publicly, at his own initiative."[68]

Inasmuch as Fontana was absent, it was probably Tripepi who guided the discussion, which, in any event, was not as simple as it appears in Cicognani's note.
In the Archive of the Congregation of the Index, there is another document describing this meeting —a memorandum prepared by Cicognani for the cardinal
prefect, Serafino Vannutelli, informing him about the debate in greater detail. It
is a manuscript folio written in the distinctive calligraphy and style of Cicognani,
who, as previously noted, was in the habit of using only the right half of the
page for his personal summaries, reserving the left side for notes.[69]

The text of this memorandum is short but interesting enough to reproduce
here in full:

> There was a heated debate in the Preparatory Congregation over Leroy's work.
> The rationales pro and con were calibrated by the Consultors with great pru-
> dence. It was first proposed to conclude with a *moneatur* [warning] to the author.
> The current resolution was because the doctrine of the book (or, better, evolu-
> tionism) is opposed by science and faith. Evolutionism is condemned by true *Onto-*
> *logical* and *Empirical* science. In ontology it is demonstrated that the essence of
> any object is an immutable *type*, that is, *incapable of any change* [evoluzione],
> *whether toward the greater or toward the lesser.* In empirical science there is *Hy-*
> *bridism*, which maintains living species distinct. Therefore, the evolutionist sys-

tem is impudent and anti-Christian, particularly when applied to the body of man, inasmuch as the Fathers and Scripture have a language for the formation of man's body that proves that it was formed immediately by God.[70]

We do not know with certainty what the rationale was for this memorandum. The prefect of the Index, like the rest of the cardinals, was informed of the results of the Preparatory Congregation in the customary way: the "informative sheet" attached to the letter of convocation of the General Congregation. In this case a General Congregation was convoked for Friday, January 25, at 9.30 in the morning in the Apostolic Vatican Palace. The notice of the convocation included the results of the Preparatory Congregation held eight days prior. The conclusion of the Leroy case was reproduced with the same words that Cicognani had used in the Diary of the Congregation.[71] But, on this occasion, perhaps owing to the complexity of the case, Cicognani decided to provide the cardinal prefect with supplementary information, or perhaps, after having described to him orally what had taken place, the cardinal himself solicited a short written summary. Either way, this supplementary information indicates that in the meeting of the consultors some new event occurred that was not clearly reflected in the overly formulaic customary documentation.

What could such an event have been? Cicognani's memorandum for the cardinal prefect indicates that a "heated" debate took place. Although it is just an educated guess, we can imagine who the protagonists were. It is possible that Domenichelli, author of the first report, which was in Leroy's favor, insisted that there was nothing contrary to faith in Leroy's book. Domenichelli had in his favor the new report by Fontana, who said the same thing. And, reading Tripepi's report, one could have concluded that he, after asserting that Leroy's doctrine was unacceptable, always recognized that it had not been officially defined as such, and that some authorities, like Cardinal González, had declined to condemn it, even though considering it erroneous.

Perhaps Cicognani also subscribed to this position, at least initially, as indicated by the memorandum he had written several months before, on the occasion of the first examination, where he demonstrated a disposition favorable to Leroy. It is certain that his position, as secretary of the Congregation, was administrative, obliging him to faithfully report the results of the discussions, as the second memorandum, with a verdict against Leroy, shows. But Cicognani had also been a consultor of the Index before he was named secretary, so his role was not only administrative.

Although Tripepi was strongly opposed both to evolution and Leroy's version of it, his final position was, in some sense, less harsh than the decision finally reached in the Particular Congregation. In effect, Tripepi had offered two possibilities: prohibit the book or warn the author, inviting him to retract. He favored the second, more benign outcome, out of consideration for the author and his order. The final decision of the consultors stated that "the doctrine, as it is found in the book, should be proscribed" and the author invited to retract.

A New Report: Enter
Buonpensiere

The key to understanding the progressive hardening of the proposals of the consultors is contained in another document from the Archive of the Index. This is another (the fourth) report on Leroy's book, signed by the Dominican Enrico Buonpensiere. What is surprising is that this report was written *after* the Preparatory Congregation, for it was signed on January 21, 1895. This was clearly not one of the official reports of the Congregation, because those were always printed for distribution to the consultors and cardinals. This is a short, handwritten document, eight pages long.[72]

Why did Buonpensiere write this report? What use was made of it? The explanation is found right at the beginning, where Buonpensiere writes: "Reverend Father: I place before you in writing all that I said orally about the work of Father Leroy in the Congregation that Your Most Reverend Father presided over on the morning of the 18th [*sic*] of this month."[73] We can thus deduce that Buonpensiere was one of the consultors most active in the debate and that his position was completely opposed to Leroy. He presented new arguments that appear to have been decisive in the fashioning of the final agreement among the consultors.

Buonpensiere argues that evolutionism is, beyond any doubt, a completely false theory from the scientific perspective, because it contradicts "ontological and empirical" science:

> Evolutionism, as all Catholic philosophers teach, is resolutely condemned by *onto-logical* and *empirical* science. In *ontology*, the *essence* of any object is an *immutable type*, that is, *incapable of any change* [evoluzione], *whether toward the greater or toward the lesser.* In empirical science there is an inexorable law of *hybridization*, which maintains living species distinct, in such a way that from the pairing of two organisms belonging to *different species* no fruit will be obtained, or else such fruit is totally *infertile*.[74]

This is close to the wording that Cicognani used in the memorandum for the cardinal prefect and reflects the importance he now attached to Buonpensiere's opinion. It may well be that Cicognani asked Buonpensiere to commit to writing the arguments he originally expressed in the debate.

Hybridization

For Buonpensiere the law of hybridization was definitive. It showed, from an empirical perspective, that species are immutable. Therefore, how could there be evolution? If species arose by evolution, why were there no intermediate species, the links that join two species that directly succeed each other? Consequently, Buonpensiere continues: "evolutionism, opposed by all philosophical science, cannot even be called a *hypothesis:* it is a simple, more or less Platonic, *desideratum* of materialism."[75]

Buonpensiere seems to have thought that evolution would have to come about by a mechanism similar to hybridization, which would give rise to the simultaneous existence of various stages of the same species. But if from the pairing of two species there is either no progeny or progeny that are sterile, how could the evolution of species occur?

Consultors who found against the theory of evolution typically did so on the basis of Scripture and Church tradition, focusing on a very few, narrowly defined issues such as the origin of Adam, or whether Adam's body, along with his soul, had to be created directly, immediately, and simultaneously by God, or whether his body might have been previously prepared to receive a soul by a natural process like evolution. Virtually the only scientific argument ever introduced by anti-Darwinians was hybridity. Such arguments typically conflated two separate phenomena: the sterility of crosses (when the cross of two species does not yield progeny) and hybrid sterility (when the progeny of such a crossing, when bred with another, is sterile). In either case such reproductive isolation was thought to be both a proof of the existence of species and a proof of the permanence, that is, fixity, of species.

In *Origin of Species,* Darwin notes that the received view among naturalists is that "species, when intercrossed, have been specially endowed with the quality of sterility, in order to prevent the confusion of all organic forms." On the other hand, to judge by the evidence from botany, the sterility thereby produced is so different in degree that "for all practical purposes it is most difficult to say where perfect fertility ends and sterility begins." Animal hybrids seem more sterile than those of plants, and some hybrid varieties are more fertile than their parents.[76] Thomas H. Huxley had held that natural selection could not be established as a *vera causa* of the origin of species until artificial selection had been shown capable of producing "physiological species," that is, varieties of a species that were cross-sterile.[77] Darwin took Huxley seriously and performed a series of spectacular experiments on di- and trimorphic plant species—plants that had flowers of two or three distinct morphological types. What he found was that such flowers could be fecundated by pollen from an opposite type, and surely, Darwin suggested to Huxley, he didn't think that there were two species growing on the same plant.

Darwin had also observed that there were no hard and fast rules regarding intersterility: hybrid progeny were sterile in some plants, vigorous in others. "Taking a general review of the facts now given of the infertility both of first crosses & hybrids—we see a most insensible gradation from absolute sterility to high or perfect fertility."[78] This observation has been amply documented since Darwin's time. For example, ice age glaciers divided many European species into two geographically isolated populations. When the glaciers receded and these populations came into contact again, some pairs could not interbreed and were considered distinct species, whereas some merged easily. And in some cases intersterility was only partial.[79]

Hybridization, as understood by nineteenth-century breeders, referred to the crossing of individuals that differed from one another in terms of taxonomy as then understood. Plant breeders did in fact hybridize different species. The problem comes from an overly facile assumption that what works for individuals can be applied to whole populations. When one refers to hybridization in natural populations, the reference is usually to interbreeding of closely related species that share the same habitat but are reproductively isolated (sympatric or sibling species). But this meaning is clearly not what was intended by the term as used in the polemics of the 1890s, where the concept of species was typological, not populational. Moreover, even though reproductive isolation among, particularly, animal species is a fact and biologists frequently still define a species as a population bounded in this respect, that does not mean that the (genetic) contours of such populations are fixed. Evolutionary species change requires only limited genetic plasticity within the breeding group to be able to adapt to a new niche. It is that dynamic of population genetics that gives rise to new species.

Darwin, who at first thought there must be an adaptive purpose in such sterility (that it was helpful to the establishment of newly emerging species), retreated to a view that held that intersterility was not affected by natural selection and was the result of incidental causes. Although there was a distant reflection of some of Darwin's views in the standard explanation of hybrid sterility in Catholic apologetics, the most immediate source was Armand de Quatrefages, a French biologist and correspondent of Darwin who held that it was a way to prevent confusion and chaos in the natural world. "Infertility among species," he wrote (in *Darwin and His French Precursors*), "has, in the organic world, a role which is almost analogous to gravitation in the sidereal world. It preserves the zoological and botanical distance among species, as attraction maintains the physical distances among the stars."[80] John Zahm argues that the exceptions to hybrid sterility found by Darwin's experiments show that exaggerated insistence on hybrid sterility leads to absurdities like declaring the flowers on dimorphic

species to be independent species. For Zahm, hybrid sterility is simply an epiphe-
nomenon of the extreme sensitivity of the reproductive systems of many organisms, so finely tuned that they can fail for a variety of reasons. Zahm rejects the
"indefinite immutability" of species as proposed by Quatrefages and his school,
an explanation that assumes too much. The many instances where scientists are
unable to determine whether certain related groups are varieties of species,
species of a genus, and so forth ought to lead to the realization that the case for
fixity in the natural world has been overstated.[81]

Conclusions of Buonpensiere

Once he has established the total falsity of evolutionism according to "science,"
Buonpensiere arrives at a logical consequence:

> Consequently, the impudence of those who attempt to harmonize evolutionism
> with Revealed Doctrine is reckless and anti-Christian. This is what Father Leroy
> does in the work cited, where he holds that it is not contrary to Genesis but agrees
> with it. He asserts that the body of Adam, instead of having been formed immedi-
> ately by the work of God from moist earth, may have been a kind of matter organ-
> ized in a form more or less close to the human type, through the law of evolution,
> under the impulse of the First Cause.[82]

Moreover, according to Buonpensiere, Leroy's intent is also anti-Catholic and
contrary to the established doctrine of the Church, inasmuch as Scripture and
the fathers use language totally contrary to his. The following pages teem with
learned references: Saint Thomas, Albertus Magnus, Peter Lombard and Gregory the Great, Bede, and Clement of Alexandria. These texts are very similar to
those adduced by Tripepi and the other consultors: they affirm that God is the
only creator, and any other causality is excluded. Nor are the scriptural texts
that Buonpensiere mentions original. His principal argument is very similar to
Tripepi's: although natural causes may, at some level, act along with the action
of God in the creation of animals and plants, the creation of man has but one
efficient (and therefore immediate) cause, that of the Holy Trinity.

Buonpensiere's report is completely opposed to Leroy's work, and his conclusion and recommendation, which must be identical to what he had proposed
four days before in the Preparatory Congregation, are the harshest of those we
have described:

> If, therefore, evolutionism is contrary to Science and Faith, the impudence of Fa-
> ther Leroy has been *truly reckless*, both for defending such absurdities, and for pre-
> tending he believes it is not opposed to Revelation. Thus my recommendation is:
> *Leroy's work should be condemned, or, to act with more benignity, suppressed; and,*

meanwhile, let the author be warned. For me, it would be better that this work be placed on the *Index of Prohibited Books,* in accordance with the rigor of justice.[83]

There is every indication that this report played an important role in the final decisions. In his memorandum for Cardinal Vannutelli, Cicognani used Buonpensiere's ideas and even his exact words. This indicates that at least Cicognani, secretary of the Congregation, identified the results of the Preparatory Congregation with Buonpensiere's arguments.

Here we might pause to comment on the interpretation of these events proposed by Barry Brundell. Referring to Cicognani's memorandum (which, we have noted, he considers to have been written by Vannutelli), Brundell sees in one of the first expressions in this text—that evolutionism is condemned by "ontological and empirical science"—proof of the influence that the Jesuits of *Civiltà Cattolica* had over the decisions of the Congregation of the Index:

> The first evidence of the CC [*Civiltà Cattolica*] group's involvement in these proceedings is found in the very wording of the memo of the cardinal prefect, specifically in the fact that it reflected articles in CC that had been written by Fr Francesco Salis Seewis, SJ, and which Fr Brandi, SJ, who was very active in Rome at the time and who had earned the rebuke from Pope Leo XIII already mentioned, cited as primary references for CC's "demonstration of the incoherence of evolution theory with traditional Catholic teaching." CC had been arguing for decades that the doctrine of evolution is "contrary to science and faith," so much so that this was almost a refrain in their writing, but only Salis Seewis had used the terminology "ontological" and "empirical" in the discussion of evolution theory; one does not find these terms in other Catholic writings on the subject. The fact that these very terms are found in the cardinal's memo is a firm pointer to their origin in CC and to the influence of the CC group on the outcome of the Leroy investigation.[84]

Brundell seems not to know of Buonpensiere's report that we just considered, because expressions in Cicognani's memorandum are taken from this report; thus, they originated in arguments used by Buonpensiere in the Particular Congregation and were then put into writing afterward, in his report of January 21. Furthermore, the articles by Salis Seewis that he mentions are later, not appearing in *Civiltà Cattolica* until 1897. Also, his assertion that only Salis Seewis had used the terminology "ontological and empirical" in the discussion of the theory of evolution is puzzling, because that terminology does not appear in the articles that Brundell cites.

The Second General Congregation,
January 25, 1895

The General Congregation met eight days after the Preparatory, on Friday, January 25, 1895. It was quite well attended, especially in comparison with the preceding one, for ten cardinals attended. The five cardinals who had been present at the previous General Congregation, in which Leroy's book had been discussed the first time, were also present on this occasion: namely, cardinals Vannutelli (prefect), Mazzella, Parocchi, Verga, and Granniello. Another three cardinals attended the General Congregation for the first time, having all been named members of the Index a few weeks before, on December 14, 1894: cardinals Steinhuber, Galimberti, and Segna.

The Jesuit Andreas Steinhuber (1825–1907) became cardinal *in pectore* on January 16, 1893, and his appointment was made public only on May 18, 1894. Steinhuber had entered the Society of Jesus in 1854, just after his ordination as a priest. He had been professor of philosophy and theology at Innsbruck, rector of the German-Hungarian College in Rome, and consultor of some Roman congregations, the Holy Office among them. Later on he became prefect of the Index.

Luigi Galimberti (1836–1896) was a Roman and had been nuncio in Austria from 1887. On January 16, 1893, he was created cardinal, and on June 25, 1894, he was named archivist of the Apostolic See.

Francesco Segna (1836–1911), after having been professor of dogmatic theology in Sant'Apollinare in Rome, had occupied various positions in the Roman congregations and in the nunciature of Spain. He was created cardinal at the consistory of May 15, 1894. Since June 20, 1893, he had occupied the post of assessor of the Holy Office, a position of great importance, because it made him the principal collaborator of the cardinal secretary of the Holy Office. This made Segna one of the principal authorities on doctrinal questions in the Roman Curia. Somewhat later, in 1908, he became prefect of the Congregation of the Index.

Thus, of the ten cardinals who participated in this meeting, four were prefects of the Congregation of the Index: Mazzella held the post from 1889 to 1893; Vannutelli from 1893 to 1896; Steinhuber from 1896 to 1907; and Segna from 1908 until 1911.

Two other cardinals participated in this session: the Jesuit Paul Melchers (1813–95) and the Capuchin Ignacio Persico (1823–95). The former was the eldest of the cardinals present, having just reached ninety-two. He had been archbishop of Cologne for almost twenty years, more than half of which had been passed in exile, after he had spent several months in prison owing to his opposition to the German *Kulturkampf*. In 1885 the pope accepted his resignation as archbishop of Cologne, but called him to Rome and made him cardinal. At the age of nearly ninety, he became a Jesuit. Ignacio Persico had been a bishop in India and in the United States, and apostolic delegate in Canada and Ireland. He had been cardinal from January 16, 1893, and was prefect of the Congregation of Indulgences and Relics.

The discussion on Leroy's book occupied the fourth place on the agenda of the meeting. As in the other cases, we know the results of the Congregation from the annotations of Secretary Cicognani in the Diary and from his personal memorandum for the audience with the pope that always followed the Congregation. The Diary note is, as was the custom, a tight synthesis:

> About no. 4°, nine said that it had to be *Proscribed,* but the decree did not have to be published. It would be communicated only to the Father General of the Order who, in the name of the Congregation, might indicate public retraction to the author; because the doctrine that he propounds in the book, according to the judges of the Roman Congregations cannot be taught or sustained; and that he withdraw the copies in the best possible way. One said: "It does not have to be condemned, but only warned via the Father General."[85]

This brief note indicates that complete unanimity was not achieved in the decision of the General Congregation. The final decision of the majority of the cardinals reflects a hardening with respect to what we saw in the second series of reports: it is said now that the book's doctrine "cannot be taught or sustained." But it also reflects indecision: the decree need not be published. The cardinals see a danger in Leroy's ideas but decide not to prohibit the book.

The decision was simply to have the author of the book retract his opinions, as if under his own initiative. The formula employed seems to indicate, moreover, that the retraction be asked of him based on the authority of the judges of the Roman congregations, so as not to involve the pope (although counting on his approval). It is not clear why the decision refers to "Roman Congregations" in the plural, inasmuch as the entire procedure and the decision were solely those of the Congregation of the Index.

Perhaps the most significant point, however, is the dissenting opinion of one of the cardinals present who, up to the last minute, held that, even though it was appropriate to warn the author, there was not sufficient reason for a condemnation or prohibition of any type. Who was the dissenter? Thanks to the memorandum that Cicognani prepared for his audience with the pope, we know with certainty. This time Cicognani prepared his memorandum after the General Congregation.[86] It is three pages long and begins with the words: "Audience of January 26, 1895. Informative Page of the General Congregation of January 25, 1895." On the second page Cicognani writes the results of the General Congregation with almost the same words he had used in his Diary. But, in noting the dissenting opinion, Cicognani wrote the name of the cardinal in question in the margin: "Em. Segna."

This identification is important, because of Cardinal Segna's position as assessor of the Holy Office. The Holy Office was the congregation whose mission was to look after Catholic doctrine, and it had the authority to decide whether a doctrine could be accepted or should be rejected, and in what measure. The mission of the Congregation of the Index was simply to examine, and in some cases prohibit, those books which sustained doctrines already prohibited or that were considered dangerous. Their point of reference was, besides the definitions of solemn Magisterium of the popes and councils, and of ordinary Magisterium, whatever the Holy Office judged or decreed about erroneous, false, or dangerous doctrines. The fact that the assessor of the Holy Office voted against condemning Leroy's book is highly significant.

We have some more indications about proceedings in the General Congregation in another document from the archive. Among the material relating to this second General Congregation is a copy of the report that Domenichelli wrote for the first Congregation.[87] This copy was used by the prefect, Cardinal Vannutelli, during the meeting of January 25, and it has some handwritten annotations made by the secretary, Cicognani, and by Vannutelli himself. When he gave the copy to Vannutelli, probably a few days before the second General Congregation, Cicognani added on the last page what the decision of the previous General Congregation had been. The prefect wrote two more notes on the first page, perhaps after studying the material, or even during the discussion itself in the General Congregation. In the right-hand margin he noted, in four short lines: "Caverni / 1878 / ? p. 27 / ? Tripepi." This is a reference to Tripepi's report, which on page 27 mentions the prohibition of Raffaello Caverni's book *About the New Studies of Philosophy.* Caverni's name is written in bold strokes, as if in a particular moment; perhaps when Tripepi presented his report, or during the discussion, Vannutelli had suddenly recalled there already existed a book that had been prohibited because of evolutionism.

Vannutelli's second note was perhaps written at the end of the General Congregation, for it records what the final decision should be. But it might also indicate some of the indecisions and changes of mind that occurred during the discussions. In the left-hand margin of the first page, Vannutelli writes (sideways): "Condemn—have the Decree written—do not publish—tell the [Master] General to order public retraction (since already prohibited) / withdraw the copies and [illegible]."[88]

These words could be the first draft of the Congregation's final decision, written by Vannutelli during the same meeting. The decision seems to have been modified. When first he wrote: "have the Decree written," perhaps the prevailing opinion was to publish the condemnation of the book, as Buonpensiere had proposed. Vannutelli would have noted "have it written," because it was customary, in the case of Catholic authors, not to publish the decree immediately, but rather first ask the author to accept the prohibition, in which case the words "he laudably submitted" were added to the decree. As debate continued, possibly due to the intervention of Cardinal Segna, it was decided not to publish the decree, when Vannutelli wrote between two lines: "do not publish."

The final verdict of the Congregation was a kind of compromise. The book was condemned, but the condemnation was not made public (meaning there was no "decree" as such). The task of rectifying his opinions through a public retraction was left to Leroy, who was also to do what he could to withdraw from bookstores the copies of his book that were still for sale. In this way, without an official condemnation, it was made known that Leroy's book had been judged negatively by the competent authority.

As we will see, the text of Leroy's retraction mentioned neither the Congregation of the Index nor the Holy See; nor did it specify which "competent authority" in Rome had judged the work negatively. As a result, one might have thought the authority was his order, the Dominicans. On the other hand, it did mention the cause of the retraction: that his thesis "has been judged untenable, above all for that which refers to the human body, which is incompatible both with scriptural texts and with the principles of sound philosophy." If the Congregation of the Index had published its decree, the reasons would not have been indicated; in this sense, the letter of retraction was more explicit and had greater impact than the publication of the decree would have had.

On the day following the General Congregation, January 26, 1895, Secretary Cicognani was received in audience by the pope. In the Diary of the Congregation, Cicognani wrote that the pope "deigned to confirm the resolutions of the Most Eminent Fathers, approved the Decree and ordered it published."[89] Just as the cardinals had decided, the decree, which was published two days later, did not include Leroy's book.[90]

Leroy's Retraction

The Congregation's decision, confirmed by the pope, was then to be communicated to the master general of the Dominicans, Andreas Frühwirth. Cicognani did not delay. On Wednesday, January 30, he noted in the Diary that he had, by resolution of the members on January 25, forwarded the decision to the master general.[91]

Perhaps because he wanted to deliver the bad news personally, or thinking that direct contact between Leroy and the Index authorities would be appropriate, Frühwirth asked Leroy to come to Rome. Leroy himself recalled the event in a letter two years later.[92] Once the news had reached him, Leroy, "somewhat stunned by the blow," decided to retract immediately and wrote an open letter to the French daily, *Le Monde*. The letter, dated in Rome on February 26, 1895, appeared in *Le Monde* shortly after, on Monday, March 4:

Rome, February 26 [1895]

To the Editor:
When Darwinism began to cause a stir, I thought it my duty to study that doctrine from which our enemies hoped to be able to derive a great advantage against the teachings of the faith. When I studied it in depth, it seemed to me that not everything in it was reprehensible. Even in the interest of religion and to better combat error, I believed that what should be done was to separate the chaff from the wheat, with the objective of making what was plausible in the system of evolution serve in the defense of revealed truth.

I have devoted various writings to the analysis and defense of this theory, most importantly a book titled *Evolution Limited to Organic Species*, published in Paris in 1891 by Delhomme and Bréguet.

Now I learned that my thesis, examined here in Rome by the competent authority, has been judged untenable, above all for that which refers to the human body, which is incompatible both with scriptural texts and with the principles of sound philosophy.

As a docile son of the Church, resolved above all to live and die in the faith of the holy Roman Church and, obeying higher authorities in this moreover, I declare that I disallow, retract, and repudiate all that I have said, written, and published in favor of this theory.

Furthermore, I declare that I wish to withdraw from circulation, insofar as I am able, what remains of the edition of my book on *Limited Evolution*, and also prohibit its sale.

In the hope that you might publish this act of retraction in your excellent newspaper, I beg you, Mr. Editor, to accept the expression of my religious respects.

Fr. M.-D. Leroy, O.P.[93]

Although Leroy did not say so explicitly, both the tone and contents of his letter suggest the Holy See's intervention. It almost gives the impression of having been dictated, to ensure that everything that had to be said was expressed with precision: the book's title and publication data, examination by the competent authority, the reasons for the repudiation of the book, the author's good intentions, the motives that led him to write the book, and the complete retraction of everything having to do with the subject. It looks like the Vatican authorities had preferred not to involve themselves directly in an official document and had used, instead, Leroy's retraction to make it known that they disapproved of the attempt to explain the origin of the human body by the theory of evolution.

Beyond this letter, no other documents were known about the Leroy case until

1998. Everything else was filed in the Index archives and was known only to those who had participated in the activities of the Congregation. One of these documents is a press clipping of Leroy's retraction. On the sheet of paper on which the clipping is glued, Cicognani wrote: "Retraction of Father Leroy. Decreed by their Eminences the Cardinals in the Congregation of the Index of January 25, 1895," and he added the reference to the newspaper.[94] On March 21, Cicognani took note of the retraction in his Diary: "Father Leroy retracted in the newspaper *Le Monde* of March 4 the doctrine expounded in his book *L'évolution*, submitted to the Decree in a praiseworthy way, and reproached the book."[95]

Some time after, Cicognani notified the pope that Leroy had retracted. The exact date is uncertain, because the relevant document is a memorandum by Cicognani, which in this case is not dated. There he wrote: "Audience with his Holiness. / 1° Submission of Leroy, Fiolchini and Angelini..."[96] The audience must have taken place in April, because elsewhere on his list he refers to a petition written on March 29.

Leroy's Doubts

The letter of retraction is clear. Leroy was informed of the opinion of the authorities, and he obeyed. He was a conscientious priest; he was not a rebel, nor did he wish to be. Moreover, he was treated as a trustworthy person and allowed access to a critical report on his book. But there is a lingering question: was he really convinced he was wrong?

We know the answer, which is: not completely. His letter of retraction was dated in Rome on February 26, 1895, and was published by *Le Monde* on March 4. In those days, Leroy was trying to understand the meaning of the condemnation of his book. He was living in Rome, in the General House of the Dominicans, where Cicognani, the secretary of the Index, also lived. Leroy had been permitted, with express authorization from the pope, to read one of the reports, the one written by Tripepi.[97] On March 7, Leroy was still in Rome, where he wrote a long letter of fourteen pages to Cardinal Serafino Vannutelli, prefect of the Congregation of the Index.[98] The letter is preserved in the archives of the Congregation: it begins "Most Eminent Lord," which is the etiquette proper to a cardinal, and he spoke in detail about his case, which shows that it was directed to the cardinal prefect of the Index, although he does not name him.

Leroy knew what type of arguments had been debated in the Congregation, and in this letter he attempts to clarify them, not only out of personal interest (as he states) but also for the good of science and of the authors who, because they share his point of view, could believe themselves affected by the prohibition decreed against him; and even, he adds, for the good of religion, because it is to its benefit if its official representatives are well informed about the matters before them. He begins begging the reader's pardon because he is going to make some observations on the decision made on his book. And he adds that he will refer only to the theory of evolution, leaving aside the problem of the human body.

If he has understood, Leroy says, his thesis does not oppose Scripture directly but only indirectly: sound philosophy shows it is false, so it cannot be regarded as a true interpretation of Scripture. But, what sound philosophy is meant? The scholastic, no doubt, because it is used by the Church. The rationale must be that species are different and irreducible; therefore they cannot be transformed into others. But, Leroy objects, philosophy cannot decide this matter, because it is a factual question in the realm of science. Philosophy can evaluate whether scientific observations and conclusions are authentic, but it cannot decide *a priori* that organic species cannot change into others. In his book, Leroy continues, he includes proofs that species can be transmuted; he will summarize them now, limiting himself to generalities about the nature of species and the problem of hybridization. Leroy mentions three scientists (Louis Agassiz, Henri Milne-Edwards, and Edmond Perrier) who attest that the differences between neighboring species are small ones. Besides, the variability of races exceeds these differences. All of this leads him to conclude that, according to "physiology," neighboring species do not possess irreducible essences. He asks the cardinal to consider his arguments carefully and to compare them with those provided him by others, so that the voice of science might be heard.

In this long letter, Leroy only argues that, for those who claim that the evolution of species is impossible for philosophical reasons, science shows that it is possible; and so he concludes that evolutionism is not false, and that it can be harmonized with Scripture. He abstains completely from saying anything about the origin of man.

The Excuses of a Jesuit

In this same month, Joseph Brucker, the Jesuit editor of *Études* and a perennial critic of Leroy, brought out a collection of articles on Scripture, previously published in journals, especially in *Études*. In a preface, Brucker explains the changes that he has made to the articles, now in book form. One of them is of interest here. Brucker comments that some of his articles had originally been somewhat polemical, although, in his opinion, always strictly scientific and within the bounds of courteous discourse. In any event, he has now tried to expound his ideas in a more impersonal way. He has not omitted a discussion of the doctrines, but he has depersonalized it, without naming any living Catholic author, in hopes of avoiding the least suspicion of animosity toward persons whom he respects, without sharing their ideas.[99]

Although Brucker does not mention him, Leroy is implicitly present in this comment. Chapter 4 of the book is titled "The Bible and Transformism" (pp. 215–53), in which he combines the two articles on the subject originally published in *Études* in April and May 1889. In the May article, on the origin of man, he had mentioned Leroy and criticized one of his assertions. The same text appears in the book (on p. 243), but the reference to Leroy is gone.

In any case, there was another matter that must have preoccupied Brucker. It had only been a short while since Leroy had been called to Rome, from where he had published his letter of retraction, a public declaration that his ideas, especially on the origin of man, had been found unacceptable by the competent authorities in Rome. Leroy's polemic with Brucker and the Jesuits of *Études* was well known, and Brucker himself had ended his review of Leroy's book stating that he was not going to press for a review of Leroy's book by the Congregation of the Index in Rome. This happened in 1891. Brucker (as we have seen) had written: "We will never press for the anathema of the Index to fall on a book which is a serious work, displaying a conscientious search for the truth, even though it seems to stray a bit from the straight path." But now, four years later, in 1895, Rome had intervened: it was not known who was behind the move or why, but it was an authority sufficiently strong to cause Leroy to make a public retraction in a Parisian newspaper. Appearances could have pointed to the Parisian Jesuits of *Études* as the cause of Rome's intervention.

There must have been comment in ecclesiastical circles: Brucker's preface says as much. Without naming anyone he says:

> My conclusions have the value of my arguments; I never dreamed of imposing them on anyone except through conviction. I am no "inquisitor," whether with a "mandate" or without one. I am a theologian who writes with the permission of ecclesiastical authority and that of my superiors, but with my exclusive responsibility. I simply draw attention to doctrines that themselves have provoked discussion and require censure, because they are asserted publicly, at times with errors, at times without permission—I don't know. I do not condemn anyone; I have no mandate for that; I have never asked nor advised, directly or indirectly, any authority to act against the authors whom I have criticized. I confess that the ingenuity of people who are dragged along by the anti-Jesuit tide makes me laugh when they attribute to me and my brothers an influence which I have never had, let alone on decisions made in Rome.[100]

Brucker's book was published in 1895. Was this preface written before or after Leroy's retraction? We know it was after, because Brucker dates it: Paris, on the feast of the Annunciation of the Virgin, March 25, 1895—three weeks to the day after *Le Monde* published Leroy's letter of retraction. It could have been a coincidence, but one would have to go to great lengths not to connect this preface with Leroy's retraction. How could Brucker's insistence in disassociating himself from any kind of denunciation or condemnation related to the authors he had criticized be explained in any other way?

If we presume, as would seem logical, that Brucker was telling the truth, it appears unlikely that Leroy's polemic with Brucker, with other Jesuits, and with other anti-evolution Catholic authors would not have influenced the officials in Rome who judged Leroy's work. Evolutionism was making inroads among Catholic authors, and Leroy was the champion of those who sought to harmonize it with Christianity.

The Fifth Report:
Angelo Ferrata

On February 2, 1897, two years after his retraction, Leroy wrote a letter to the cardinal prefect of the Index from Paris. Just as on the last occasion, he does not use the prefect's name, but rather begins "Most Eminent Sir" and uses a similar complimentary close ("of Your Eminence"). The text of the letter is now in the Index archives:

Paris, February 2, 1897

I have the honor to make known to you the following facts. In February 1895 I was summoned to Rome by the Reverend Father Frühwirth, Master General of our Order. When I arrived, I found out that my book on *Limited Evolution* was rejected by the competent authority. There were two roads open to me: either renounce my book publicly, or rewrite my work completely. A bit stunned by the blow, I decided to retract.

Some time later, an article in the journal *Études religieux* attacked me. I thought I should respond, but my situation imposed upon me the duty to have my reply reviewed precisely in Rome. The Jesuit Fathers of the journal criticized my thesis on the body of man above all; I changed it, in the direction that Father Cicognani, secretary of the Index had indicated to me; and I strengthened it with the authority of Saint Augustine and Saint Thomas. The Reverend Father General of our Order, whom I informed of it, had it examined by two Roman theologians who found nothing to reprove. Nevertheless the Reverend Father General informed me that the examiners and he himself did not think it opportune to embark on a polemic with the fathers of the Society of Jesus on this matter. But, he added: rewrite your work, taking into account the advice you have received; that would be the best response.

I have followed his advice; the work is rewritten, but the Reverend Father General does not think that he should assume responsibility for the revision; because the case is in the hands of the Congregation of the Index, I resubmit to it.

Therefore, I humbly beg your Eminence to please let me know if you will have my book, duly corrected, examined by whomever you may choose, so that, if it is the case, I might prepare a copy worthy to present you.

I pray your Eminence receive the homage of all the respect that I profess,

Your most humble and very obedient servant

Fr M. D. Leroy, O. P.

Paris, 94 rue du Bac[101]

The prefect's reply must have been positive, because the following month, on March 13, 1897, Leroy wrote the secretary, Cicognani:

Paris, March 13, 1897

Most Reverend Father,

I have the honor to send you the copy of my book on *Limited Evolution*, corrected in accordance with your instructions. I have tried to comply as best I can, adding the clarifications demanded by your excellence. The Consultors of the Index to whom you may wish to undertake the scrutiny will, after examining it, be able to say whether I have faithfully observed your recommendations.

I pray you receive, my most Reverend Father, the homage of all my respectful feelings.

Fr. M. D. Leroy, O.P.[102]

According to the acts of the Congregation, Leroy's book arrived on March 26, 1897. It is recorded as a "little book" (*libellum*), not to show disdain, but because it was a kind of draft, in which new handwritten pages were mixed in with pages of the old, printed version. One of the consultors, the Augustinian Angelo Ferrata, was charged with its review, as noted in the Diary of the Congregation for March 26, 1897: "Father Dalmace Leroy of the Order of Preachers submitted a *libellum* requesting that he be granted permission to publish again *L'évolution restreinte*, prohibited on January 15 [sic], 1895, completely corrected and emended. The Prefect kindly agreed, and ordered the Secretary to give the book to one of the consultors; it was given for examination to Father Angelo Ferrata of the Order of Saint Augustine."[103]

While Ferrata was reviewing the book, Leroy wrote a letter on April 12, 1897, to Tommaso Maria Granello,[104] recently named commissary of the Holy Office, congratulating him on his appointment. He mentions that he is in touch with the Congregation of the Index in order to obtain authorization to publish a new edition of his book, corrected in accordance with the observations made to him in Rome by the competent authority. He adds that things were moving slowly. He asks Granello's pardon for raising confidential issues, "but the kind and very fraternal way you welcomed me during my disgrace authorizes me, or so it seems to me, to do so." Evidently, Leroy sought to advance his case with this letter and to enlist the mediation of the new commissary of the Holy Office. In a postscript, to make his quest even clearer, he says: "Right now, the manuscript is in Rome, no doubt submitted to the examination of the Consultors of the Index." Granello sent the letter to the Congregation of the Index, where it came to rest in Leroy's dossier.[105]

Just as Leroy had supposed, his manuscript was being examined by the consultors of the Index. It did not get lost in a pile of papers and was the object of careful attention. But the result was still negative. Angelo Ferrata's report was eight pages long and was dated June 16, 1897.[106] Ferrata began by stating that, following his instructions, he had carefully read the changes Leroy had made in his book, and added: "I have found nothing that could change the judgment made of this work. The changes consist of additions which are intended above all to confirm or develop evolutionary theories."

The rest of the report contains the following observations on the text sent by Leroy:

—Leroy says that in Rome the *representative* of the competent authority assured him that "neither the Bible nor the Church teaches anything concrete on the subject of fixity of species."

—Leroy, after citing Zahm in his favor, says that Zahm, in a work titled *Bible, Science and Faith,* which has been translated into French, declares himself openly favorable to evolutionism, *all the elements* of which are found in Saint Augustine, and he adds that when Zahm came to Rome, he was most kindly welcomed by the Roman pontiff, who rewarded him with the title of doctor.

—Leroy observes that the scholastics thought species to be immutable but that this has been corrected by science, like so many other errors. When it was thought that the days of the biblical narrative of Creation were of twenty-four hours, people tended to conceive an immediate and instantaneous Creation, but now that they are interpreted as epochs, a way is opened to evolutionism. Miracles should not be invoked when it is not necessary.

The report includes a few more observations on Leroy's ideas, for example, about matter and form in general, and form in the animal kingdom in particular, and concludes:

What I have said is enough to demonstrate that the author has changed none of the book; indeed, he has added more arguments, including some eccentric ones, to confirm his theory. Therefore, in my opinion, the judgment already made should stand firmly, the work should be proscribed, and its publication cannot be permitted.[107]

Rome, Santa Maria del Popolo
June 17, 1897
Father Angelo Ferrata, Augustinian

The finding could not have been more negative. The Diary of the Congregation for June 19, 1897, reports:

> Regarding Father Leroy's book. In view of the fact that Father Ferrata's report is not sufficiently evenhanded and complete, the Prefect determined that the same book be returned so that another consultor might examine it, and the Secretary, putting into action the wish of the Prefect, entrusted it to Father Enrico Buonpensiere of the Order of Preachers.[108]

Buonpensiere knew Leroy's work, having written a very negative report on it two years before. He seemed to be an appropriate person to determine whether the new version corrected the defects of the previous one.

The Sixth Report: Buonpensiere Again

Buonpensiere's first report is short, focused on a philosophical issue (ontology) and a biological one (hybridity). His second report is long (fifty-six pages, divided into forty-six numbered paragraphs) and mainly addresses the theological implications of Leroy's views. It is signed and dated in Rome, Minerva, August 12, 1897, and it contains an evaluation of the ten chapters of the new volume that Leroy sought to publish.[109] Most important are his comments on chapters 1 (on Scripture, tradition, and Magisterium) and 10 (on the origin of Adam's body). His comments on the other chapters are cursory by comparison.

The first chapter considers evolution in the light of Scripture, tradition, and the teachings of the Church (§§4–17, pp. 4–24). Leroy writes that it is a completely certain fact that the Church had never defined nor taught anything regarding evolutionism. This he had already stated verbatim in the edition of 1891 (p. 33). According to Buonpensiere, Leroy goes too far and confuses his wish with the facts. Even conceding what he says, it does not seem reasonable to conclude that there is freedom to embrace any opinion *in the scientific field* (Buonpensiere's emphasis); especially when there is some relationship with dogma, one would tend toward the soundest doctrine, that most accepted by theologians. Besides, Buonpensiere adds, it is not true that the Church has said nothing:

> —Buonpensiere says that, "even granting that the Church may not have expressed its opinion on evolutionism with complete *dogmatic precision*, it is nevertheless true that the Church has *persistently* shown its repugnance to this doctrine. It has shown it with the condemnation of the works of Darwin" and also with the condemnation of the books of Niccola Marselli, Pietro Siciliani, and Leroy himself.[110] Niccola Marselli (1832–99) was a professor of history and a deputy in the Italian Parliament. Two of his works were listed in the *Index* by the decree of June 27, 1881: *The Origins of Humanity* and *The Great Races of*

Humanity.[111] In both works, Marselli writes from an atheistic and materialist perspective, denying the spirituality of the human soul and the value of religion. He presents Darwin's theories as one of several hypotheses positing the purely material origin of humanity.[112] The physician Pietro Siciliani (1835–85) was professor of philosophy and pedagogy in Florence and Bologna. On December 5, 1881, two of his pedagogical works (one of which was on the teaching of religion) were placed on the *Index.* Then, on April 3, 1882, six more of his books were condemned. Some of these included occasional references to Darwinism in relation to various psychological and sociological issues. The evolutionary content was not directly studied. According to the consultor who examined the works, Giuseppe M. Granniello, a Barnabite, his "philosophical system" was "atheist and materialist."[113] From our perspective, it is interesting that such relatively minor figures were known to Buonpensiere as Darwinians, suggesting the equivalence in his mind of Darwinism, atheism, and materialism, and the apparent interchangeability of such terms.

—Leroy thinks what happened in the case of Galileo will come to pass with evolutionism. Buonpensiere comments: "I think otherwise, because in the case of Galileo: first, there was no *persistence* in the condemnation of the system that he advocated concerning the movement of the Earth and the immobility of the Sun, and it is known that those who supported Galileo's ideas were not harassed by the Church tribunals; second, the condemnation of Galileo was related to the way in which he defended his position, more than the substance of it. By contrast, we see the opposite in the case of evolutionism, inasmuch as the Church not only has condemned rigid evolutionism in the works of those who seek to derive man from the monkey, both in body and in soul, but also with the condemnation of Leroy's book it has shown that it opposes viewing Adam's animal side as an evolution from any animal" (p. 6).

—Buonpensiere asks himself: could the Congregation of the Index modify its own determination on evolution restricted only to organic species, *excluding man in regard to both soul and body?* His answer is, "I do not know: but, for my part, I am against any concession to the evolutionists, whatever the scientific explanation may be." Motives apart, the system is not and cannot be, at the most, anything more than a hypothesis: "Rather than lowering my voice before a scientific hypothesis, I would urge with all equanimity that the hypothesis abandon its pretensions and cede the field to the command of revelation" (pp. 6–7).

Regarding Scripture, Buonpensiere says that Leroy at least should have shown that the literal meaning of Genesis is not at odds with evolutionism, but he cites only Vigouroux, Farges, and Brucker to reach his desired conclusion, as if there

was not a pleiad of dissenting exegetes. Moreover, Buonpensiere rejects Vigouroux (and Leroy) who "have said nothing about the particular way in which
plants and animals are produced." He adds that he cannot accept the conclusion
that Leroy draws, that is, that Scripture does not condemn limited evolution, and
he criticizes Leroy's assertions based on a very literal interpretation of Scripture.
According to Leroy, the Bible, just like the Church, teaches nothing on the way in
which the Creator has formed species. Leroy adds that "A representative of the
competent authority in Rome gave me this surety." Buonpensiere comments: "I
do not understand, or better, I do not know how true Leroy's assertion is" (p. 19).

With respect to tradition, according to Leroy, the Church fathers lacked data
on evolution, in the realms both of science and of revelation; hence they were
uncertain and produced various interpretations. To support this view, he cites
Valroger, Farges, Brucker, and Zahm and concludes that everyone agrees that on
this point there is no theological tradition, "and I venture to repeat, the Fathers
are not competent on this matter." Buonpensiere acknowledges that Leroy displays erudition, but also conceptual confusion. For example, it is well known
that Saint Augustine differed from other fathers in the way he understood the
production of plants and animals, but he never thought of evolution, not even of
a limited kind. The notion that the fathers were not competent judges should be
rejected as rash, even though it is true that deductions they made on science
based on Scripture could be wrong. And here Buonpensiere adds: "But it remains to be demonstrated in the present case whether the conclusions of the
Church Fathers or that of the moderate evolutionists accord with the truth. And
so long as the contrary is not clearly demonstrated, it seems very just to me that
Catholics should follow the understanding of the Fathers, based on what Scripture says" (p. 23).

Leroy had characterized the first pages of Genesis as a primitive epic poem, a
patriarchal psalm. To Buonpensiere, "such expressions strike me as robbing Genesis of any historical value; and such a position clashes violently with the opinion of the Catholic Church" (pp. 23–24).

With respect to chapter 2, on limited evolution and philosophy (§§17–21, pp.
24–29), Buonpensiere says that it contains many philosophical misconceptions;
to show the weakness of the foundations on which Leroy seeks to support his
opinion (p. 24), Buonpensiere identifies his views that all the "brutes" form one
sole species and that the notion of species is a product of our minds, and then
discusses these views and several more in the light of scholastic philosophy.
When Leroy cites as a fact in his favor that evolution explains the extinction of so
many species, Buonpensiere replies that this can be explained by catastrophes
occurring in the early epoch of humanity. When Leroy says that proof should be
sought from evolutionists, but it should also be sought from those who advocate
"separate creation" of the various species, Buonpensiere responds that the latter

have in their favor the Church fathers, the entire Church, and the invariability of living species.

Turning to chapter 3, on the current notion of "physiological species" (§§22–29, pp. 29–40), Buonpensiere observes that the text must either be new or completely rewritten, inasmuch as it is handwritten (p. 29). Here are discussed philosophical problems that, in Buonpensiere's view, would incite a store of protest against Leroy if published (p. 34). With a certain irony, Buonpensiere says that the chapter's conclusion is "precious" and deserves to be transcribed. Leroy compares the development of the individual with the evolution of the species, with the difference that the constituent elements remain united in the first case and separate in the second. Buonpensiere takes it as given that this is unworthy of comment (p. 40).

Buonpensiere comments briefly on chapters 4–9, but focuses closely on chapter 10, on the origin of the body of the first man (§§36–44, pp. 44–55). This is the central point. Leroy asks whether the human body can be considered a product of evolution and says it cannot. Following the provincial Council of Cologne, theology, Saint Thomas, the Church fathers, and Genesis, Leroy writes: "no, man's body is not derived from animals; it is the direct product of divine power through the infusion of the rational soul [*par l'infusion de l'âme raisonnable*]." Buonpensiere underlines "par l'infusion de l'âme raisonnable," underlining "par" three times, and adds that he does so because, in his opinion, there lies the kernel of what distinguishes Leroy from many other Catholic authors. Leroy does not properly accept that Adam's body was formed immediately by God; rather the "immediate" action of God consists only of the infusion of the soul, while preexisting living matter is converted into the human body.

To determine the orthodoxy of this view, Buonpensiere cites another passage where Leroy explains it: "As for Adam, God intervened directly, created a human soul, then infused or breathed it [*soufflé*] into an animal, and he gave to this soul the extraordinary power of instantaneously transforming this organism into a human body."[114]

> From everything said up to here, it is quite clear that Scripture, understood according to the doctrine of the Church Fathers, excludes, with the exception of the efficiency of God and the passivity of the *humus,* any other action by secondary causes in the formation of the first man. And it is very probable that this interpretation of Genesis, given by the Fathers, is in accord with Catholic truth and doctrine [*ad Catholicam Veritatem et Doctrinam*], as Albertus Magnus and Suárez say explicitly. (p. 53)

Buonpensiere adds that, if to this were added the canon of the provincial Council of Cologne, "I do not see how a Catholic writer could defend the system of limited evolution, even though limited to Adam's body only" (p. 54). The council declared that evolution is contrary to Scripture and to the Catholic faith, and this is substantially the same evolution of which Leroy speaks (p. 54). Leroy's interpretation is that the council only condemned *spontaneous* evolution, which is atheistic because it excludes divine intervention, and did not condemn an evolutionism that admits the divine plan and the action of the Creator. But Buonpensiere says that this argument is weak, because everything that happens in brute nature in accordance with its laws and with the general motion of God comes about by spontaneous change (*spontanea immutatione*) (p. 55).

Buonpensiere's conclusion and recommendation are predictable:

> I have examined all these reasons of Leroy's before and, insofar as Philosophy is concerned, I have written that he greatly strays from the doctrine of Saint Thomas. As far as the fit between limited evolutionism and Divine Revelation is concerned, I have declared that this system, especially as it refers to the formation of Adam's body, is erroneous and cannot be sustained.
>
> Therefore I judge the book, as I have it before me, unworthy of printing. Even more, since in this volume Father Leroy has inserted many *totally new* points, that strike me as quite dangerous—for example, the reference to the union of the soul with the body and the generation of organisms. Perhaps it would be useful to advise the author of this through his superior. (pp. 55–56)

Reply to Leroy

Buonpensiere submitted his report right after finishing it. It is dated August 12 and, in spite of the great heat that normally envelopes Rome at this time of year, two days later, August 14, 1897, the following note appears in the Diary of the Congregation:

> Father Buonpensiere, to whom the examination of Father Leroy's book was entrusted, submitted a very detailed report, where he concludes that the aforementioned father cannot and should not be granted permission to publish the book *L'évolution* etc., even though he has corrected it, because the doctrine contained in it cannot be sustained, and continues to be worthy of censure. Therefore, the Prefect charged the Secretary to warn Father Leroy through the Father General of the Order of Preachers to abstain in the future from any publication of this book, etc. etc.[115]

In the letter that Leroy had sent from Paris the previous March 13, accompanying the corrected book, there is a handwritten annotation: "He was informed that the Congregation could not approve his book, and that he cease insisting on it."[116] Owing to a later annotation in the Diary of the Congregation, we know that on August 18, 1897, the secretary communicated the decision of the Congregation by letter to the master general of the Dominicans, attaching the conclusion of the consultor.[117]

Leroy obeyed. But he did not change his ideas, and in the years immediately following he expounded them in public.

Leroy Insists on the
Origin of Man

Leroy was a loyal Dominican who had experienced the difficulties attending the restoration of his order in France and the successive actions of the civil authorities against Catholic religious orders. He wished for the well-being of his Church and his order. Nevertheless, he was unconvinced by the arguments made to him. He continued to think that evolution and Christianity could be harmonized, and he himself had proposed a solution for the most delicate aspect, the origin of the human body.

He was not the only Catholic cleric who thought this way. He had published his book with all the requisite permissions, listed at the beginning. Two theologians vouched that the book contained nothing contrary to faith or morality, and both indicated its utility to readers. Finally, the license to publish was signed by the relevant authority. Father Gardeil, of the *Revue Thomiste,* saw nothing inappropriate in Leroy's book, and Leroy's response to the critiques in *Études* was published in the same journal. The consultor of the Index who had written the first report, Domenichelli, had proposed that no measure be taken against the book, and the secretary of the Index had been on the verge of endorsing that proposal.

In 1898 Leroy published a six-page letter to Albert Farges in the form of a small pamphlet, which included explicit authorizations from J. Monsabré and V. Scheil, and with the *imprimatur* of R. Monpeurt,[118] in order to respond to an article by Farges on the origin of man. Leroy tried to clarify his position. Once again he repeats his thesis: evolution had prepared the substrate destined to receive the rational soul directly from God, a position that could be reconciled with the doctrine of the Church and Scripture. He states that this opinion conforms to a Church doctrine according to which God directly and uniquely made Adam's body, unassisted: to see that this is so, one must bear in mind Saint Thomas's doctrine on the "human compound."[119] For further clarification, Leroy references Cardinal González's chapter on the anthropology of Saint Thomas in his *History of Philosophy.*

Leroy's chain of reasoning is this: every material substance is composed of form and matter. Form expresses the nature of the substance. Two substantial forms cannot coexist in the same substance. The rational soul is the form of the human body in an immediate and exclusive way: this is not open to debate, because it is a teaching of the Ecumenical Council of Vienne. Therefore, he who infuses the form (in this case, the human soul) in the material substrate directly and immediately is he who forms, directly and immediately, the body of man, whatever the origin of the substrate. Therefore, one can hold that the body of the first man was formed immediately by God and that divine action fell upon a substrate prepared by evolution.

When Leroy published this letter in 1898, it was clear that his ideas on evolution and Christianity had not changed nor would they change. In the *Revue Thomiste* of 1899 there is a review signed only with the initials D. L. Just reading it we recognize it as written by Leroy, because it expounds the same ideas on the origin of man we have just summarized. Indeed, years later, Leroy is identified as the review's author in the journal's cumulative index.[120]

The book reviewed is *L'origine des espèces*, by the Sulpician priest J. Guibert, superior of the Seminary of the Catholic Institute of Paris. Leroy congratulates him for publishing a book on so important a subject, the more so since Guibert was a competent individual whose orthodoxy could not be doubted. Moderate evolutionism is presented as a hypothesis but, at least it is a sensible one, supported by scientific data. Leroy devotes the second part of his review to the origin of man. He is in complete agreement with Guibert on the divine origin of the soul. Regarding the human body, Leroy remarks that two opinions are found among Catholics. The first, that God made the human body directly, is the position formulated with great vigor and erudition by Brucker (whom Guibert follows). The second is that evolution could prepare the substrate upon which God infused the rational soul to make the human body. Leroy comments that this opinion, so long as there is no final pronouncement of the Church on the subject, is considered acceptable by Cardinal González and is the one that he argues, stressing the substantive differences between animals and human beings.

Leroy Tries Again

Leroy still held out hope that the judgment against his work might be revoked. From the Diary of the Congregation we know that he again solicited authorization to print a new, revised version of his book. Thus an entry for November 21, 1901, reads: "Father M. D. Leroy of the Order of Preachers asks that he be permitted to prepare a new, corrected edition of his prohibited book, whose title is *L'évolution restreinte aux espèces organiques*."[121]

We do not know how Leroy made this petition. However, yet another petition, written a bit later, survives. This is a letter dated in Paris, December 20, 1901,[122] addressed to the secretary of the Congregation, insisting on the same theme:

7, rue de la Chaise
Paris, December 20, 1901

Most Reverend Father:

A while ago I was officially informed that this was a propitious time to solicit a review of the judgment against my book *L'évolution restreinte* by the Congregation of the Index. The priest who transmitted this message to me said that my request had to be sent via Father Esser, secretary of the Congregation. Not having had the honor of meeting you, I have strictly followed the advice given me. Today, when the occasion presented itself to offer my wishes for a good Christmas and new year, I take the opportunity to call your attention to my case and to thank you beforehand for whatever interest you might have.

In the Diary of the Congregation it is noted on December 22, 1901, that the secretary had received the letter from Leroy urging him to act on the matter.[123] The reply was negative. A Diary entry for January 7, 1902, states that the secretary has replied to Leroy's petition in the negative.[124] Here we find a summary of the case, with the most important dates. A draft of the secretary's response, written in Latin, is also preserved (Leroy had written him in French). It is interesting, because it makes clear that Leroy's statement did not agree with the facts available to the Congregation:

Rome, January 7, 1902

What you assert in your letter to the Congregation of the Index, that is, that you were ordered either to retract or else to completely correct your book on evolution, does not correspond to the acts of this Congregation. Rather, the book was specifically and absolutely prohibited. Notwithstanding this, through a special kindness shown you when in 1897 you asked the Congregation whether you could correct the book, you were permitted to display the corrections you had made. But, not only were they insufficient, they also proved once again that your book needed correction, and you were again denied permission to publish the corrected version. Bearing all this in mind, the Congregation determined to give the following response to the new request, in which the preceding petition and denial were not even mentioned: Just as was decided.[125]

A Diary note for January 13, 1902, records that Leroy had humbly accepted the Congregation's decision regarding his prohibited book.[126] Leroy's letter of the same date reads:

Paris, January 13, 1902

Most Reverend Father:

I acknowledge receipt of the decision of the Congregation of the Index that you had the honor to address to me in response to my request relative to the reedition of my book on *Limited Evolution*. Of course, I accept it as a docile son of the Church.

Now let me tell you, Most Reverend Father (I am speaking to Father Esser and not to the Secretary of the Index). I want to know whether it is the theory of evolution itself that is rejected, or only the manner in which I have treated it. I have reason to believe that it is mainly that manner that caused the rejection. My most sincere desire would be that it were so, for the honor of the Church, because many persons, philosophers, scientists, and theologians could not understand the condemnation of an opinion which perhaps they do not share, but which they view as inoffensive and within the domain of physiology. Therefore, in the probable case that the blame is only mine, I was prepared to make whatever corrections or deletions are asked of me, above all that which has provoked a polemic.

The letter that you had the kindness to send me before the official decision made me hope that it still were possible to present the desired clarifications, with a different title and in another format, for example, that of journal articles, avoiding as much as possible the controversies that have caused, I believe, the charges of which I am the object.

Please receive the homage of all respectful and devoted sentiments in Our Lord and the Holy Virgin.

Fr. M. D. Leroy, O. P.[127]

Echoes of the Retraction

Leroy's retraction was known widely in the ecclesiastical world, frequently cited as the most outstanding example of the reaction of the Roman authorities to evolutionism.

The Spanish Dominican Juan González de Arintero (1860–1928), also an exponent of limited evolution, wrote a work titled *Evolution and Christian Philosophy*, in a planned series of eight volumes, of which only a *General Introduction*, of 194 pages, and volume 1, titled *Evolution and the Mutability of Organic Species*, of 559 pages, were published, both in 1898.[128] In the *General Introduction* he mentions Leroy several times. Arintero argues that evolution shows God's perfection better that the separate creation of species and, to make the point, he cites some words from the first edition of Leroy's book: "The genesis of the organic world, through the mediation of natural agents, requires infinitely more genius than direct Creation. I have no doubt about the difference between a watchmaker who makes a precision watch, and an inventor who creates a machine capable of producing the same watch by itself; the inventor is 100 yards beyond the bench worker."[129]

This quotation demonstrates, clearly, that Arintero knew Leroy's work, which he admired, and that both made the case that evolutionism, far from hiding God's action or making it unnecessary, demonstrates it even better than do special creation and perfect adaptation. On the same page Arintero cites the Marquis de Nadaillac who, in spite of being opposed to transformism, wrote: "A certain school is accustomed to say that the Catholic Church condemns the transformist doctrine. This is wrong: there are Catholic scholars who have supported it and do so with determination. It is enough to mention Father Leroy, of the Order of Preachers, and Dr. Maisonneuve, professor of the Catholic Institute of Angers."[130]

Arintero also mentions Leroy's opinion that evolutionism will follow the same course as Galileo: that, as the exaggerated emotions of both sides are calmed, perhaps a priest who did not fear to predict a bright future for this idea will be thanked. Arintero states that, at the present moment (1898), many Catholic writers are favorable to evolutionism, thanks in great part to the publications of Leroy, Gardeil, and many others. And he provides original texts of the letters of Lapparent and Monsabré that Leroy reproduced at the beginning of his book.[131]

Still, Arintero knew that Leroy had been censured by the authorities. Although he praises Leroy's book and notes that the second edition corrects various defects of the first, he adds:

> But both in this version and the previous, there are a number of exaggerations. In the first place, Leroy sets out to defend evolution, without defining its limitations, and so seems to propose an excessively advanced theory, lending to the data and conclusions a general scope that they do not have. It appears that he recognizes no limits to mutability within each of the two plant and animal kingdoms, excepting man only.[132] And even in the case of man, Leroy substantially adopts Mivart's hypothesis; for even though he says he rejects it for the immediate formation of man's body by God, he reduces this act of formation to the simple infusion of the rational soul. This differs from Mivart's opinion only in the wording he uses. Thus it is that that interpretation has attracted censures that I believe are just.[133]

Arintero followed the opinion of Zeferino González, who accepts the possibility of an evolutionary process leading to the human body, but with a stipulation: God made the human body immediately at the same time he created the soul, but by appropriating matter organized by evolution. In this way the literal sense of the biblical text and its traditional interpretation could be saved. Thus he thought Mivart's stance, shared by Leroy, was exaggerated. He also says that a reason for not accepting Mivart's opinion is that his works had been listed in the *Index of Prohibited Books,* although in a footnote he admits that he does not know for what reason they were listed: "I must nevertheless confess that we do not know what the principal motive of the condemnation of these works was; perhaps it was something other than the theory mentioned. But I suspect it must have had some influence."[134]

In reality, the only work of Mivart placed on the *Index* was an essay in which he expounds a scarcely orthodox opinion about the punishments of hell. The confusion of Arintero, a notably erudite figure, illustrates Catholic media's misinformation about the action of Church authorities in Leroy's case, through a lack of data. Leroy's retraction did not identify the authority that had judged his doctrine and in what exact terms it had done so. A work of Zahm that also urged the reconciliation of evolutionism and Christianity was, at just that time, the object of complex negotiations in the Vatican, as we will see. To add to the confusion, the *Index* also listed a work by Mivart, the Catholic champion of reconciliation, but this work (really, three articles) had nothing to do with evolution. Cardinal González, who considered Mivart's approach to evolution exaggerated but who refused to condemn it as heterodox, had died recently. At this juncture it was not easy to determine what, exactly, was the position of the Church authorities.

This became abundantly clear in the case of John Zahm's book, *Evolution and Dogma*, published in 1896. There, Zahm cites Leroy with approval. In 1898 the Jesuit priest Salvatore Brandi wrote:

> When Professor Zahm wrote this, he did not know that the good father Leroy, who was summoned to Rome in February 1895 to be informed, had already "repudiated, retracted, and condemned" his theory in an authentic document signed by him. Therefore, he could not be cited then, nor can he be cited now, in favor of that theory. Neither can he be praised as having "nobly defended it with valid arguments," without causing a grave injury for him [Leroy] and the *competent authority* that sought his formal retraction.[135]

Brandi next reproduced the text of Leroy's letter to *Le Monde*. It seems likely that the letter was not known in America. Yet it is still striking that Zahm, a very competent, well-informed scholar, with broad international contacts, who had lived in Rome for twenty months beginning April 1, 1896, was unaware of Leroy's retraction. This underscores the complexity of the situation at that moment.

Until 1998 Leroy's retraction and its attending circumstances have been mentioned frequently, but always with much brevity and a certain air of mystery, completely logical if one recalls that the only available document was Leroy's brief letter to *Le Monde*. The resultant confusion dissipated only after the opening of the archives of the Holy Office in that year. Nevertheless, studies have continued alluding to Leroy almost schematically. An exception, however, is an article by Jacques Arnoud on Leroy and the problem of the origin of man, in which he notes that the reasons behind the summons of Leroy to Rome were obscure.[136]

The declassification of the Vatican documents, which included a full account of the Leroy case, has offered the possibility of lifting that obscurity. However, new confusions have since appeared, only now they are presented as guaranteed by the archival record. Two examples illustrate the point.

A New Legend: Leroy
on the Index?

The first is an article in *Living Tradition*, where material from the Archives of the Congregation of the Doctrine of the Faith was used, but in a somewhat disconcerting way.[137] The author, Brian Harrison, says that he has worked in this archive and cites documents that are found only there, but he makes some mistakes. Thus, on various occasions he speaks of meetings and consultors of the Holy Office when he refers, in reality, to meetings and consultors of the Index. Another error is much more serious and affects the core of the subject. Specifically, the author states that the books of both Mivart and Leroy on evolution were included in the *Index of Prohibited Books*, which is false.

To justify his assertion, the author cites a four-volume theology text book published by Valentín Zubizarreta in 1926.[138] This book is written in Latin. In footnote 5 two books by Mivart, *Lessons from Nature* and *Genesis of Species*, are mentioned, with the comment: "Haec opera relata sunt in Indicem librorum prohibitorum" (These works were brought to the *Index of Prohibited Books*). In footnote 6 he mentions *L'évolution restreinte* by Leroy and says: "Opus relatum est in Indicem librorum prohibitorum, 1895" (The work was brought to the *Index of Prohibited Books*, 1895), adding that Leroy had hardly learned that his thesis had been reproved by Rome when, with great humility, he retracted by means of the letter in *Le Monde*.[139]

Apparently all appears in order, except for a small detail of Latin. In Latin, *relatum* and *relata* are the neuter past participles, singular and plural respectively, of the verb *refero, referre, retuli, relatum,* meaning "to refer, take to, submit or remit to someone." In translation, the passage from Zubizarreta says that the works mentioned were "taken to" or "submitted to" or "denounced to" the Index: it does not mean they were added to the *Index of Prohibited Books,* and in fact they were not. The truth of the matter is easy to ascertain. A perusal of the successive editions of the *Index* from the period during which the works of Mivart and Leroy were published reveals that, of the two authors, only Mivart's "Happiness in Hell" appears, and that title has nothing to do with evolution. Moreover, these editions have always been publicly accessible.[140]

From this erroneous material and other data that he expounds in his article, Harrison extracts this broad generalization:

In the first place, we are in a position to correct a widespread popular perception about the history of the Church's relations with science. It is commonly held that while the Vatican notoriously blundered in the seventeenth century by condemning Galileo and proscribing all his works propagating the Copernican worldview, Rome "learned the lesson" from having "burnt her fingers" during the first great outburst of tension between traditional faith and modern scientific theories, and therefore "prudently" abstained from intervening with similar condemnations the next time around, when evolution became the new bone of contention, even though many theologians were shrilly calling for Darwin's head on a plate. Instead, it is not uncommon to hear statements to the effect that the Catholic Church "has never had a problem with evolution."

In fact, the record shows great similarities between the initial Vatican responses in both historic controversies. As Galileo was called in and rebuked by the Holy Office, so were Fr. Caverni and Fr. Léroy. As, in the seventeenth century, works defending the Copernican system were placed on the *Index of Forbidden Books*, so, in the nineteenth, were works defending human evolution—by Caverni, Mivart, Léroy (and possibly others). The main difference seems to have been that, for whatever reason, these anti-Darwinian censures emanating from Rome never received nearly as much publicity as the Galileo case.[141]

This conclusion is quite inaccurate. Mivart and Leroy's books on evolution were never listed in the *Index of Prohibited Books*; therefore, there is no need to look for "whatever reason" to explain why this fact was not publicized. Harrison corrects the "popular perception" by introducing new inaccuracies, presented as a consequence of his study of archival documents, when in reality they are due to an inaccurate translation from Latin of a work published in 1926 that, patently, has nothing to do with the archive. We have seen that the Congregation of the Index decided to proscribe Leroy's book but without publishing the decree, and, for this reason, the book did not appear in the *Index*. Rather than correcting an error, Harrison is creating a new myth.

Harrison wants today's Church to adopt a negative stance toward evolution and uses archival material to attempt to show that, in the past, this stance was much harsher than what is generally thought. This could explain the imprecision of his assertions.

Leroy and the Jesuits of 'La Civiltà Cattolica'

The second example of the use of the Index archives is found in a scholarly article published by Barry Brundell in a specialized journal, to which we have already referred.[142] Brundell contends that toward the end of the nineteenth century there took place a change in Church policy with respect to evolutionism, and that this change was provoked especially by the Jesuit editors of *La Civiltà Cattolica*. He aims to base his thesis on documents from the Archives of the Congregation for the Doctrine of the Faith and claims that, thanks to the opening of the archive, he can now propose for the first time an explanation of events inexplicable till now: above all, the conflicts of Leroy and Zahm with the Roman authorities.

Brundell devotes scant space to Leroy, less than two pages, and his reconstruction of the case is faulty because he has overlooked important documents.[143] According to Brundell, when the denunciation arrived, Leroy's book was examined by consultors, who completely rejected the accusations; but, as we have seen, in this phase it was examined by only one consultor, Domenichelli, who did not completely reject all the accusations. Brundell says it must have been hoped that the case could be wrapped up at this moment, but the opposite occurred: the Congregation submitted the book to its own scrutiny, which he supposes to be an unusual procedure. But, in fact, the normal procedure was followed: the reports of the consultors were always discussed in the so-called Preparatory Congregation, attended by the consultors and the secretary of the Congregation. Anything else would have been abnormal. Brundell does not allude to the existence of these Preparatory Congregations, which are not the same as General Congregations where the cardinal members of the Congregation of the Index evaluated the reports sent up by the Preparatory Congregations.

Brundell's reconstruction of the events does not represent what really happened. He is surprised that the prefect of the Congregation decided to broaden the study, extending it to the second edition of Leroy's work, and asks what the causes of this change might be. But there was no new examination of this type. From the start, it was the second edition that was scrutinized, which was logical because it was a corrected and expanded edition, which already existed. The secretary of the Congregation, in person, had received a copy, perhaps sent by the author himself. Brundell attributes this nonexistent change to pressure from persons who did not belong to the Congregation and who could bring pressure to bear on its members. Hence, the core of his thesis: those who applied the pressure were the Roman Jesuits, almost all of them associated with *La Civiltà Cattolica*.

The first piece of evidence in favor of this thesis, according to Brundell, is that the memorandum written by the cardinal prefect reflects articles published by the Jesuit Francesco Salis Seewis in *La Civiltà Cattolica*. Nevertheless, we saw that the memorandum in question was not written by the cardinal prefect but by the secretary, and that the four articles by Salis Seewis that Brundell mentions were published in 1897, two years after Leroy's retraction appeared in *Le Monde*. Thus, the argument collapses. Brundell believes that Roman Jesuits led the charge against Leroy while ignoring the equally strong opposition by the Parisian Jesuits of *Études*. The Roman Jesuits might have played a role in the Leroy case, but Brundell does not document his contention, while introducing new confusions in a matter much in need of objective clarification.

A Paradigm

The case of Leroy's book can be considered a paradigm of the interventions of the Vatican authorities regarding evolutionism. The beginning was provoked by a denunciation that originated casually with a Frenchman with scant knowledge of theology. The successive reports show that the Vatican officials followed the established procedures, as in any other issue, without any fixed purpose. A first report was quite favorable to Leroy. In the meeting of the consultors, Leroy's book met some obstacles, coming at least from consultor Buonpensiere, who later on confirmed his frontal opposition to evolutionism in writing. A serious debate took place when the report was discussed in the meeting of the cardinals, and two new reports were commissioned. Cardinal Mazzella took part in this meeting, and we can safely presume that he was inclined against the book, as he had strongly opposed evolutionism in a published textbook widely used at the time. In order to explain why new reports were requested, there is no need, therefore, to presume any influence external to the Congregation. Curiously, the secretary Cicognani had written a proposal that no action be taken against the book. Surely it was written before the meetings led to commissioning new reports, but it clearly shows that he was not seriously opposed to evolutionism.

The new reports were less favorable to Leroy. One of them did not recommend prohibition of Leroy's book. Nevertheless, the other was more severe and, even though it did not clearly advocate prohibition, it concluded that it was necessary to take some measure—for instance, to order the author to withdraw the book from sale and to retract his theory. In the new meeting of the consultors there was a heated debate, probably due to the intervention of Buonpensiere. He was asked to write a report summarizing his expressed position, and in this new report the recommendation was clearly for the prohibition.

Finally the cardinals decided to forbid the book but not to publish the corresponding decree. Instead, Leroy would be asked to make a public retraction, which he did in a brief letter published in the Parisian newspaper *Le Monde*. In the following years, however, Leroy tried to obtain permission to publish a new version of his book. This provoked a new report. The prefect of the Congregation was not satisfied with this fifth report and again ordered Buonpensiere to produce a new, sixth report, which was completely negative to Leroy's new version of the book. Therefore, Leroy was not authorized to publish it.

The main points discussed were the interpretation of the Holy Scripture, the evaluation of evolutionism, and the special intervention of God in the origin of the human being. The discussions and incertitudes clearly show that no official doctrine existed in the Catholic Church regarding the issue of evolutionism. This was obviously well known by the officials and authorities in the Vatican. The outcome of Leroy's case contributed to a similar decision taken shortly afterward in the case of a book written by the American priest John Zahm.

Americanism and Evolutionism

John A. Zahm

In 1896 the American priest John Zahm published a book in which he argued the compatibility of evolutionism and Catholic doctrine. Zahm extended his thesis to the origin of man (that is, of the human body). The Congregation of the Index decided to condemn the book, but it did not publish the corresponding decree. The matter was then carried into the public arena, because both Zahm's supporters and his adversaries waged a long struggle with the Roman authorities, all documented in numerous letters. Zahm's problems were complicated owing to his relationship with the leaders of the influential Americanist movement. Documents in the Holy Office archives shed new light on both problems.

Priest, Scientist, Writer

John Augustine Zahm was born on June 11, 1851, in New Lexington, Ohio. His mother, born in Pennsylvania, was the descendant of an English general. His father had emigrated from Alsace to the United States. John thought of becoming a priest and, after contacting Father Edward Sorin of the Congregation of Holy Cross, enrolled at the College of Notre Dame in South Bend, Indiana, where he arrived on December 3, 1867.[1]

The Congregation of Holy Cross had been founded in France, soon after the French Revolution, by the priest Basil Moreau (1799–1873). Edward Sorin (1814–93), a priest of the congregation, together with seven other brothers of the order, arrived in Indiana in 1842 and founded a college, which in time became the University of Notre Dame. John Zahm was a member of the congregation, in which he held a series of important posts, and played a prominent role in the development of the university.

Zahm studied for the priesthood at Notre Dame. After graduating in 1871, he became a novitiate of the order the same year. On June 4, 1875, he was ordained. Under the influence of Joseph Carrier, science professor and director of the university science museum, Zahm became professor, and later chairman, of the Science Department at Notre Dame. He oversaw the construction of a modern, well-equipped building for science instruction and was also named vice-president of the university.

In 1892 Zahm published his first book, *Sound and Music*,[2] which was followed by others on the relationship between science and religion, some of which were also published in French, Italian, and Spanish: *Catholic Science and Catholic Scientists* (1893),[3] *Moses and Modern Science* (1894),[4] and *Bible, Science and Faith* (1894).[5] In recognition of his achievements, Pope Leo XIII awarded him an honorary doctorate in 1895. In 1896 Zahm published three new works on science and religion: *Scientific Theory and Catholic Doctrine*,[6] *Science and the Church*,[7] and the book that provoked problems with the Vatican, *Evolution and Dogma*.[8]

Evolution and Dogma was published in February 1896. At the same time Zahm traveled to Rome, although there was no connection between those two events. The post of procurator general of the Congregation of Holy Cross had fallen vacant, a position that required residence in Rome. Father Gilbert Français, superior general of the congregation, proposed two names. One of them was Zahm, who was the choice of the superior general. On February 10, 1896, the provincial council of Notre Dame examined the superior general's request and responded affirmatively. On March 4 Zahm was surprised to receive the notice of his appointment. He seems not to have been terribly enthused about moving to Rome, but he was obedient and off he went. He resided in the Eternal City for twenty months, from April 1, 1896.[9]

When he arrived in Rome, Zahm was forty-six years old. There he had the opportunity to interact with many figures in the ecclesiastical world. He was received cordially by Pope Leo XIII and by many cardinals and successfully negotiated a number of projects put forward by his congregation. He was also a member of an informal group of American clerics who pushed an ecclesiastical program for the United States ("Americanism") that met some resistance in Rome.

On July 22, 1896, Zahm wrote to his brother that the Italian and French versions of his book *Evolution and Dogma* were in press and would be published soon—and that he foresaw adverse reactions.[10] In fact, *La Civiltà Cattolica* quickly ran an extremely critical review written by the Jesuit, Francesco Salis Seewis.

After living in Rome a little less than two years, Zahm was summoned to Notre Dame in January 1898 because the provincial of the Congregation of Holy Cross had died. Zahm was elected as the new provincial and never returned to Rome. What he did not know is that, while he was still in Rome, in November 1897, his book had been denounced to the Congregation of the Index. When the Roman authorities decided to condemn the book, Zahm was already in the United States. He resumed his activities at Notre Dame, supervising the construction of buildings and a library, until his tenure as provincial was over in 1906. After-

John Zahm and Theodore Roosevelt in South America. By permission of the Archives of the University of Notre Dame.

ward he continued to write and also made several long trips to South America, one of them in 1913, in the company of Theodore Roosevelt (president of the United States from 1901 to 1909), which was recorded in several books written under the pen name of H. J. Mozans. Zahm died in Munich, Germany, in 1921.

'Evolution and Dogma'

The American edition of *Evolution and Dogma* is 461 pages long and is divided into two parts: the subject of the first (pp. 13–202) is the theory of evolution and its origin and development in the course of history; it concludes with two chapters in which proofs and difficulties of the theory are examined. The second part (pp. 203–438) is about the relationship between the theory of evolution and Catholic doctrine. Zahm analyzes and criticizes the materialist and agnostic interpretations of evolutionism and defends theistic evolutionism. The remaining chapters of the second part cover the origin of life, the origin of man, and teleology.

The second part is the nucleus of the book. The central idea is that the theory of evolution is compatible with Catholic doctrine, so that it can be distinguished from nonscientific ideological interpretations, such as materialism and agnosticism. Evolution, as a scientific theory, leaves room for divine action in the creation and ordering of the world, as it also does for the special intervention of God in the creation of the human soul. At the end of the book, Zahm asserts that evolution is not about the creation or origin of things, but rather the way in which things are formed, once the universe has been called into existence by divine omnipotence. Evolution requires Creation and is based on it:

> Evolution, then, postulates creation as an intellectual necessity, for if there had not been a creation there would have been nothing to evolve. . . . And for the same reason, Evolution postulates and must postulate, a Creator, the sovereign Lord of all things. . . . But Evolution postulates still more. In order that Evolution might be at all possible it was necessary that there should have been not only an antecedent creation *ex nihilo,* but also that there should have been an antecedent involution, or a creation *in potentia.* To suppose that simple brute matter could, by its own motion or by any power inherent in matter as such, have been the sole efficient cause of the Evolution of organic from inorganic matter, of the higher from the lower forms of life, of the rational from the irrational creature, is to suppose that a thing can give what it does not possess, that the greater is contained in the less, the superior in the inferior, the whole in a part. . . . But this is not all. In order to have an intelligible theory of Evolution . . . we must also believe that creative action and influence still persist, that they always have persisted from the dawn of creation, that they, and they alone, have been efficient in all the countless stages of evolutionary progress from atoms to monads, from monads to man. (pp. 431–32)

Zahm's conclusion is that evolution ennobles our ideas about God and man. He also views human evolution positively: "As to man, Evolution, far from depriving him of his high estate, confirms him in it. . . . It teaches, and in the most eloquent language, that he is the highest term of a long and majestic development, and replaces him in his old position of headship in the universe, even as in the days of Dante and Aquinas" (p. 435).

With his clear, elegant, and equanimous style, Zahm declares his support of evolutionary theory. Although he alludes to its hypothetical character, at the same time he states that it has solid evidence in its favor, and he replies to the arguments that were typically adduced against it. Moreover, he presents it as an ancient doctrine, supported even by Aristotle and Saint Thomas Aquinas.

Saint Augustine and Saint Thomas, Precursors of Evolution?

Here, Zahm's enthusiasm carries him away. He presents the ancient Greeks as precursors of evolution, especially Empedocles, who formulated an idea similar to Darwin's natural selection, as Aristotle confirms. For Zahm, Aristotle, more so than Empedocles, should be considered the father of the theory of evolution. But here he is at least partly mistaken. He cites a passage where Aristotle expounds the idea of Empedocles (Aristotle, *Physics*, bk. 2, ch. 8), as if Aristotle had approved it; but, through inadvertence or confusion, he does not say that, in this same passage, Aristotle dismisses and criticizes Empedocles' idea (pp. 25–28).

It is no less surprising to learn that Thomas Aquinas is a father of evolutionism. Early in the book, Zahm says that Saint Thomas had formulated principles of causality pertaining to living organisms that have great value for the interpretation of evolutionism (pp. 29–30). This is not difficult to see, because the distinction and complementarity between the causality associated with divine action (the "First Cause") and the causality of created things ("secondary causes") is basic to the philosophical understanding of evolution. But later on he introduces Aquinas as an authentic precursor of evolutionism, stitching together texts where Saint Thomas comments on the creation of the world (pp. 284–93).

On the other hand, as Zahm observes, Saint Augustine could indeed be considered a precursor of some evolutionary ideas—in particular, his theory of "seminal reasons." Zahm uses texts of Saint Thomas to similar ends, to interpret the biblical narrative of Creation in accordance with evolutionism. The argument was weak, but Zahm uses it to conclude that "theistic evolution" is a Catholic idea developed and taught by the most eminent saints and doctors of the Church. "Theistic evolution" is a process in which God participates as creator and ruler of the world. His conclusion is clear and explicit: "there is nothing in Evolution, properly understood, which is antagonistic either to revelation or Dogma; that, on the contrary, far from being opposed to faith, Evolution, as taught by St. Augustine and St. Thomas Aquinas, is the most reasonable view, and the one most in harmony with the explicit declarations of the Genesis narrative of creation" (p. 300).

Moreover, according Zahm, this is the way to avoid anthropomorphism: there is no need to attribute to God actions, such as modeling clay with his "hands" to form the body of man, which have an anthropomorphic flavor. Along the same lines, Zahm continues: "Theistic Evolution, in the sense in which it is advocated by St. Augustine and St. Thomas, excludes also Divine interference, or constant unnecessary interventions on the part of the Deity, as effectually as it does a low and narrow Anthropomorphism. Both these illustrious Doctors declare explicitly, that 'in the institution of nature we do not look for miracles, but for the laws of nature'" (p. 304).

Playing with Fire

"Derived creation" or evolution, Zahm clarifies, does not require continuous miracles. Rather, God has acted "indirectly," through natural laws that He himself has put in place and continues to maintain with his providence, and this is not what we ordinarily call "miracles." Zahm criticizes Darwin and other naturalists, when they depart the realm of their science, owing to their lack of clarity on these issues. He then repeats his conclusion, adding an important nuance:

> From the foregoing pages, then, it is clear that far from being opposed to faith, theistic Evolution is, on the contrary, supported both by the declarations of Genesis and by the most venerable philosophical and theological authorities of the Church. I have mentioned specially St. Augustine and St. Thomas . . . but it were an easy matter to adduce the testimony of others. . . . Of course no one would think of maintaining that any of the Fathers of the Church taught Evolution in the sense in which it is now understood. They did not do this for the simple reason that the subject had not even been broached in its present form, and because its formulation as a theory, under its present aspect, was impossible before men of science had in their possession the accumulated results of the observation and research of these latter times. But they did all that was necessary fully to justify my present contention; they laid down principles which are perfectly compatible with theistic Evolution. They asserted, in the most positive and explicit manner, the doctrine of derivative creation as against the theory of a perpetual direct creation of organisms, and turned the weight of their great authority in favor of the doctrine, that God administers the material universe by natural laws, and not by constant miraculous interventions. As far as the present argument is concerned, this distinct enunciation of principles makes for my thesis quite as much as would the promulgation of a more detailed theory of Evolution. (pp. 312–13)

So Zahm does not claim that Saint Augustine and Saint Thomas had conceptualized an evolutionary theory in the modern sense. He claimed only that their principles are compatible with this theory and provide support for it. On essential points, in Zahm's view, everything was already present in Aristotle, Gregory of Nyssa, Saint Augustine, and Saint Thomas (pp. 382–83).

But Zahm was playing with fire. In an epoch of philosophical and theological crisis that witnessed a renaissance of Saint Thomas expressly promoted by Pope Leo XIII, those who supported Neo-Thomism had nothing good to say about the theory of evolution, which was frequently used as a weapon in favor of materialism and agnosticism. They must have been disgusted by seeing Saint Thomas used to support a theory they deemed contrary to Catholic doctrine.

Arguments for and against
Evolutionism

In the two last chapters of the first part of his book, Zahm examines the arguments in favor of evolution (ch. 7, pp. 84–139) and the objections commonly raised against it (ch. 8, p. 140–202). The principal difficulties can be reduced to three and are the same that are raised today: that there are no observed examples of changes from one species into another, that the fossil register is imperfect, and the result of a crossing of distinct species is infertile. Even recognizing there are difficulties, Zahm is still favorable to evolution. He is convinced that, over time, evolutionists will win the debate. At the least, evolution is a highly probable theory (pp. 68, 75, 134–35, 388).

Zahm also presents the classic philosophical objection: that the evolution of species is impossible because species are fixed (pp. 313–19). Here he limits himself to stipulating that species could be fixed in a metaphysical and logical sense, but experimental research must determine which species of real living organisms are fixed. Zahm cites the French Dominican Dalmace Leroy, who had also written in favor of the compatibility of evolution and Christianity. On this occasion he shows that, even though early he had cited the first edition of Leroy's work, he was also familiar with the second (*L'évolution restreinte aux espèces organiques*, 1891), because he refers the reader to chapter 3 of this work (p. 317). The allusion to Leroy was certainly not going to help matters, because Zahm's book was to be examined in the Congregation of the Index a short while after it had judged Leroy's work negatively.

According to Zahm, the fact that the doctrine contrary to evolutionism, that is, the theory of special creation, had brought science to a dead end explains nothing, and by its own nature it was an obstacle to progress (p. 419). Zahm's blunt way of expressing himself could not possibly have pleased those theologians who favored "special creation."

Zahm asserts that there is no middle ground between creation and evolution, so we must all be either creationists or evolutionists (p. 75). This way of speaking, which persists in our times, is infelicitous. The terms that are really opposed are not creation and evolution, creationism and evolutionism, but rather evolution or evolutionism on the one hand and "special creation" on the other. Special creation asserts that God from the beginning created the various living species exactly as they exist in the present, or that God was creating directly, in different epochs, the new species that were populating the world. This is the only issue standing in the way of admitting that some species arise from others by successive transformations. On the other hand, evolution is not opposed to the creation of the universe or to the action of God who is present in the being and activity of all creatures. What is unfortunate is only Zahm's way of expressing himself in some passages. In others, Zahm strives to dissipate the misunderstandings that then as now make creation oppose evolution and to assert roundly that, far from an opposition between evolution and divine action, evolution supposes continuous creation and divine action in nature: one could even say that this is one of the central ideas of his book.

The Origin of Life

The fifth chapter of the second part of Zahm's book is an analysis of the origin and nature of life, beginning with a consideration of spontaneous generation (pp. 320–23). Zahm asserts decidedly that the Church fathers and the scholastics had no problem with spontaneous generation. Ancient Greek and medieval scholars supposed that the lowest forms of life had originated spontaneously from the earth or from decaying organic matter.

Zahm writes that, from the scientific point of view, very little is known about the origin of life; indeed, it lies beyond the competence of science, just as does the origin of matter (p. 327). He is concerned to show that, in the hypothetical case that some day life might be produced artificially, even that would not pose any doctrinal problem, because the Church fathers had accepted spontaneous generation (p. 331). Zahm concludes that spontaneous generation without God is very hard to conceive, but with God it poses no major problem (pp. 338–39).

His treatment of this subject is somewhat confused, which is unsurprising given the state of scientific knowledge at the time. Yet those who were trying to create a new school of scholasticism (Neo-Thomism) quickly zeroed in on Zahm's discussion. Francesco Salis Seewis, a Jesuit on the staff of *La Civiltà Cattolica*, immediately wrote four articles on spontaneous generation, attempting to clarify what Aristotle, Aquinas, and the scholastics had to say about it.[11]

The Origin of Man

Zahm next discussed the origin of man (ch. 6, pp. 340–68), a key issue whenever the compatibility of evolution and Catholic doctrine was debated. Following Catholic theologians, Zahm states that the human soul is created immediately by God. Most of the chapter has to do with the origin of the body, in an attempt to answer the question: can a Catholic believe that the body of the first man came about through evolution from other animals?

Zahm reviews some solutions proposed by other Catholics. He lays out the theory of St. George Mivart, who had answered that question in the affirmative in his book on the genesis of species, published in 1871. Zahm emphasizes that even though there were some who tried to get Mivart's book condemned, they were not successful; in fact, the pope conferred a doctorate on Mivart. He mentions that when Saint Thomas Aquinas speaks of this subject, he admits that the angels could have played some part in the formation of the body of the first man, and concludes that, similarly, one can admit that other creatures also could have participated, which is equivalent to presenting Saint Thomas's idea as favorable to the possibility of the origin of the human body through evolution. He observes that, although difficulties remain, the origin of the human body by evolution is compatible with the principles of Saint Augustine and Saint Thomas. He also mentions the theory of the Spanish cardinal Zeferino González, who went so far as to assert that most of the difficulties disappear if it is admitted that the evolutionary origin of the human body was complemented by a kind of final touch applied by God, a special act through which the animal organism is prepared to receive the infusion of the human soul. Nevertheless, Zahm says that this complement proposed by Cardinal González is unnecessary, and that if one admits the evolutionary origin of the human body, it is more logical and coherent to do without it.

The origin of man was, without doubt, the central point of the debates. It was generally held, as a teaching of the Bible, that there must have been a direct and special divine act to have formed the body of the first man. That seemed to fit the biblical story, the special dignity of the human being, and the special providence that God invests in man.

Some Dangerous Friends

In order to show that his ideas could be accommodated by Catholic dogma, Zahm placed special emphasis on the important support lent to them by similarly inclined Catholic authors. On repeated occasions Zahm refers to the International Scientific Congresses of Catholics that were held precisely in those years. When Zahm published his work in 1896, the first three (Paris, 1888 and 1891; Brussels, 1894) had already taken place. Zahm quotes lectures by Monsignor d'Hulst in the 1891 Congress and Duilhé de Saint Projet in that of 1894. He quotes Cardinal Zeferino González, who, referring to Mivart's theory on the evolutionary origin of the human body, had stated that he would not condemn this theory, because only the Church is competent to assess this kind of proposition and determine its compatibility with Scripture (pp. 361–63).

No one doubted the orthodoxy of Cardinal González. But the same cannot be said of Monsignor d'Hulst, of Mivart, and of various other Catholics who presented their opinions before the International Scientific Congresses. By invoking them, Zahm wanted to lend solidity and verisimilitude to his ideas, but the strategy backfired. These quotations, more than quieting things down, increased the general alarm, because they were evidence that ideas favorable to evolutionism were spreading.

Something similar happened with the authority cited next, the Dominican Leroy, who held a privileged status for Zahm. He praises Leroy as a partisan of Mivart's theory, while expressly claiming that his work, besides having the official permissions granted by ecclesiastical censors, was itself supported by authorities as distinguished as the scientist A. de Lapparent and the theologian Monsabré (pp. 363–64). Evidently, Zahm was not aware of Leroy's public retraction, which had taken place in 1895, when Zahm had finished preparing his book for publication. It is illogical that Zahm would have referred to Leroy as an authoritative support for his ideas had he been aware of the latter's problems.

Evolution and Teleology

"Teleology, Old and New" (pp. 369–77) is the title of chapter 7 of the second part of Zahm's book. It was necessary to cover this topic. Up to this point Zahm had argued that evolutionism is compatible with Creation and divine action in the world. But he had said nothing about one of the principal objections. As he wrote at the beginning of the chapter

> But there is yet another objection against Evolution, which, by some minds, is re-garded as more serious than any of the difficulties, heretofore considered, of ei-ther philosophy or theology. This objection, briefly stated, is that Evolution de-stroys entirely the argument from design in nature, and abolishes teleology, or the doctrine of final causes. In the case of Darwin, for instance, as we learn from his "Life and Letters," he had no difficulty in accepting derivative in lieu of special creation, but when it came to reconciling natural selection and Evolution with teleology, as taught by Paley, he felt that his chief argument for believing in God had been wrested from him entirely. (p. 369)

William Paley (1743–1805) was an Anglican minister who wrote very influen-tial books on Christianity and nature. In 1802 he published his *Natural Theology; or, Evidences of the Existence and Attributes of the Deity: Collected from the Appear-ances of Nature*, in which he expounds his famous "argument from design" as a proof of the existence of God. For Paley, the best evidence for design in nature is what he called "perfect adaptation"—that is, God created each plant and animal species perfectly adapted to its environs. It is a proof of God's existence because, just as we conclude from observing how a watch works that it had to have been designed by an intelligent craftsman, a watchmaker, with even greater reason, living creatures, which are considerably more complex than a watch, give evi-dence of the necessary existence of God as author of nature.[12]

Darwinism, Zahm notes, had done away with this kind of argument. The or-der that we observe in the world is simply the result of an accumulation of favor-able variations. In the struggle for existence, those organisms with favorable variations will tend to survive in greater numbers than those with unfavorable variations. Over the long term this process, natural and blind, will produce pop-ulations of organisms ever better adapted to their environments. Natural order is not the result of a plan but of a complex play of blind forces. The ancients saw in the natural order of things a way to recognize God's purpose in creating and governing the world; evolutionism sought to do away with this illusion.

Zahm states that evolutionism deprives the classical "argument from design" of meaning, but in its place it substitutes a purpose that is much richer and more interesting than the ancient one. It is not difficult to see why. The unfolding of evolution over millions of years, beginning with matter in a primitive state and ending with enormously sophisticated organisms, supposes the existence of a set of potentials that are realized little by little, in such a way that the process as a whole would be unintelligible if there were no such grand plan.

In this line, Zahm rejects antifinalist philosophies that seek to base themselves on evolution, citing in their favor authors who see in evolution a new way to prove, in a more profound way than before, the existence of purpose in nature and of a higher plan.

The Marquis de Nadaillac's Review

Reviews of Zahm's book appeared in short order. In the *Revue des Questions Scientifiques* for July 1896 appeared one signed by the Marquis de Nadaillac.[13]

From the beginning the review is very laudatory. Zahm has admirably summarized, he says, everything written on the subject from antiquity to our own times. Zahm favors evolutionism and sets out to show it agrees with the doctrine of the Church, of which he proclaims himself a faithful disciple and energetic defender. Evolution must be distinguished from its atheist, agnostic, pantheist, or materialist interpretations and shown to be a magnificent witness of God's presence throughout nature. Nadaillac does not oppose these assertions, but he has serious doubts about several of Zahm's statements and about the scientific value of evolutionism.

Thus, after summarizing what Zahm says about the presumed anticipations of evolutionism by the Church fathers and Saint Thomas, Nadaillac comments: "The conclusions of this learned theologian evidently exceed his premises. It strikes me as difficult to tease evolution, as we currently understand it, out of the words of the philosophers of antiquity or the doctors of the Church, and we would not know how to find in their theses any support for the doctrines of Lamarck or Darwin, and even less for those of the neo-Darwinians or neo-Lamarckians" (p. 231).

Nadaillac is completely opposed to the possibility of abiogenesis (the origin of organisms from inorganic matter): "Without doubt science will make enormous gains. . . . But we would not know how to accept, whatever new power science might bring to mankind, that life could be produced through physical, chemical or mechanical combinations. On that day man would be equal to the Creator, something no Christian can accept" (p. 232).

Nadaillac then considers the origin of man and mentions Mivart's and Zahm's concept of the infusion of the human soul. But he says he is not competent on this subject, stating that from the scientific perspective, which is his own, the question is insoluble. Nevertheless, he recalls that Mivart's hypothesis, although it caused a great scandal, after twenty-five years people still support it, while the Holy See, guardian of the faith, has not thought it worthwhile to condemn. He recalls that Mivart, after having proposed his theory, was awarded a doctorate by Pope Pius IX. Nadaillac further alludes to Cardinal González's version of the origin of man, and agrees with Zahm that the modification he proposes (direct, partial production of man's body by God) is unsatisfactory and has little to do with evolutionism (pp. 234–35).

Nadaillac thus shows appreciation and respect for Zahm and holds his thesis to be compatible with Catholic doctrine. But he sees great difficulties in the acceptance of evolution as a true doctrine, at least as the question stood at that moment: it lacks proof. Then, speaking as a scientist, he presents objections to evolution, one after another.

An important objection is that we do not understand the processes by which evolution happens, the weakest aspect of the new doctrine. The result is proclaimed without having understood the causes. Moreover, Nadaillac reviews a whole series of problems: the impossibility of species being transformed by generation, the problem of the infertility of hybrids, the uncertainties surrounding presumptive explanations based on natural selection, the role of the environment, and the inheritance of acquired characteristics. He concludes that evolutionists themselves agree that they still have not identified an explanation of the process of species transformation (pp. 239, 242). This was a common theme in the age when Darwinism was "in eclipse."

In spite of these objections, Nadaillac strongly asserts the new vistas opened by evolutionism contain nothing contrary to Christian doctrine (p. 238). His analysis is respectful and objective, as he concludes:

> Doctor Zahm's book has great merit . . . but, if I may be permitted to say so, it takes too much as truths already acquired what are hypotheses that ought to be proven. . . . If I am not very disposed to accept the conclusions of the evolutionist school, neither can I reject them in any absolute way. In Scotland, the jury, besides the habitual replies, has the right, without pronouncing on the deed itself, to respond *not proven*. As of now, that is my only conclusion, and I do not doubt it is the same for all those who approach these studies without bias and with the sole desire of arriving at the truth. (pp. 245–46)

The Review in the 'Dublin Review'

Another review—favorable to Zahm—appeared in the *Dublin Review* in October 1896, signed by F. David, pen name of the Franciscan David Fleming, who worked in the Vatican Curia (in the Holy Office, as a matter of fact) and was acquainted with Zahm and his friends.[14]

After praising Zahm's usual clarity and skill, Fleming fixes his attention on specific issues, the first of which is spontaneous generation. Fleming points out that Saint Thomas seems to accept it with some reluctance, and that on this point, as on many other difficult ones, he changed his mind frequently. But in spite of this and the fact that the Church fathers and the scholastics were wrong on this point (as Zahm had also said), Fleming admits that Zahm's reasoning was correct when he concludes

> that it cannot be looked upon as against faith to hold that in the beginning the Creator infused into matter the activity necessary for the origination of organic life to be evolved and differentiated in the course of ages in accordance with the laws which He had laid down, and that furthermore, that should any future researches necessitate the admission of "abiogenesis," we should be following in the footsteps of great Catholic thinkers and theologians in the past, in loyally accepting it. (p. 248)

Fleming also endorses Zahm's notion of "derived" or mediated creation, that is, the formation of certain organisms from others. He says that this point has been developed in his country by Mivart and others for a number of years, and that "The point of this argument is, that if the Schoolmen had all the facts before them which we have, they would not have the slightest hesitation in admitting that all the organic species at present existing are the outcome not of direct and immediate, but of derivative and mediate creation" (p. 248).

Fleming also agrees with Zahm that pantheism, materialism, atheism, and agnosticism, which seemed to be inseparably linked to evolutionism, can be dissociated from it. That dissociation removes a source of the strong opposition to evolution by distinguished clergymen. Moreover, the true idea of evolution is singularly beautiful and harmonizes perfectly with divine guidance of the world, because it is more perfect to create beings capable of contributing to the production of others, than to create everything directly. Fleming also agrees with Zahm's views on teleology (pp. 248–49).

Fleming argues (as did Zahm) that we can reject the individual theories of Lamarck, Darwin, and others, but we cannot reject evolution itself as contrary to either reason or faith. This does not mean that we must accept it: evolution must first be proved. In Fleming's opinion: "It seems to us that, as matters stand at present, the theory of Evolution has passed from the state of being *merely possible* to the state of *probability.* . . . It is quite evident to us that there is no incompatibility between Evolution and Theism, or between Evolution and Spiritualism—I mean, the doctrine holding the spirituality of the soul and its separate creation" (p. 250).

On the relationship between the evolution of plants and animals and the biblical narrative of Creation, Fleming cites the Jesuit Hummelauer as an exponent of the general opinion, which Fleming accepts, according to which there is no such conflict.

Fleming's final point, which is always the culmination of such discussions, is the origin of Adam. Zahm, Fleming recapitulates, takes the same line as Mivart, Leroy, and, with some nuancing (to which Zahm and Fleming both allude), Cardinal Zeferino González. The philosophical rationale, for which he cites Saint Thomas, leads Fleming to conclude that no objection can be made on either metaphysical or physical grounds. With regard to revelation, Fleming reasons that Genesis does not support either immediate formation or the evolution of Adam's body. The Church fathers have no teachings in this regard. With respect to the creation of Eve, Fleming mentions the opinion of Origen and of Cardinal Cajetan, adding that they were not censured by the Church. He concludes: "In our humble opinion, and we speak under correction, we are not compelled by any *principle* of theology or exegesis to insist upon the strictly historical and scientific nature of the account of the creation and formation of Eve. . . . The practical unanimity of interpreters on the point does not place the matter beyond all respectful and reverent inquiry. We gladly leave the matter to the authority of the Apostolic See" (p. 254).

At the end of his article, Fleming rates Zahm's work as "admirable and useful," and adds some small observations. But the review is completely favorable to each and all of the principal theses of Zahm's book, from a priest who, repeatedly throughout the article, is disposed to follow the infallible guidance of the Church.

'La Civiltà Cattolica' 's Review

The Italian edition of Zahm's book was published the same year as the American (1896), and, as we have seen, in January 1897 it was reviewed in the Roman Jesuit journal, *La Civiltà Cattolica*.[15] On the book's cover it is stated that Zahm was named doctor of philosophy by Pope Leo XIII, was currently professor of physics at the University of Notre Dame, and that the Italian translation had been authorized by the author.

The reviewer, the Jesuit Francesco Salis Seewis, first notes that *La Civiltà* had always given due praise to Zahm's works and laments having to criticize his new book. To forewarn the reader, he says, it is enough to know that, in an appendix, Zahm asks whether a good Catholic can be an evolutionist, to which he responds affirmatively, quoting the words of a Protestant, Sir William Gladstone. Salis Seewis says that an Italian priest would never use such a citation, but that the quotation reflected both the spirit and writings of a certain school of reconciliation, although misconceived, which had taken shape in Europe and America in the past several years. The quotation also reflects some well-known Catholic journals, outside of Italy, that had praised Zahm's work, asserting that *Evolution and Dogma* contained nothing that was not true and that its publication was the most important event of 1896. Salis Seewis comments that this chorus of elegies marks 1896 as a bad year for the countries in question, inasmuch as he doubts that these elegies were shared by the bishops of those places, who are the legitimate custodians of sound Church doctrine.

Salis Seewis did not lower his tone in what followed. Not all the commentary had been favorable he notes, citing a critical review in an English magazine, according to which Zahm's book is so bad that if he had written against evolution, instead of favoring it, he still would have been viewed as a disgrace to Catholic literature. How so? Salis Seewis says that each of the several times *La Civiltà* had written about evolution, it had taken care to analyze it in a rigorous way:

> We have always judged it only from a scientific perspective. Our question has always been this alone: *What value does evolutionism have in the court of positive science and logic?* And having considered the objections that have been raised and never resolved, those which its partisans, Zahm included, admit, we have concluded each time that evolutionism can only be deemed a tissue of vulgar analogies, arbitrary suppositions, not sustained but proven wrong by the facts, fantastic adages and subterfuges that render indecent the seriousness of science. (p. 202)

But, Salis Seewis laments, there are but a few independent spirits. Too many Catholics (always outside of Italy) have blessed this idol as if it were a saint and, whether to justify themselves or to please others, kneel before it; they do not cease to assert that evolution is not contrary to dogma, and that one can be a good Catholic and an evolutionist at the same time. As if the principal and true impediment to evolutionism, Salis Seewis states emphatically, arises from the fear of clashing with the Bible! Absolutely not! he continues; it arises from the scientific defects of the system. Of course, it is required not to twist the words of the Bible capriciously, as some Catholic evolutionists have done. Among the various meanings that sacred texts might have, good Catholics, if there is nothing defined by authentic Magisterium, may choose that which best coincides with scientific evidence, avoiding fantastic interpretations; this is what the doctors have always said:

> When evolutionism has passed its examination in the court of sciences, and emerges with the patent of a system founded on evident principles, logical deductions, and positive facts, only then will it be worthwhile to consider it in the light of Revelation. . . . But so long as it presents nothing more than the hope of future

demonstrations . . . while it continues reasoning as if logic had nothing to do with it . . . while things are as they are, we hold that it is useless to introduce this scientific failure into the sacristy and it is astonishing that there are Catholics rushing to do so. . . . This is why we have also kept the question on the grounds of science; that its followers have tried to take it into theological terrain where they have no business, seems to us not only dangerous, but even insidious. (p. 203)

In the conclusion of his review, Salis Seewis alludes to Zahm's intent, which, no doubt, is honest, as his previous works, which have done much good, demonstrate. He laments that his goodwill has on this occasion been dragged down by an ill-conceived reconciliation that led him to concede too much to the enemy, in hopes of peace more than of justice and truth. And he also laments that, with the Italian edition, evolutionism, which till now had scarcely made inroads among Italian Catholics, and less still had been promoted publicly by a priest, could easily drag along the naive, believing they can trust to the author's fame. Salis Seewis ends by disrecommending the book as unfortunate and prejudicial for anyone lacking the will, the means, and the time to form an independent opinion.

The International Scientific
Meeting at Fribourg

The fourth International Scientific Congress of Catholics took place in Fribourg, Switzerland, from Monday, August 16, through Friday, August 20, 1897.[16] After the two first meetings, held in Paris in 1888 and 1891, and the third, held in Brussels in 1894, Fribourg was a step forward in every sense. It signaled greater integration of the Latin and German worlds, an important step in internationalization. The presiding officer was a German professor from Munich, and 38 percent of the 192 papers presented and published were in German, together with a good representation in other languages, including Italian, English, Dutch, and Spanish. For the first time, the participants were allowed to submit papers in their national languages. Attendance exceeded that of the previous congresses, and the number of subscriptions to the published papers reached 3,007. The acts were published in eleven volumes with a total of 3,610 pages.[17]

Cordiality was a keynote at this meeting. The University of Fribourg, a Catholic institution of recent foundation, had made a notable impact on the life of this small and picturesque city, whose residents lodged attendees in their houses and threw themselves into details of the meeting amid a holiday atmosphere.

Pope Leo XIII had personally set the parameters of these meetings in a letter written to Monsignor d'Hulst on May 20, 1887: the congresses were to debate scientific questions as an aid to theology, but without dealing directly with theological issues, which required a different setting.[18] Nevertheless, problems arose on various occasions when topics related to evolution came up in the scientific or anthropological sections.[19] At the Fribourg Congress a new section devoted to biblical exegesis was included, as a kind of tribute to the founder of these congresses, the recently deceased canon Duilhé de Saint Projet. In spite of all precautions, protests occurred, and some participants were openly irritated.[20]

Zahm, who had already attended the Brussels Congress of 1884, came to Fribourg from Rome in August 1897 and lectured on evolution and finalism, a subject he had treated in *Evolution and Dogma*. The basic idea was the same: that evolutionism, far from abolishing purpose in nature, reveals a deeper, richer, and more interesting purpose than had previously been thought.

It was an optimistic speech, with a certain air of risk. The *Revue des Questions Scientifiques*, edited by Belgian Catholics, printed a complete text,[21] but preceded it with two notices. The first was that the editorial board left responsibility for the ideas expressed wholly to the author, pointing out that these differed notably from the ideas on evolutionism that had many times been expounded in that journal. Second, the journal recommended to its readers several works on evolution and teleology: the three last chapters of a book by Pierre Janet on final causes, which are a defense of the existence of purpose in nature; the articles of Salis Seewis in *La Civiltà Cattolica* on spontaneous generation, critical of evolution; and David Fleming's favorable review of Zahm's book in the *Dublin Review.* Finally, the journal added details on various aspects of Zahm's text.

Zahm's talk was calm and direct. He openly espoused evolutionism against those who backed "special creation," and he argued in favor of an allegorical reading of the Genesis narrative, not a literal one; he asserted that the theory of special creation was "completely discredited" and was "unworthy of the least consideration as a practical hypothesis to guide the researcher in the study of nature and her laws." He spoke of Darwin' *Origin of Species* as a "great work." And with various arguments he claimed that an adequate analysis of evolutionism shows that this theory is not contrary to purpose in nature, but favorable to it.

The voice of Americanism, which was about to play an important role in Zahm's case, was also heard in Fribourg. Monsignor Denis O'Connell, who had been rector of the American College in Rome, was a close friend of Zahm and an influential figure in ecclesiastical circles, in both Rome and the United States. His exaltation of American values in his lecture was heard by some as a critique of the Church in western Europe. There were immediate echoes in France. Zahm's pro-evolution stance was increasingly drawn into the battle between Americanists and Europeans.

Evolution and Dogma
in the Holy Office

A short time after the Italian edition was published, *Evolution and Dogma* was discussed in the Congregation of the Holy Office. The extant documentation is very brief (and perhaps incomplete) and seems to indicate the proceedings soon ceased. No official document was published, and, insofar as we know, the episode was not known to anyone other than Holy Office insiders. It was not mentioned during any of the subsequent events (the denunciation and examination by the Index, campaigns for and against the publication of the decree of condemnation, or commentaries in journals and theology textbooks).

The document source that provides the details of this discussion consists of a short series of manuscript notes in the Archive of the Holy Office preserved along with a copy of *Evoluzione e dogma*.[22] A particularly significant fact is that the book, which bears the seal of the Library of the Holy Office, was unread: only the first fourteen pages, up to the beginning of the introduction, and some of the last pages containing the indexes have been opened, while the remaining pages are uncut, as if the book had just arrived from the bindery.

The section of the archive in which it is filed (Censurae Librorum) includes all materials related to the study of the books examined; these were frequently voluminous, as, for example, in the case of Mivart. Zahm's book is only wrapped in two sheets of paper. The inside sheet probably came together with the book. It has the title "Doctor Zahm and his scientific works" and it includes a short biographical note about Zahm and various texts intended for publicity. Here someone marked, with thick red and blue pencil strokes, two references to John Ireland, the American archbishop who was a friend of Zahm and who was considered to be the leader of the Americanist movement. The two references are about the relationship between Catholic and Protestant doctrine, not about evolution.

It is the second sheet of paper that interests us here, because on it are written a series of dates and notes that permit us to follow the chain of events. The first note states simply:

Saturday, January 16, 1897

The meeting of the cardinals took place in the Vatican. Let Zahm's book, *Evolution & Dogma*, be purchased and brought to the Particular [Congregation] to designate a Consultor.[23]

The Italian edition had appeared several months before, probably in late summer. But the date of this first note is interesting, because it coincides with the publication of the review by Salis Seewis in *La Civiltà Cattolica*. We cannot be sure if there was any connection between these two events. The note seems to refer to the previous General Congregation, that of Wednesday, January 13, before the journal was published. But at this point the journal had already been read in the office of the Secretariat of State. The negative finding by Salis Seewis could have already reached the ears of some of the cardinals. The petition, in any case, could not have been especially "official," because it does not appear in the Decreta S. Oficio, where the decisions taken in each Congregation were registered.

Some later annotations indicate that the matter was proceeding slowly. It was not until early May that a report on the book was sought from one of the consultors, the Jesuit Michele De Maria (who was also a consultor of the Index). But then comes the stipulation: "But first let us ask the Congregation of the Index for everything they have related to this issue."[24] The same note appears in Decreta S. Oficio, although curiously the dates do not coincide, because it appears only in the Congregation of May 6 (that week the Congregation of cardinals took place on Thursday, rather than Wednesday, the usual meeting day).[25] It is possible therefore that the May 1 note reflects the decision of the cardinal secretary or the assessor, formalized in the next Congregation.

Two final notes complete the available information. One written on May 7 says: "The Father Secretary of the Index awaits the official letter to transmit forthwith the documents relating to evolution." And four days later there is written, simply, "To the Index." On May 11, the assessor of the Holy Office asked the secretary of the Index to send him "everything found in that Congregation concerning the evolutionist scientific system, including books, report, and decrees."[26]

The Holy Office archival records stop at this point. There is no further indication that the scrutiny continued. The consultor De Maria did not examine Zahm's book: no report on Zahm appears under the name of this consultor in the card catalog of the Holy Office, which is usually quite complete. As noted, the pages of the copy acquired for this purpose remained uncut. Nor does the Congregation of the Index preserve any trace: the May 11 letter was not registered in the Diary, as was habitually done with letters. A petition of this type could not be ignored. But when some months later Zahm was denounced to the Index, no one recalled the examination pending in the Holy Office, although five of the seven cardinals present in the Congregation of May 6 were also members of the Index.[27] Two of them, Vannutelli and Parocchi, attended the Congregation that decided the case. The fact is that the Holy Office did not go forward. From November of that year on, the issue of evolutionism was examined only in the Congregation of the Index.

The Denunciation of Zahm's Book,
November 5, 1897

In the nineteenth century, many American Catholics were Irish or German im-migrants. Among the former there was a distinct tendency to Americanize while the latter frequently formed tightly knit, German-speaking communities. This particular cleavage was reflected in the American Catholic hierarchy and was exacerbated because the "Americanists" tended to be liberal both in religion and politics, while the Germanophiles were more conservative. Zahm was aligned with the Americanist faction.

Zahm's book was denounced to the Congregation of the Index by Archbishop Otto Zardetti. Born in Saint Gall, Switzerland, in 1847, ordained in 1870, he was a professor in the Seminary and canon in Saint Gall. In 1881 he went to the United States where he held various positions in an area where there were many German Catholics: he was professor of theology in Saint Francis Seminary in Milwaukee, Wisconsin, from 1881 to 1886, then vicar general of Northern Min-nesota from 1886 to 1889, and first bishop of the new diocese of Saint Cloud, Minnesota, from 1889 to 1894. The diocese of Saint Cloud was home to numer-ous German Catholic immigrants. In spite of his Italian surname, Zardetti was culturally German and was closely associated with the head of the American Catholics' conservative wing, Archbishop Corrigan of New York.

In 1894 Zardetti was named archbishop of Bucharest, Romania, but at the end of one year, in 1895, he resigned, probably for reasons of health, and went to live in Rome where he held a few minor posts, as consultor of two congregations of the Vatican. He died in Rome in 1902.

When he denounced Zahm's book to the Congregation of the Index, Zardetti was already living in Rome. Zardetti's charge is very complete and fills eight manuscript pages.[28] The first page is a letter in Latin, dated and signed in Rome on November 5, 1897, addressed to the cardinals of the Congregation of the In-dex, in which he says that, after having carefully examined Zahm's book, he had found it worthy of censure (Zardetti enclosed a copy of the Italian edition, which had already been published). The five pages following contain a summary of the ten principal points that Zardetti judged worthy of censure, and on the last two pages he transcribes the letter from *Le Monde* two years before, where Dalmace Leroy retracted his support for a doctrine similar to that contained in Zahm's book. The ten points extracted from Zahm's book are:

1. Evolution proposes a concept of God nobler than what is obtained by think-ing of a God who experiments with prime matter and only obtains results after many tries, a view that is, in addition, anthropomorphic.

2. The whole book is a defense of how a Catholic can be an evolutionist.

3. He supports the evolution of man's body, asks if there is anything against that either in dogma or in metaphysics, and replies no.

4. Analogy and congruity require acceptance of the evolution of man's body.

5. He cites Mivart in his favor as having been approved by the Holy See, which is untrue. Mivart's article on the punishments of hell was condemned, and after this Zahm proposes him as a theologian of recognized ability.

6. The animal origin of the human body is in perfect harmony with Saint Augustine and Saint Thomas. This is refuted in the attached article from *La Civiltà Cattolica*.

7. He also cites the authority of Leroy, whose book was judged untenable and incompatible with scriptural texts and with the principles of sound theology, and who retracted his views in a letter of February 26, 1895: a copy is attached. Zahm was unaware of the judgment of the competent authority.

8. The last chapter is completely favorable to evolution as an ally of religion. Special creation explains nothing and impedes progress.

9. Evolution does not degrade man but rather ennobles simians.

10. He recommends his thesis as in agreement with Scripture, the Church fathers, and eminent Catholic scientists like Monsabré, d'Hulst, Leroy, Lapparent, and Mivart.

Zardetti adds some additional details: the theory promoted by Mivart, Leroy, and Zahm has been refuted theologically in every detail in Cardinal Mazzella's treatise *On God the Creator;*[29] at the recent Congress of Fribourg there was active propaganda for the ideas of Zahm and evolutionism; and, finally, Cardinal Satolli, in his book *On Habits*, says that evolution lacks proof and is repugnant to the principles and conclusions of metaphysics and natural science.

Zardetti enclosed a pamphlet composed of the four articles that Salis Seewis wrote, between July and September 1897 in *La Civiltà Cattolica*, on spontaneous generation in ancient philosophy, Saint Augustine, Saint Thomas Aquinas, and Suárez,[30] and also the very negative review of Zahm's book in number 1118 of *La Civiltà*, analyzed previously.

Buonpensiere's Report,
April 15, 1898

Zardetti's denunciation was clear, complete, and ordered. It followed the normal procedure of the Congregation of the Index. The first step was the preparation of a report by a consultor, who in this case was the Dominican Enrico Buonpensiere, who had already intervened in the Leroy case on two occasions, writing very negative reports, as we have seen. On this occasion he drew up a report that was also completely negative, on the compatibility of evolutionism with Christianity. Dated April 15, 1898, in the Dominican House of Santa Maria sopra Minerva, where the abjuration of Galileo had taken place 265 years before, this report fills fifty-three printed pages (plus the text of Zardetti's denunciation).[31]

At the beginning of his report, Buonpensiere summarizes Zardetti's denunciation and transcribes two long paragraphs from the *Civiltà Cattolica* review of Zahm's book. He praises Zahm as an active, well-intentioned priest, versed in physics and metaphysics, but, agreeing with *La Civiltà*, he says that Zahm puts his good intention in the service of a bad cause. In any event, the praises of Zahm were exhausted early on. Buonpensiere transcribes several paragraphs of the book's introduction, where Zahm states that the Church has not formulated any theory on the origin of the world and its inhabitants and, as a result, no Catholic is obliged to follow any concrete theory so long as there is no evidence supporting it. Zahm alludes to criticisms leveled, in the name of religion, against Copernicus, Newton, and various theories of geology, adding that it is now evolution's turn and that it is shocking that some have not absorbed the lessons of the past. Buonpensiere comments that such opinions demonstrate Zahm's lack of familiarity with metaphysics, theology, and history (p. 4). Zahm's competence is questioned from the start.

Pros and Cons of Evolution

Buonpensiere devotes scant space to the history of evolutionary ideas and arguments for and against—the first part of Zahm's book. Inasmuch as Zahm makes no doctrinal assertions in these chapters, their analysis is the shortest part of the report (pp. 5–12). Buonpensiere says there is nothing worthy of comment in chapters 1 and 3–6. In chapter 2, Zahm presents the ancients, the Church fathers, and medieval authors who contributed to the bases of evolution, claiming that Aristotle should be considered the father of evolution. Buonpensiere says that one can judge the truth of this assertion by noting that Zahm erroneously attributes to Aristotle a doctrine of Empedocles that Aristotle cites in order to criticize it. Buonpensiere is right: As we have noted, Zahm has taken the reference to Aristotle lightly, or at second hand, and misses Aristotle's critique of the evolutionary ideas of Empedocles. From here on, Buonpensiere continues, we can imagine how much truth there might be in Zahm's attributions of evolution to Albertus Magnus and Thomas Aquinas. Buonpensiere is right enough here: Zahm's enthusiasm led him to present ancient and medieval figures as precursors of the modern theory of evolutionism (pp. 5–7).[32]

Chapters 7 (evidence in favor of evolution) and 8 (difficulties) fill more than a hundred pages of Zahm's book, but Buonpensiere spends only four summary pages on them. He limits himself to transcribing a few passages from chapter 7, preceded by the following observation:

The author sets out to prove the *reality* of Evolution, not with metaphysical arguments but rather on observation and the correct interpretation of the facts of nature. It is worth transcribing a few paragraphs, so as to see what Zahm's *famous demonstrations of the reality* of evolution are. They are simple assertions, and arbitrary explanations of biological phenomena that, with the same ease and certainty that Zahm proposes them, can be perfectly refuted by others, inasmuch as they lack both philosophical and traditional theological arguments. (p. 8)

This dialogue of the deaf is surely at the heart of many misunderstandings. A mentality accustomed to use metaphysical reasoning to study the physical world could have difficulty crediting the hypothetico-deductive method that is so frequently applied in the natural sciences. As for chapter 8, Buonpensiere only makes a brief comment, followed by a long quotation from Zahm that Buonpensiere seems to regard as Zahm's admission of the failure of evolutionism:

In chapter VIII, the Author tries to resolve the *Objections to Evolution,* but with scant success, to the point that he feels obliged to confess that "all the theories of Evolution connected to the above-named factors . . . involve numerous and grave difficulties, which, so far, have not been satisfactorily answered." Nevertheless, here is the general conclusion of *Part One* of the work: "The lack of this perfected theory, however, does not imply that we have not already an adequate basis for a rational assent to the theory of organic Evolution. By no means. The arguments adduced in behalf of Evolution in the preceding chapter, are of sufficient weight to give the theory a degree of probability which permits of little doubt as to its truth. Whatever, then, may be said of Lamarckism, Darwinism and other theories of Evolution, the fact of Evolution, as the evidence now stands, is scarcely any longer a matter for controversy." (p. 12)[33]

Buonpensiere did not think it necessary to say anything more on this subject. His report does not provide an adequate idea of Zahm's arguments and he dwells too much on the difficulties. He disposes of chapter 8 in three lines, referring to a presumed statement by Zahm confessing his failure. But Zahm had devoted more than sixty pages of his book (pp. 140–202) to discuss three of the most common objections to evolution (lack of evidence of species change, the imperfection of the fossil record, and the infertility of hybrids), to which he responds in great detail (pp. 143–62, 162–82, 182–93). Buonpensiere's report does not allude to this extensive analysis, and so evolution is presented as an arbitrary theory that lacks evidence.

The reference to Zahm that Buonpensiere presents as a confession of failure has little to do with what Zahm really said: he considers that taken together the arguments are conclusive, although he is also careful to point out the current limitations on explanations of the mechanisms of evolution. Zahm distinguished between evolution itself (as a fact based on the convergence of a number of arguments) and the mechanisms of evolution, about which there was no consensus. Buonpensiere does not appear to have grasped the distinction.[34]

Two Different Perspectives

The second part of the report (pp. 12–43) on the compatibility of evolution and theism and the origin of man is much more extensive. Buonpensiere keeps up the attack. According to Zahm, he says, evolution and faith are compatible, but the only proof he provides is the opinion of a few contemporary authors. But in this chapter (part 2, chapter 1 of his book) Zahm does little more than explain the plan of the following chapters; so Buonpensiere's critique doesn't make much sense. In the next two chapters, Zahm offers an intelligent critique of agnosticism and Haeckel's monism; Buonpensiere says nothing about agnosticism, but remarks that Zahm's journalistic language does not constitute an adequate scientific refutation of Haeckel.

On evolution and theism (Zahm's next chapter), Buonpensiere introduces a personal observation, whose scope is difficult to evaluate. He says that Zahm poses the problem poorly, because everyone agrees that God did not create everything immediately out of nothing: he created primitive matter and formed the rest (with the exception of the human soul), making use of the potential of previously created cosmic matter. According to Buonpensiere's complicated reasoning, the disagreement between evolutionists and anti-evolutionists is whether the potential for developing the first plants and animals, and the first man, out of cosmic matter was active or only passive, and if it were partially active for some organisms, whether it was principally or instrumentally active: "With these distinctions and subdistinctions, and with still others soon to be introduced, problems multiply and the basic theorems of evolutionism and anti-evolutionism form themselves into two opposing sides, as anyone can observe" (p. 15).

Evolutionists, Buonpensiere continues, speak of an active principal potential in secondary causes. Anti-evolutionists do not agree: in the chaotic matter initially created by God, there was only passive potential—the potential to be activated by mineral and animal substantial forms. Such passive potential was activated by the word of God, in the first days of Genesis, with the multiple forms of minerals, plants, animals, and men, just as Moses describes it. (Buonpensiere seems to find Zahm's book scientifically useless because it does not deal with philosophical issues.)

Buonpensiere then moves on (pp. 16–28) to examine the positions of Saint Augustine and Saint Thomas, to show that Zahm has erred by presenting them as precursors of evolution. Augustine's "seminal reasons" do not refer to any evolution but to a passive potential deposited in prime matter from the beginning, which God activates later on. Thus, Buonpensiere concludes, to attribute evolutionism to Saint Augustine is to defame a saint. Neither was Aquinas an evolutionist. Although it is difficult to determine exactly what Augustine meant by "seminal reasons," it is not difficult to rebut Zahm, who exaggerates Augustine's stance somewhat and Aquinas's greatly. Moreover, these appeals to authority are minimally relevant to the problems posed by evolution.

Man and Monkey

The origin of man is the focus of Buonpensiere's principal conclusion. Zahm had written that the human soul does not owe its origin to evolution. Although this is not a defined dogma, it has been a Catholic doctrine almost universally taught since the days of the apostles. The issue, therefore, concerns the body only. He acknowledges the difficulty and even the impossibility of finding the missing link between simians and men, but says that, if evolution takes place on other scales, it is logical to believe it acts here as well. Nothing in either dogma or metaphysics opposes this. In his favor, Zahm cites Mivart and Saint Thomas, who stated that angels could cooperate in the formation of Adam's body; we might as well believe in the cooperation of other creatures as links in evolution. Rejecting this line, Buonpensiere attempts to show (pp. 34–40) that Saint Thomas does not accept angelic cooperation: he expressly explains the immediate creation by God of Adam's body and soul at the same time.

Zahm states that a Christian is free to accept moderate evolutionism and says that this doctrine has not, until now, been condemned by any of the congregations, and mentions Cardinal González and Leroy as favoring this view. Buonpensiere comments: "anyone can understand the value of this last argument . . . therefore, I will dispense with refuting it" (p. 40); he refers the reader to Leroy's retraction. (Recall that the Congregation of the Index, to whom his report is addressed, had recently decided, with Buonpensiere's participation, to condemn Leroy's book, although the corresponding decree was not published.)

In section 31 of his report (pp. 40–41), Buonpensiere summarizes the basic canon of the hermeneutics that Catholic evolutionists used to interpret Scripture: they distinguish between revelation and its interpretation, in such a way that the interpretation (and, therefore, traditional interpretation) can change. Buonpensiere says that this can be interpreted benignly but adds that in such a way one would not distinguish between the unanimous opinion of the fathers and the personal opinion of some private author: "It is because of this rashness that, in the event that no measure be taken against the present work [by Zahm], I would deem it opportune to warn the illustrious author about the ambiguity of the expressions he uses in biblical exegesis" (p. 41).

It is interesting that Buonpensiere, in spite of constant criticisms of Zahm's work, admits that it is possible that the final decision of the cardinals of the Index might be *dimittatur*, which here can be translated "no measure be taken against." This seems to indicate that, in spite of the anxiety that evolution provoked, there was no clear criterion for making a decision. Buonpensiere leaves the origin of man here, although he will return to it, more forcefully, in his concluding remarks.

Natural Purpose

Zahm discusses purpose in nature in chapter 7 of his book. He argues that evolution overthrows ancient ideas about purpose in nature but also provides a basis for a much more interesting and profound understanding of this purpose and, ultimately, of God's plan that orders nature. As we have already seen, that was the subject of his Fribourg lecture at the 1897 Catholic Science Congress. Buonpensiere summarizes this chapter in a brief comment: "Not of interest" (p. 42).

Buonpensiere treats chapter 8 with similar indifference: Zahm asserts that evolution is an ancient idea and was taught by Aristotle and Saint Thomas; he waxes poetic on future evolution; he predicts that in a not-too-distant future evolution will be accepted by the doctors of the Church as an ally of religion; he claims to have resolved objections but only supplies ostentatious erudition.

In section 34 of his report, Buonpensiere provides a synthesis of evolutionism as Zahm presents it: God created the matter of the world in a chaotic state, and along with it primordial forces and motions. He subjected this matter to a law of ascendant progression, in order to make the cosmos. From natural forces there arose plant life, which from the hand of divine providence developed little by little until animal life was produced and, with the passage of enormous periods of time, anthropomorphic animals. God summoned one of these beings, infused the rational soul, and made Adam. Adam's body thus was not formed directly and immediately by God, but indirectly, through the forces that God himself infused in matter from the first moment of Creation.

Buonpensiere observes that in the final analysis, if we think of man from an evolutionary standpoint, it is difficult to conceptualize him as created in God's image. Zahm attempts to show that his limited version of evolution is not opposed to Catholic dogma, but his arguments are simple assertions, frequently lacking explanations, ranging from error to sophism.

Adam's Body and Catholic Doctrine

Toward the end of his report, Buonpensiere presents a most unusual proposal: he formulates his own theological thesis, attempts to demonstrate it, and proposes that the Congregation condemn the contrary doctrine. To begin, he writes his thesis in capital letters:

I THINK IT IS CATHOLIC DOCTRINE TO STATE THAT GOD HAS MADE ADAM, IMMEDIATELY AND DIRECTLY FROM THE MUD OF THE EARTH. (p. 45)

To support this thesis, Buonpensiere considers the interpretation of the words of Genesis, based on the hermeneutic rule that the obvious, natural sense of biblical words should not be abandoned unless it leads to absurd conclusions, which does not happen in this case. He follows with seven pages of citations from the Church fathers and theologians, including Peter Lombard, Albertus Magnus, Thomas Aquinas, Alexander of Hales, Durand, Duns Scotus, Saint Buonaventure, and Suárez, all of whom—he concludes—agree with his thesis, and some of whom explicitly teach that this is a doctrine of Catholic faith: "Now then, the unanimous agreement of the Church Fathers and Scholastic Theologians in matters of Faith and Customs bears certain witness to Catholic Dogma; just as Melchor Cano, among others, instructs. . . . Thus the truth of the conclusion presented above is sustained" (p. 51).

Further support is found in two conciliar canons. One from the provincial Council of Braga I (A.D. 633), canon 13, which condemns Manichaeans and Priscillians, according to whom the creation of "all flesh" is not the work of God but of bad angels. And another of the provincial Council of Cologne (1860), which teaches that "The first parents were created [conditi] directly by God. Therefore, we declare as contrary to Sacred Scripture and to the faith the opinion of those who are not ashamed to assert that man, insofar as his body is concerned, came to be by a spontaneous change [spontanea immutatione] from imperfect nature to the most perfect and, in a continuous process, finally [became] human." Buonpensiere only adds: "It is unnecessary to say anything about the value of these two canons" (p. 52).

Buonpensiere's Proposal

In the four last sections of his report, Buonpensiere sets out his conclusions.

> Reverend Father Zahm's book, leaving aside other theological ambiguities, is a continuous apology of a doctrine contrary to the truth of the Catholic faith: it does not seem to me susceptible to emendation, for it would have to be done over from the beginning. Therefore, it merits proscription. After the condemnation inflicted on Father Leroy's book, to leave without censure Zahm's book, which teaches the same theory, does not seem to me to conform to the rules of justice. (p. 52)

Still, Buonpensiere continues, for the same reasons that led the Congregation not to publish the condemnation of Leroy's book, limiting itself to warning him and requesting that he withdraw his book from sale and issue a brief retraction, it should display the same charity in Zahm's case.

Buonpensiere ends his report proposing that the evolutionary origin of Adam's body be officially condemned by the Church:

Nevertheless, inasmuch as it is necessary, once and for all, to let Catholic natural-ists know publicly that it is not permitted to teach that Adam's body may not have originated immediately from the mud of the earth but comes from an anthropo-morphic brute, I propose that, as was done on other occasions, this Holy Congre-gation condemn the following proposition, or another like it, that is: "God did not make the body of Adam immediately from the mud of the earth, but out of the body of an anthropomorphic brute, which had been prepared to be produced by the forces of natural evolution from lower matter." (p. 53)

The Congregation of the Index approved the way it would act with respect to Zahm, but it did not condemn any concrete proposition. A condemnation of this type is found rather as falling within the competence of the Holy Office. As we will see, some consultors proposed, precisely, to ask the Holy Office if trans-formism contradicts divine revelation. There is no record that this question was ever posed.

The Assessment of the Italian Translation, and Something More

Buonpensiere's report is dated April 15, 1898. In a note from the archive we learn that, inasmuch as this report was based on the Italian edition of the book, the prefect of the Congregation of the Index, Cardinal Andreas Steinhuber, wanted another consultor to ascertain whether the Italian translation reproduced the original English faithfully and that, in the event there were serious, substantive variants, that he give his opinion on the original. This task fell to the consultor Bernhard Doebbing, a German Franciscan fluent in both languages (he had once been a philosophy professor in Cleveland, Ohio), who on June 10, 1898, re-ported that the two versions were virtually identical. Once that was established, the cardinal prefect was content to rest with Buonpensiere's report. This note is printed and is filed with two more pages, also printed, containing the report by Doebbing, who points out some minor differences between the Italian transla-tion and the English original: for example, a few sentences and some words ad-dressed to Italian readers added by the translator, "who sings the glories of evo-lution," and which are not found in the original text. Doebbing writes that he has compared the two texts diligently, and that the Italian translation is a very faithful one: even more, it is a literal translation, insofar as language permits.[35]

Doebbing's annotations are reproduced in Buonpensiere's report. There are paragraphs in English, written in the margin, wherever quotations from the book used by Buonpensiere do not perfectly agree with the English original. In some instances the changes have some importance, and he calls attention to Doebbing's not having noticed them. Doebbing, however, had reached the correct conclusion, which shows that the Congregation attempted to act with honesty and rigor.

The circumstances of Doebbing's report reveal a more important piece of information. In the Diary of the Congregation, a note about the prefect's charge explains how the copy of the original English version of Zahm's book had been obtained. Cardinal Satolli had a copy, which he gave to the secretary of the Congregation of the Index. But Satolli also had an English original of another work, *The Life of Father Hecker,* a key Americanist text that was also under examination by the Congregation of the Index at this time, and which had been given to the consultor Eschbach in its French translation. Both originals were given to Doebbing for comparison.[36]

Francesco Satolli had been a professor in the United States, and from 1893 he was, besides, the first apostolic delegate of the Holy See in the United States. He knew the principal American clergy. He was named cardinal in 1895 and then returned to Rome. In those years he had already declared himself in favor of the "conservative" wing of American Catholicism and against the "liberal" wing of Zahm and the paladins of Americanism. The French translation of *The Life of Father Hecker* was the catalyst that led to the crisis of Americanism. We have already noted that evolutionism and Americanism converged in Zahm's case, and now we see that the two themes were under study by the Congregation of the Index at the same time, and that Cardinal Satolli furnished the Congregation with the original American editions of the two books that were the focus of the two controversies. This convergence decisively influenced the way events unfolded.

The Preparatory Congregation,
August 5, 1898

As was customary, once the report had been prepared, it was examined in the Preparatory Congregation, which on this occasion was held in the full Roman summer heat, Thursday, August 5, 1898, in the Palace of the Chancellery, one of the principal palaces of Rome. The Congregation of the Index had its seat in some rooms in this palace.

At the meeting, besides the Dominican Marcolino Cicognani, who was secretary of the Congregation, and Alberto Lepidi, another Dominican who was master of the Sacred Palace, eleven consultors were present: Giuseppe Pennacchi, Alphonse Eschbach (rector of the French Seminary), Franz Xaver Wernz (Jesuit), Davide Farabulini, Enrico Buonpensiere (Dominican, rector of Santa Maria sopra Minerva and presenter in the case), Teofilo Domenichelli (Franciscan), Pie de Langogne (Capuchin), Angelo Ferrata (Augustinian), Bernhard Doebbing (Franciscan, rector of the College of Saint Isidore, in Rome), Arcangelo Lolli (regular canon), and Rafael Merry del Val.[37] Six of those present had participated in the meetings in which Leroy's book was examined: three had been in the two Preparatory Congregations (Cicognani, Buonpensiere, and Domenichelli), and the other three in the second (Pennacchi, Eschbach, and Wernz).

This Preparatory Congregation reveals that differing opinions were expressed in the meeting. In the archive there is a draft of three summaries of the results of the deliberations.[38] The last one coincides with the summary recorded in the Diary:

> Bearing in mind that the author does not directly espouse the doctrine of evolutionism in his book, rather he expounds it with some support, and that this doctrine is not opposed to any defined dogma nor does it impugn the faith, three favored that *the author only be warned.*
>
> Inasmuch as this doctrine does not agree with the verdict of Holy Scripture, nor with the doctrine of the Church and the Church Fathers and Theologians, five declared themselves *in favor of the condemnation of the work and the publication of the Decree:* four, with *prior admonition of the Author;* one, *absolutely.*
>
> Since the Holy See already made clear its position on this doctrine when it condemned the work of Father Leroy, one said: *that the book should be prohibited and the Decree published in accordance with the norm and a warning to Catholic writers.*
>
> Finally, another said: that he should not be condemned now; in the meantime, the Holy Office should be asked to consider and resolve the doubt: does transformism contradict divine revelation? Afterward, this Sacred Congregation might proceed against the book.
>
> Only one was in favor of the report of the Consultor.[39]

Eleven votes were recorded. What happened to the other two? The manuscript minutes show that some consultors had changed their minds. One version of the minutes is written in a hand that is not that of Secretary Cicognani and is quite different from the rest. There we read that

> one abstained; another, upon not finding in the work any error about defined dogmas, also abstained from emitting any judgment; three favored prohibiting the work, and suggested that meanwhile the Holy Inquisition [Holy Office] be asked to proclaim its judgment regarding whether the doctrine of transformism can be accepted and taught; five thought that the work should be prohibited and the decree published; three opined that the work should not be prohibited now and that, meanwhile, the author should be severely warned that in the future he not adopt these and other new and pseudo-philosophical opinions which in our times depart from Catholic doctrine on the origin of the world.[40]

Thus, thirteen votes were actually recorded, and there were also two abstentions: it is possible that the final draft of the report did not record the abstentions; or it may reflect a preliminary round of voting.

Cicognani wrote another report. Because it is filled with deletions and interpolations, its contents seemingly represent a draft intermediate between the earlier report and a final one, or perhaps a second round of voting.

As we have just seen, it is said in these reports that evolutionism is not opposed to any defined dogma of the faith. Even those who favored prohibition said only that it "was not in agreement" with Scripture, tradition, theologians, and the doctrine of the Church. Others said openly that it was not opposed to any dogma, or based their vote on other reasons, or abstained. Moreover, the proposal to ask the Holy Office to pronounce on evolutionism was initially supported by three attendees, which shows a clear awareness that up to that moment the Holy Office had not pronounced.

**The General Congregation,
September 1, 1898**

With the end of the hot Roman August, the results of the Preparatory Congregation were debated at the General Congregation of September 1. Besides Secretary Cicognani and Lepidi, master of the Sacred Palace, six cardinals participated in the meeting: the prefect Andreas Steinhuber, Serafino Vannutelli, Lucido Maria Parocchi, Francesco Segna, Girolamo Maria Gotti, and Raffaele Pierotti.[41]

Three of them had participated in the two General Congregations in which Leroy's book was examined: Serafino Vannutelli, who at the time was prefect of the Congregation and now was prefect of the Congregation of Bishops, a very important post; Parocchi, who also occupied two important posts: secretary of the Inquisition or Holy Office, and vicar general of the pope for Rome; and Pierotti, not yet a cardinal, who had participated in his role of master of the Sacred Palace. Two others had taken part in the second General Congregation only: Steinhuber, who now was prefect, and Segna, who had voted in favor of Leroy. Only Gotti had not participated in either of those Congregations. Six cardinals who had taken part in the Leroy case were not present; some of them had died.

As was the custom, the place of the meeting (in a designated room in the Vatican), who attended, and the results of the deliberations were all noted in the Diary of the Congregation. With respect to Zahm's book it was decided:

> That it be prohibited, but that the Decree not be published until it was known (via the Father General of his order) whether he wanted to submit and repudiate his book; that he would also be warned via the same Father General that he notify Alfonso Maria Galea, the Italian translator of his book, of the Decree of prohibition, and so that in the future he not publish books that treat theological and religious matters without the prior censorship of his Ordinary and religious Superior.[42]

As was also the practice, this summary furnishes no account of how the meeting developed. The memorandum prepared by the secretary for his audience with the pope is somewhat more explicit:

Audience of September 3, 1898.

Holy Father,
The Sacred Congregation of the Index which took place on the first day of the present month of September was quite interesting. All the cardinals who belong to our Congregation and were in Rome participated, that is, the Most Eminent Parocchi, Serafino Vannutelli, Gotti, Segna, Pierotti, and Prefect Steinhuber.[43] Six books were judged, after examination and serious discussion. . . .

The fourth is Father Zahm's book, *Evolution and Dogma*. It was denounced by Monsignor Zardetti and criticized by *La Civiltà Cattolica*. The report was written by Father Buonpensiere, Dominican, on the basis of the translation from English by a certain Alfonso Maria Galea, and concluded in favor of the prohibition, giving his reasons. Because Zahm supports the system of evolution not only for plants and the lower animals, but also for the body of man: that is, man can be genealogically kin to some unknown simian or monkey species, and that in such a genealogical affinity there is nothing opposed to metaphysics and Dogma. To support his thesis the author instructs that the opinion of the origin of Adam's body from a simian or monkey is in perfect harmony with the principles adduced by the two great luminaries of the Church, Saint Augustine and Saint Thomas. All this is totally false, as the Consultor established when he concluded that Father Zahm, in addition to so many theological ambiguities, consistently argues in favor of a doctrine contrary to the truth of the Catholic faith. But since Galea's translation was the version studied, to assure us that it was in conformity with Zahm's original, Father Doebbing was charged with comparing the two, which he found to be in perfect agreement. Based on these materials and after a long and serious debate, the Cardinals decreed the prohibition, but that the Decree not be published until Father Zahm, through the mediation of his General, makes an act of submission.[44]

The cardinals' long discussion produced a decision that was seemingly unanimous, inasmuch as no dissents are mentioned. This decision was stronger than that of Buonpensiere, who had said the book deserved to be prohibited but proposed to act in a gentler way, as had been done with Leroy and for the same reasons. The cardinals, however, decided to prohibit the book. For the moment the decree would not be published to give Zahm the opportunity to accept the decision, so that it could be stated in the decree that the author had submitted in a praiseworthy manner, but it was also decided to publish the decree without conditions: if Zahm failed to submit, it would be published just the same, but without the clause stating Zahm had accepted the decision, which would be worse for Zahm.

The final steps of this stage were the usual ones. On September 3, the pope received the secretary in audience, approved and confirmed the report, and ordered the decree published. The decree refers to the list of books prohibited in the session of September 1: seven are listed but Zahm's book is not one of them. The decree is dated September 3 (the day of the audience with the pope), and on September 5 it was published in Rome in the usual way.[45]

The publication of the prohibition of Zahm's book was delayed pending receipt of his submission via the general superior of the Congregation of Holy Cross. Everything now appeared in order, with only a few customary steps remaining, simple ones in the case of a good Catholic like Zahm, who quite recently had been procurator of his congregation in Rome. Now, however, is when the really complicated part of his case began.

The Decree That Was
Never Published

Secretary Cicognani communicated to the general superior of the Congregation of Holy Cross what the cardinals had decided about Zahm's book and gave him permission to communicate the decree to Zahm and to ask if he wished to submit.[46] On September 18, the general superior, Gilbert Français, who was to become one of the key players in the complex negotiations just under way, wrote from the General House of the Congregation of Holy Cross, in Neuilly-sur-Seine (eight kilometers from Notre Dame de Paris), the following letter to the cardinal prefect of the Congregation of the Index:

Congregation of Holy Cross
General House
Neuilly-sur-Seine, 30 Avenue du Roule
September 18, 1898

Eminence,
I have received, through a letter from the Secretary of the Sacred Congregation of the Index dated this September 10, notification of the decree condemning the work *Evolution and Dogma*, written by Father Zahm, one of our members in the United States.

I immediately transmitted a copy of the decree to him, inviting him to make an act of Christian submission to this supreme decision, and to comply with all the obligations which follow from it.

I am convinced that he will do it in the full spirit of faith and with all humility. As for myself, Eminence, I am entirely disposed to work everything in accord with the tenor of the decree, the holy laws of the Church, and the duties of my position.

With deep respect, I am Your Eminence's most humble and obedient servant in Our Lord

Gilbert Français
General Superior of the Congregation of Holy Cross[47]

Zahm responded in submission, and Gilbert Français so communicated to the cardinal prefect in a letter written in Rome, dated the following November 4. Up to this point everything was normal and expected. But that same day, Français wrote another letter to the cardinal, in which he asked that the decree not be published. The two letters can be consulted in the Archive of the Index.[48] The first reads:

Eminence,
As I had the honor to write you several weeks ago, I transmitted to Father Zahm a copy of the decree of the Sacred Congregation of the Index which contained the condemnation of his book, *Evolution and Dogma*. I have received his reply, and it is just as I hoped. I am expressly authorized to say, in his name, that he submits with no doubt, with the greatest respect and with all his heart, to the judgment made on his work.

I have the honor to be, with the deepest respect, etc.

Gilbert Français
Superior General of the Congregation of Holy Cross
Rome, November 4, 1898
19, via dei Cappuccini

The second letter is similar to the first, but with an important addition: a plea that the decree prohibiting Zahm's work not be published:

Eminence,
Just as it was agreed, I have the honor to send you, in writing, Father Zahm's submission to the decree of the Sacred Congregation of the Index, which condemns his work *Evolution and Dogma*. I have scrupulously translated the expression of his sentiments, as he wrote them to me.

I can add in my own name and in his that we are wholly disposed to do all that is prescribed for us, to repair that which ought to be repaired in this matter.

Again I dare to solicit the great goodness of your Eminence, that you might stop the publication of the decree. Your Eminence knows America well enough, the juxtaposition of the different races found there, and their present state of mind, to be able to think that such publication perhaps could harm the life of Father Zahm, who is a truly worthy priest, and a man of religion whose great influence only has need of direction for it to have great utility.

Moreover, such publication, besides being for myself and my Congregation certainly a cause of authentic suffering, it could create serious difficulties of administration.

In these circumstances, Father Zahm has ceased to follow presumptuous and rash ideas. It is also beyond doubt that he possesses real qualities as an administrator, and that, in this area, his cooperation could not be more precious to me.

Since I must return to France Monday morning, I humbly ask the blessing of Your Eminence, and I beg you accept the expression of my most respectful sentiments in Our Lord

Gilbert Français
Superior General of the Congregation of Holy Cross
Rome, November 4, 1898
19, via dei Cappuccini

Indeed, the publication of the decree would have been a powerful blow to Zahm, who had returned from Rome to the United States to take up the post of provincial of his congregation, and also for the superior general and the Congregation of Holy Cross, which was active at that time in several countries.

November 4, 1898, was a Friday, and Français said in his second letter that on Monday he would return to France from Rome. The same Friday, in the Diary of the Congregation it is noted that Zahm, having been apprised of the prohibition of his book, submitted in a praiseworthy manner and repudiated his book, in accord with the decree of the Congregation.[49] The next Diary note states that three months had passed and the decree had still not been published. Even more: it says that on February 3, 1899, the pope personally gave instructions to the cardinal prefect of the Congregation of the Index not to publish it:

> In the audience with the Cardinal Prefect . . . the Pope ordered that the publication of the Decree of September 1, 1898, in which Father Zahm's book *Evolution and Dogma* was prohibited, be suspended, even though the Author submitted in a praiseworthy manner and repudiated the work, until Father Zahm, who will soon come to Rome from America, can be heard.[50]

Zahm did not go to Rome, and the decree was not published. Documents found in other archives show that both Zahm and Français applied all available means to avoid the publication of the decree. Moreover, they were not alone: they were abetted by come clerics involved in the cause of Americanism, who carried out many complex negotiations that reached the pope himself. Other clerics sought the condemnation of both Zahm's book and Americanism. To fully understand Zahm's case, it is necessary to look into these complex circumstances.

Americanism and Its Opponents

Americanism was not a movement, properly speaking, but rather an attitude that initially sought to "Americanize" the Catholic Church in the United States. In the nineteenth century, many American Catholics were European immigrants, and it was often difficult for the established population to view them as authentic Americans. Prejudice was particularly aimed at poorer immigrants, especially Irish, Italians, and, somewhat later, Poles. Because of the Irish potato famine, a severe agricultural failure that was historically the last subsistence crisis in western Europe, some 780,000 Irish Catholics arrived in the United States. Immigration of German-speaking Catholics was also large. The Catholic Church in the United States was considered a church in a missionary country. For that reason it was administered through the Congregation for the Propagation of the Faith (Propaganda Fide in Latin, thus its abbreviated name: Propaganda). Catholics of German origin wanted to have their own schools and hierarchy, whereas the Irish supported a complete Americanization of the Catholic Church.

The Americanists also had their own positions with respect to the social question (most of the immigrants were workers), Catholic and public education (they favored the integration of both), and some religious issues, and they placed special emphasis on everything related to democracy and the separation of church and state. At the end of the nineteenth century there was tension between the Americanists, sometimes called "liberals," and those Catholics described as "conservative." The liberals generally supported modernizing the Church to adapt it to the ideas of the epoch. However, they fully accepted the faith and the discipline of the Church. This ideological context is relevant to Zahm's case through his association with the liberal faction.

The leader of the Americanist group was John Ireland.[51] From 1875 he was bishop of St. Paul, Minnesota.[52] In 1884 that diocese was elevated to a metropolitan archdiocese with five suffragan dioceses, one of which was that of St. Cloud. The first bishop of the latter, named in 1889, was the Swiss cleric Otto Zardetti who later was to denounce Zahm's book to the Congregation of the Index in Rome. Zardetti knew Ireland personally and had firsthand familiarity with the Americanist current and its principal representatives.

Ireland wanted to Americanize the Church, conceding great importance to American values and adapting the Church to the social and political context of the United States. This is the period in which the American nation began to play a role on the world stage. In 1898 the conflict between the United States and Spain led to drawing Cuba and the Philippines, two Catholic countries, into the orbit of the United States. Ireland and the Americanists favored the diffusion of American values in the world, although at the same time they tried to mediate with American administrations to work for better conditions for the Catholics of the newly independent countries of Cuba and the Philippines.

James Gibbons had been a bishop in North Carolina from 1868 when he was thirty-four years old—at the time, the youngest bishop on the entire Catholic world. Amiable and effective, a great patriot, he was convinced that the Church enjoyed conditions for carrying out its mission in the United States that could not have been better. He was named bishop of Richmond in 1872 and coadjutor archbishop of Baltimore in 1877. In 1888, when the previous archbishop died, he became titular archbishop of Baltimore, the principal see in the United States in that period. The pope named him president of the Third Plenary Council in Baltimore (1884) and made him a cardinal in 1886. He was the highest Catholic authority in the country, was greatly respected, and was a moderate individual, although he counted himself, nonetheless, on the Americanist side.

In 1889 the first centennial of the establishment of the Catholic hierarchy in the United States was celebrated. The Church was flourishing, with seventy-five dioceses already and no end of growth in sight. In that moment the Americanists were well situated. Cardinal Gibbons and Archbishop Ireland were outstanding figures, and two other Americanists also occupied important positions: Bishop Keane and Monsignor O'Connell.

John Joseph Keane had been bishop of Richmond, Virginia, since 1878 and was later named the first rector of the new Catholic University of America, authorized by the pope on April 10, 1887, and established in Washington, D.C. The university lay in the diocese of Baltimore, and Gibbons was its chancellor. Moreover, in 1885 Monsignor Denis O'Connell had been named rector of the American College in Rome, founded the year before to receive American seminarians studying in Rome. O'Connell had known Gibbons ever since he was a young priest. The level of mutual understanding among Ireland, Keane, and O'Connell was very high, and all three worked together in Rome in 1886 to obtain the creation of Catholic University. When he was in Rome, O'Connell was, in practical terms, an unofficial representative of the American hierarchy (especially of Gibbons and Ireland), owing to his tireless pursuit of ecclesiastical politics.[53] The liberal group had very good relations with Roman clerics, among whom the two Vannutelli brothers, both cardinals, stood out. Serafino Vannutelli was cardinal prefect of the Congregation of the Index from 1893 till 1896.

The conservative wing, led by the archbishop of New York, Michael Corrigan, was backed by three important Italian clerics who had lived in the United States and occupied influential posts in Roma: Francesco Satolli, Camillo Mazzella, and Salvatore Brandi. In 1893 Satolli was named the first apostolic delegate of the Holy See in the United States and was made a cardinal in 1895. The Jesuit Mazzella, as we have seen, was prefect of the Congregation of the Index from 1889 to 1893, when he was named prefect of the Sacred Congregation of Studies. Salvatore Brandi, also a Jesuit, was a member of the editorial team of *La Civiltà Cattolica*, was in touch with Corrigan by mail, and aggressively opposed Zahm's book.

In September 1893 a religious conference (the so-called World's Parliament of Religions) was held in the United States, which Catholics attended along with non-Catholics.[54] This matter, as well as the support by liberals of some secret workingman's societies, was among the causes of the increasing confrontation of Corrigan and his backers in Rome and the liberals. On January 6, 1895, the pope published an encyclical (*Loginqua oceam*) addressed to the American bishops. It included both great praise and but also reservations about specific issues in debate. Around this time the influence of the liberals began to decline. This same year the Vatican, responding to pressure from the conservative wing, asked O'Connell to resign as rector of the American College, which he did, and the following year Keane, at the pope's request, resigned as rector of the Catholic University of America.

Nevertheless, the activities of the liberals continued, and to some extent were internationalized. O'Connell continued to reside in Rome; Cardinal Gibbons named him vicar of the Church of Santa Maria in Trastevere, of which the cardinal was the titular. O'Connell moved into an apartment on the Via del Tritone, known as Liberty Hall, where a group of his friends used to gather. One of them was John Zahm, who arrived in Rome in 1896 as procurator of the Congregation of Holy Cross and became a great friend of O'Connell. Keane went to live at the Canadian College in Rome. O'Connell's group, besides Keane and Zahm, included some influential figures, including the Vannutelli brothers; Louis Duchesne, rector of the French School in Rome; and David Fleming, consultor of the Holy Office, who had written a favorable review of Zahm's book.

The Life of Father Hecker

Although the decisive battles were fought in Rome, France was the scene of the greatest difficulties that Americanism faced. Liberal French Catholics, who favored collaboration with the Republic, found in American Catholics a model to emulate in their relations with a democratic government. The American liberals had many points of contact with France. Now, the French translation of *The Life of Father Hecker* became the fuse that exploded existing tensions among liberals and conservatives.

Isaac Thomas Hecker (1819–88), son of poor Prussian immigrants was born and died in New York City. Although his education was quite limited, he was an original, intuitive thinker. Born into a Protestant family, he was baptized in the Catholic Church in 1844. In 1845 he joined the Redemptorists in Belgium, was ordained a priest, and returned to the United States in 1851. He worked tirelessly among the burgeoning Catholic immigrant population, while at the same time becoming a practiced, very persuasive orator, admired among Catholics and, perhaps principally, among non-Catholics. Hecker and some of his friends then left the Redemptorists and founded a new society—the first male Catholic order to be founded in the United States—known as the Paulists, who focused on preaching and the press. Cardinal Newman said that Hecker's work in the United States was similar to his own in England: Hecker devoted himself to building acceptance for Catholicism in a quite hostile society, where Catholics were viewed as foreign and antidemocratic.

After Hecker's death in 1888, his disciple Walter Elliott, a Paulist priest, wrote a biography titled *The Life of Father Hecker,* published in New York in 1891.[55] The book was translated into French, with an introduction by Archbishop Ireland and a preface by Félix Klein.[56] The book's publication in France ignited a broad polemic. The most famous critique was that of Charles Maignen a priest who wrote a series of articles in *La Vérité* beginning on March 3, 1898. He asked himself whether Hecker was a saint and replied that he was, in fact, a radical Protestant. Maignen criticized Hecker for his insistence on the Holy Ghost as an interior guide, for favoring the active virtues over the passive (humility, obedience), and accused him of denying the objective certainty of Catholic faith. Maignen collected his articles in book format, but the cardinal archbishop of Paris did not grant him permission to publish. He did receive permission, instead, from Alberto Lepidi, master of the Sacred Palace, which seemed to presage some action by the pope.

In fact, *The Life of Father Hecker* was denounced to the Congregation of the Index by Charles-François Turinaz, the bishop of Nancy, on April 24, 1898,[57] and it was examined by that Congregation, in whose archive are two printed reports by consultors, a forty-eight-page analysis by Bernhard Doebbing, dated July 5, 1898, and another of thirty-two pages by Alphonse Eschbach, dated May 26, 1898.

Doebbing examined the English original and recommended prohibition of the book. In his report's conclusion he attributes to Hecker a range of ideas. Hecker speaks repeatedly of a new era, a new phase of the Church, led by America; just as America surpasses all other nations in its social life, so, with its spirit of freedom and individuality, it will illuminate a renewed Church, under the direct, special, and abundant inspiration of the Holy Spirit. Spiritual life that until now had been practiced in the Church principally through grace, prayer, penance, the Eucharist, and other sacraments had grown weak in the development of the natural virtues; in the phase to come, natural sacraments had to be employed, that is, all earthly means for cultivating the natural virtues, in order to reinforce individual activity and the distinctive independence of America and Americans. Traditional devotion, with its tight discipline and characteristic asceticism, with evangelical counsels in monastic life, produces innocents lacking in personal energy; with the arrival of new, free, and independent activity, individuals can strive for a state of perfection under the guidance of the Holy Spirit, following the example of Saint Paul. There are no rules for discerning the spirit: one is guided by personal self-assurance and the direct instruction of God, with no place for external direction, or reducing it to the minimum. The scholastic method followed in Catholic schools is an obstacle to true science. The Paulists are the leaders prepared by the Holy Spirit. The Anglo-Saxon "race" bears within it the seed of the universal kingdom of the future. Catholicism must be unified with Protestantism through Americanism.

Doebbing's conclusion is that the little good found in this book is not new, and what is new is not good. Archbishop Ireland's preface increases the danger of deception, owing to his great influence. Not only Ireland but Klein, too, and Hecker's other followers say that Hecker still lives and will see the triumph of his ideas, for which they are actively working. According to Doebbing, liberals who want to reform the Church and save the world gather now in the name of Hecker, just as they did in the sixteenth century in the name of Luther. For all these reasons, Doebbing concludes, he must opt for the book's prohibition.[58]

But then Eschbach, the second reader, proposed that no measure be taken. He advances five reasons supporting his conclusion at the end of his report:

> 1. At the core of Elliott's work, there is no proposition which, interpreted according to the rules of the Apostolic Constitution *Sollicita* [which regulated the activities of the Congregation of the Index], is contrary to the teachings of the Church in matters of faith or morality.
>
> 2. Father Hecker founded a society of priests who are highly praised for their piety and apostolic zeal. The condemnation of his biography written by a member of that Society would be a tremendous blow, to the prejudice, as it appears, of saving souls.
>
> 3. The denunciation of this work seems in great measure attributable to political rather than doctrinal motives. To condemn it would be considered by many as a victory of the party of those who have been resistant to the Holy Father's directives to the French.
>
> 4. For reasons already mentioned, the Preface of the book, written by the priest F. Klein, should be deleted. This Preface, more than Elliott's book, had with reason provoked the *refractaires* and others.
>
> 5. Inasmuch as the Introduction is the work on an American Archbishop and, besides, does not contain any proposition contrary to faith, the response to the denunciation could be a *dimittatur* [take no measure].[59]

With respect to the third point, recall that Pope Leo XIII had encouraged French Catholics to collaborate with the Republic, but the monarchists opposed this directive: they were the very *réfractaires* to whom the report refers. Ireland and the Americanists were on the pope's side here, and for this reason the consultor introduces the reference to French politics.

We have seen that toward the end of May 1898, Cardinal Satolli passed on to the Congregation of the Index the original American editions, which he owned, of Zahm's book and of *The Life of Hecker,* and both were entrusted to the consultor Doebbing who was to prepare a report. In Zahm's case, he only had to establish that there were no important differences from the Italian version. Once established, the examination of the book followed the usual bureaucratic path, though the Preparatory and General Congregations. In the case of *The Life of Hecker,* Doebbing wrote his long report concluding the book should be prohibited. But *The Life of Hecker* was examined in neither the Preparatory Congregation nor the General, owing to an unaccustomed act. In effect, the Archive of the Congregation conserves a memorandum of the prefect, Cardinal Steinhuber, to the effect that:

1. In a dispatch of June 7, 1898, the Eminent [Cardinal] Secretary of State informed the Eminent [Cardinal] Prefect of the Congregation of the Index that it is the wish of the Holy Father that in the event the book: "Vie du P. Hecker p. Elliott" is denounced, the Congregation might not make any decree about the said book before having explored the intentions of His Holiness.

2. In the audience of September 22 the Holy Father told the Eminent Prefect that he had decided to reserve for himself the matter of the denounced book, although adding that the letters relative to this book will be sent for preservation in the Archive of the Congregation of the Index.

A. Cardinal Steinhuber
Prefect[60]

The archive also preserves a letter from the secretary of state, Cardinal Mariano Rampolla, to the prefect of the Index, dated June 8, 1898, telling him, as he had told him orally the day before, that it is the pope's will that should a denunciation of Elliott's book reach the Congregation of the Index, that "the Congregation do nothing until Your Eminence speaks personally with His Holiness."[61]

Logically, then, from June 7, and with greater reason after September 22, the Congregation of the Index took no further action. Doebbing's suggestion that the book be prohibited had no effect.

In sum, Leo XIII reserved for himself all further decisions with respect to both Americanism and Zahm. The partisans and detractors of each were the same in both cases.

Zahm and the Americanists

When Zahm arrived in Rome in 1896, O'Connell quickly brought him into his circle of friends. As Zahm wrote to his brother Albert toward the end of June 1896, "Everything delightful here. Will see the Pope again tomorrow, & a few days later will dine with one of the most prominent Cardinals of Rome. Dinners given by Cardinals are here as rare as white blackbirds."[62]

At summer's end O'Connell and Zahm made a trip to Greece, leaving Rome on Thursday, August 27. On September 13, Zahm sent a postcard to his brother from Athens to say how much he was enjoying himself. On October 2 they arrived in Sparta.[63] For Zahm, the trip to Greece marks the start of his commitment to the ideals of the liberal group. Up to this time Zahm could have been considered one of them through his friendship with Ireland and Keane, and his attitude toward science and, in a more general sense, the issue of the training and values that Catholics, especially priests, ought to have. From this point forward, however, Zahm considered himself a member of a group of clerics determined to change the way in which Catholics had traditionally treated their relations with culture and society, especially in the Old World. For Zahm and O'Connell, Greece became a symbol of the ideal of freedom, rationality, and citizenship. They and others of their circle went so far as to occasionally use in their correspondence pseudonyms taken from places they had visited in Greece: Zahm signed as "Parnassus," O'Connell sometimes as "Marathon," others as "Sparta" or "Dynamo."

On their return from Greece, O'Connell and Zahm were met with the news of Keane's removal from the rectorship of Catholic University. But this event in fact worked to reinforce the liberal group in Rome, because Keane was transferred there and joined them. The Canadian College, where Keane had lodgings, was also host to Otto Zardetti, the opponent of the liberal current who had denounced Zahm's book to the Congregation of the Index.

His first year in Rome was a peaceful period for Zahm as he adapted to his new environment and consolidated his friendship with the liberal wing. Moreover, because he approached his job as procurator of the Congregation of Holy Cross with confidence and efficiency, he had more than enough time to supervise the diffusion of *Evolution and Dogma*. He was invited to lecture on the topic in Malta, the home of Alfonso Maria Galea, who a few years before had translated Zahm's book *Catholic Science and Catholic Scientists* into Italian and was then working on the Italian version of *Evolution and Dogma*.[64] On August 23, when he informed his brother of his imminent departure for Greece, he added: "*Evolution and Dogma* is already printed in Italian & will be out in a few days. Then look out for a storm. I know it will enrage a good many here in Rome, but, *n'importe;* the die is cast, & I care not which the result may be. The book will appear in French very shortly."[65]

Zahm was aware that his book was going to make some waves, although perhaps he underestimated the scale of the ensuing storm. In another letter to his brother written after his return from Greece, he recorded the tremendous repercussion provoked both in England and Rome by a review of his book in the *Dublin Review*, written by David Fleming, whom Zahm called "the ablest theologian in England, the pope's special adviser in the question of Anglican orders." He added that he was expecting an imminent visit from Mivart, the English Catholic scientist whose defense of the compatibility of evolution and Christianity had made him famous.[66] In another letter to his brother (December 6), announcing the publication of another book (*The Church and Science*), he took note of the hostile critiques of *Evolution and Dogma* and commented: "The Controversy has fairly commenced in England, & is to break out in Rome in a few weeks. The Jesuits are already training their biggest guns on me, & you will see the result in a series of articles in *La Civiltà Cattolica*, their greatest magazine."[67]

In the course of 1897 Zahm's association with the liberal group and his involvement with Americanism both reached their high-level mark. Correspondence reflects that his friendship with O'Connell's circle continued to grow. Two important members of this circle were Cardinal Serafino Vannutelli and his brother Vincenzo, also a cardinal. Zahm's papers record various invitations to dinner, and another to spend a few days in the house of the Vannutellis, in Genazzano, near Rome.[68] Among O'Connell's papers is preserved, for example, the menu of a lunch given by Cardinal Serafino Vannutelli on March 21, 1897, signed on the back "S. Card. Vannutelli."[69] Another note in S. Vannutelli's hand is a luncheon invitation to Denis O'Connell and Zahm.[70] Zahm mentioned this luncheon in a letter written the same day to his brother Albert, in which he refers to Serafino Vannutelli as "the next Pope," because he was considered as a possible successor of Leo XIII, who had reached the age of eighty-seven.[71]

The Americanist Moment in Fribourg

Zahm and O'Connell participated in the International Congress of Catholic Scientists in Fribourg, Switzerland (August 16–20, 1897), as already noted, probably traveling there together from Rome.

The Fribourg Congress may well have been the culminating moment of Americanism. On September 18, nearly a month after the meeting ended, the *Catholic Citizen* ran a long article by F. X. Kraus, under the pseudonym "Spectator," that highlighted the positive reception given the movement at the congress. The subtitle of his article was: "'Americanism' Toasted at a Banquet."[72] Kraus judged the Fribourg Congress to have been superior to the previous ones held in

Paris and Brussels, measured not only by participation but also by the significance of the work presented. According to Kraus, four lectures merited special attention and were very well received not only for the flare with which they were presented but also because they marked the end of a long and harsh controversy. Two of these were biblical studies and the others were presentations by Zahm ("Evolution and Teleology"), which we have already described, and O'Connell ("A New Idea of the Life of Father Hecker").

O'Connell's lecture, whose aim was to present a clear and explicit exposition of Americanism, shorn of the false images spread by its opponents, created a stir in the audience. He contrasted Roman law, pagan and godless, under which the individual had no rights as such, to American law, which existed for the individual, whose rights were protected by the power and authority of the federal government. Kraus here records O'Connell's closing words, which received "long and enthusiastic applause":

> Americanism, then, in spite of all that has been said to the contrary, involves no conflict with faith or morals; it is no new form of heresy or liberalism or separatism. No, when fairly considered, Americanism is nothing else than that loyal devotion which Catholics in America have to the principles upon which their government is founded and their conscientious conviction that these principles afford Catholics favorable opportunities for promoting the glory of God, the growth of the Church and the salvation of souls in America.[73]

The importance of the Fribourg Congress for the Americanist movement can be deduced from many reports that mention, in the first place, the enthusiasm with which Zahm's and O'Connell's lectures were received by the other members of the liberal group. On September 11, even before O'Connell and Zahm had returned to Rome, John Ireland wrote Zahm from St. Paul to comment on the events of Fribourg:

> I have read with intense delight Mgr. O'Connell's letter about American triumphs in Freiburg. What a great work you & your friends are doing in Europe for the Church & for America. Here in America we are profoundly grateful.
>
> Abp Keane & myself have spent hours & hours together, talking of things in Rome.
>
> I am personally grateful to you for your friendship to Mgr. O'Connell, when he needed friendship.[74]

On October 15 John Keane wrote O'Connell from Washington, praising his Fribourg talk and the connection he made in it to Hecker.[75] Two months later, John L. Spalding, bishop of Peoria, who sympathized with the liberals, congratulated Zahm for his growing influence.[76]

Nevertheless, the Fribourg Congress proved to be a mixed blessing for the

Americanists and Zahm. It drew so much attention that it was no doubt one of the causes of the crisis that soon after was to compromise both the Americanist program and Catholic attempts to establish the compatibility between evolution and dogma.

No documents show a direct relationship between the International Congress of Fribourg and the crisis of Catholic evolutionism. But the sequence of events seems to suggest a connection between the campaign orchestrated there in favor of the positions of the liberals, including both Americanism and the defense of the theory of evolution, and the first attempts to bring the two doctrines before Vatican tribunals.

Already during the Congress the opposition of some members of the hierarchy to O'Connell's plea for Americanism was patent. The same afternoon as O'Connell's lecture, the bishop of Nancy, Charles François Turinaz, accused Hecker of exaggerating the value of individualism and of distorting the meaning of the supernatural, both coming too close to Protestant positions.[77] It would be this very bishop who, several months later (April 24, 1898) would denounce *The Life of Father Hecker* to the Congregation of the Index.[78]

With respect to evolutionism the reaction was even quicker. Zahm's book was denounced to the Index by Otto Zardetti in November 1897.[79] Among the reasons he gave for denunciation was Zahm's public promotion of evolutionism, as was apparent in Fribourg.[80]

One cannot dismiss the notion that Zardetti's denunciation may have been not his own personal decision only but rather part of a plan in which others also participated: Satolli, for one, and perhaps Mazzella, that is, the conservative wing of the American hierarchy, possible using Salvatore Brandi, a regular correspondent of Corrigan, as an intermediary. But there is no document that permits us to establish such links definitively.

The Americanist Returns to America

Zahm's returned to the United States just while the denunciation of his book was working its way through the Congregation of the Index, unbeknownst to him. The reason for his return, the death of the provincial of the Congregation of Holy Cross, resulted in a kind of emergency appointment for Zahm, but it became permanent when the following summer he was elected to be the new provincial. Everything indicates that the change was appealing to him. It was difficult for him to leave the United States, and now he was pleased to be at home. Naturally, Zahm planned to continue his fight for the Americanist cause from his new post. On February 5, 1898, he wrote to O'Connell to inform him about his new situation, and proclaiming his intent to continue with "the cause," that is, to diffuse the progressive ideal of Americanism:

What a change a month has brought about: I cannot yet realize it; it all seems like a dream. Yes, my fears were well founded. I was called home to be provincial, & here I am attending to the manifold duties of provincial. I need not tell you what a change all this makes in my plans, in my life. You know without my telling you. But possibly it is for the best. I think I can do more for "the cause" here than anywhere else. I think you will agree with me in this.[81]

That other members of the group agreed with the utility of Zahm's new position for the diffusion of Americanism is evident. Keane for one wrote so to Zahm from Rome, on January 20:

My very dear Friend:
Fr. Legrand has just communicated to me the good news of your appointment as Provincial. It was, of course, just what I expected, the right thing & the only thing for your Congregation and its excellent Father General to do. But, all the same, I hasten to send you my hearty and affectionate congratulations, and my earnest best wishes for the success of your administration of an office rendered specially important by the peculiar character of the present epoch. Providence has made "Americanism" a factor of special importance in the life of the Church today. And if, as I feel confident, under your wise and energetic action, and under the nobly broad-minded and right-minded direction of Father General, Your Congregation becomes a leading exponent of the best that is meant by "Americanism," then your Congregation will become a great power for good in the development of the future.
    You know well with what affectionate interest I shall watch your part in the march of events.
    We miss your sadly here. There is a big gap in our Tuesday meetings; and to poor dear O'Connell the loss is unutterable. But the Lord makes no mistake. Amen.[82]

The next months passed without any news about the fate of *Evolution and Dogma.* Zahm went about the promotion of his ideas confidently, in particular the publication of his Fribourg lecture in several languages. In April the English text appeared in *Appleton's Popular Science Monthly* and a French version in the *Revue des Questions Scientifiques.*[83]

The Crisis of Americanism

At this point, events unrolled at a breathtaking pace. At the end of a few months Americanism and Zahm's book were conjoined, both in the Congregation of the Index and in the maneuvering for and against each. The denunciation of *The Life of Father Hecker* (April 24, 1898) seems to have been the conclusion of a well-planned campaign. Over the preceding weeks a series of articles opposing Americanism and *The Life of Father Hecker* had appeared in the French newspaper *La Vérité*. According to Alphonse Eschbach, one of the consultors who examined this work for the Index, Turinaz's denunciation was directly inspired by the articles written by Maignen (who signed them with the pseudonym "Martel") in *La Vérité*.[84]

Immediately after Maignen tried to publish the series of articles in book form with the title *Études sur l'Américanisme: Le Père Hecker, est-il un Saint?* The articles were highly polemical, and the archbishop of Paris, Cardinal Richard de la Vergne, withheld ecclesiastical permission (*imprimatur*) to publish the book. Maignen then sought the *imprimatur* in Rome and obtained it from the master of the Sacred Palace, Alberto Lepidi. This action drew a series of protests from Keane, Ireland, and Gibbons and contributed to the tense atmosphere over the issue of Americanism.[85]

Meanwhile the case of *Evolution and Dogma* was working its way through the Congregation of the Index. Even though the Congregation worked under a rule of secrecy, some rumors must have seeped out, because in July 1898 Denis O'Connell wrote to Zahm advising him of the rumors abroad about a condemnation of evolution: "It seems as if the war were on. There is a general agreement of opinion that the Holy Office is preparing a decree against Evolution. Dave [Fleming] is the only one who says he knows nothing about it: and I imagine Dave is getting tired of the fire. They say the decree is to be of a general character."

O'Connell goes on to mention Americanism, the attacks on Elliott's book, and the controversy surrounding that of Maignen. He says that Keane requested that the publication of Maignen's work not be authorized, but received no response. Then he turns to his own activities:

> After his [Keane's] departure, I had many conversations with Lepidi. He admitted that he had fallen into a confusion mistaking "American political Americanism" for "European" Americanism, and we agreed that he would a) correct Hecker's Life just as freely as he pleased and that I would see to have a new edition prepared on these corrections b) that he would indicate to me everything he found objectionable in Americanism, and I promised to have them all author[it]atively denied.[86]

The same day, July 10, O'Connell also wrote to Félix Klein, translator and author of the preface to *The Life of Father Hecker*, describing the latest news, using almost the same words.[87] In neither of the two letters does O'Connell indicate exactly when he had the conversations with Lepidi. But everything indicates that the events followed one another quickly during June and the first few days of July.

The publication of Maignen'a book, *Le Père Hecker, est-il un Saint?* (Is Father Hecker a Saint?), was announced in *La Vérité* on May 24.[88] Archbishop Keane sent a formal protest to Cardinal Rampolla, secretary of state of Leo XIII, in a let-ter dated June 2, 1898, with the backing, as he wrote Ireland two days later, of Cardinal Serafino Vannutelli. The reply somewhat pacified Keane, because Ram-polla considered the granting of the *imprimatur* as "most regrettable," adding that neither he nor the pope had known anything beforehand. But a few days latter, Sunday, June 12, Lepidi wrote Keane to justify his conduct. The master of the Sacred Apostolic Palace asserted that his role was limited to safeguarding faith and morality, and he did not feel any obligation to stop this particular dis-cussion, even though it was controversial, because out of the debate some clarification on the question of Americanism might emerge. This letter clearly did not satisfy Keane, because Lepidi did not in any way display the concern that Rampolla had expressed. The following day, after answering Lepidi, Keane again protested to Rampolla, sending him Lepidi's letter and his own response. The sit-uation deteriorated even more the following day, June 14, when the journal *La Croix* published an article stating that Lepidi had acted with the explicit authori-zation of the pope. Klein and Keane replied, the former with a letter to the editor of *La Croix*, on June 15, and the latter with another letter to Rampolla, three days later.[89]

Meanwhile, on June 7 Cardinal Rampolla had told the prefect of the Index, Cardinal Steinhuber, that the pope had decided to reserve for himself the ques-tion of how to deal with *The Life of Father Hecker*, in the event that it was de-nounced to the Index. In fact, that had already happened but was not publicly known. As for the pope's decision, that was known only to the cardinal secretary of state and the prefect and secretary of the Index (and perhaps one or more other officials of those two bureaus).

This decision was probably triggered by the unfolding controversy. Maignen's book cast doubt on the orthodoxy of Elliott's *Life of Father Hecker*, but it was also an attack upon a very active and important sector of the American hierarchy. On July 11, Ireland also sent a letter of protest. Gibbons held off until August 26. But in the meanwhile a prominent member of the conservative group, Bishop Sebastian Messmer of Green Bay, arrived in Rome.[90] In his audience with the pope he explained that many bishops were worried by the doctrinal stance of the Americanist movement.[91]

Keane left for the United States at the end of June, leaving O'Connell in charge of the negotiations. It must have been then when O'Connell put his own strategy to work in several meetings with Lepidi. Lepidi's reaction was initially positive: he accepted O'Connell's proposal, although he asked for time to consult his supe-riors. But when they met again a few days later, his attitude had changed. The pope had reserved the whole question for himself.[92]

O'Connell, nevertheless, was insistent, dispatching letters to Lepidi in which he tried to define the nature of Americanism as he saw it.[93] To his surprise, excerpts from his letters appeared in print on August 15 in an article by Maignen in *La Vérité*, and shortly thereafter they were reproduced in an appendix to the English edition of *Le Père Hecker, est-il un Saint?*, which had been published in Rome that summer, with Lepidi's *imprimatur*. Meanwhile the controversy continued to grow, with articles in many journals, especially French ones.

On July 11, Salvatore Brandi wrote to Archbishop Corrigan of New York (with the letter marked "confidential") and told him of the "excellent reception" that Maignen's book had received among the cardinals of the Roman Curia. He said he had received warm comments from many cardinals, including Parocchi, Ledochowski, Satolli, Aloisi-Masella, Pierotti, Mazzella, Steinhuber, Ferrata, and Jacobini, and the same from the great majority of Roman prelates. Then he added: "I have it on good authority that the Life of the Fr. Hecker by Fr. Elliott has *already* been delated to the Sacred Congregation of the Holy Office by, at least, ten French Bishops. I doubt however that a decree condemning it and ordering it to be put on the *Index* will be issued in the present circumstances. What we call here the *ragione politica* will likely prevent its immediate publication."[94]

Brandi's information was not too accurate: he transforms the Index into the Holy Office, and the bishop of Nancy into ten French bishops. What information did he have? At that very moment the pope had already let it be known that he must be consulted, in person, in the event that some action be taken against the book about Hecker. It seems logical to think that, in the frequent contacts between the Jesuits from *La Civiltà Cattolica* and the Secretariat of State, someone might have revealed, as a rumor, the book's current status.

Brandi continued writing Corrigan with his impressions. On October 12, he gave a version that was only partially realistic:

> As I anticipated the *Life of Fr. Hecker* was condemned by the *Index*; but the Decree was not published and will not be published, at least, for the moment. "Americanism" will soon receive a blow in a more solemn and perhaps more telling manner in the shape of a pontifical document addressed to the Bishops of the U.S. This letter is now being prepared and it will not be long before it is ready to be made public. So much I can tell You now, of course under the strictest secret, for Your Grace's private information.[95]

The first part of the letter is surprising because, in fact, the Index had not condemned *The Life of Father Hecker*; rather, it had suspended its procedure at the express indication of the pope. *Evolution and Dogma* was under investigation, but Brandi was not aware of it. Still, it is true that the Congregation of the Index had not been inactive. And it is also true that a papal document on Americanism was in the works, although several months would pass before it was made public.

What steps were taken during the summer months? Leo XIII's decision to reserve the case for his own judgment was made in mid-September, and it is not entirely clear what transpired in the intervening months. There was talk of the existence of a special committee appointed by the pope to study the question of *The Life of Father Hecker.* O'Connell himself transmitted this rumor to Keane in a letter dated July 12, 1898, identifying Mazzella, Satolli, and Brandi as possible members of this committee. In letters written later, O'Connell identified other members: Ledochowski, Serafino Vannutelli, Domenico Ferrata, and Angelo di Pietro. Later on, he found out that Vannutelli was not involved in the matter.[96]

This was a mystery for historians. Fogarty supposed that sometime in October the pope named a special committee to investigate Americanism. This committee was to have reported to the Secretariat of State, not the Holy Office. But he admits he could find no record of such a committee in the papers of the Secretariat of State in the Vatican Secret Archive, and that the secretary of the Congregation for the Doctrine of the Faith stated that the Holy Office had not been involved in the matter. Nor did a search of the documents of other congregations yield any information.[97]

The Mystery of the Committee Resolved

Was there really a special committee charged by the pope with investigating the question of *The Life of Father Hecker* or Americanism generally? At least a partial answer can be found in the Archive of the Index, where there are some relevant documents. The folder containing the documents related to the book also holds, besides the reports already mentioned, other documents that might have originated with such a special committee. Moreover, the secretary of the Congregation, Cicognani, made some annotations in the Diary that might shed some light on the facts.

On February 6, 1899, Cicognani noted in the Diary of the Congregation that the cardinal prefect had given him a memorandum. But, realizing perhaps that this memorandum contained a very limited reference to facts that were not to be found in any other document of the Congregation, Cicognani added a long comment:

> This note from the Eminent [Cardinal] Prefect is not complete nor does it reflect what was done on orders from the Holy Father. A prior, formal denunciation of *The Life of Father Hecker,* written and published by Father Elliott, was examined by the Holy Congregation, and for this end the consultors Eschbach and Doebbing were named. In these circumstances, the pope reserved for himself the resolution of the case in the best way he can. Meanwhile he created a committee, and to examine the book he named as his own choice [Cardinal] Satolli, Rev. Cucchi, and Father Alexis Lépicier of the Order of the Servants of Mary. This committee, after a serious examination, and by written vote, informed the Pope that many inde-

fensible propositions were found in the book, which could be described as new, dangerous, erroneous, etc. etc. All this, together with the book, the Pope gave to the Master of the Apostolic Palace, Lepidi, to examine. Lepidi, by interpreting the matter benignly, weakened the commission's opposing vote [*minuit votum commissionis contrarium*]. Then the Pope ordered all papers to be given to the Eminent Prefect to inform himself and then emit his own vote. The Secretary did not know what the views or judgment of the Prefect were. He only knew that the Pope reserved for himself to give at the right time a Charter or Constitution, if it were necessary, to the American bishops about the novelties being diffused by the followers of Hecker. Still, no document of this type has been published.[98]

This document answers some of the questions. It confirms the existence of a committee, established directly by the pope to examine the book, and it identifies the members: Cardinal Satolli; Tito Maria Cucchi,[99] consultor of the Congregation of the Index; and Alexis-Henri-Marie Lépicier.[100] Cicognani states that the committee, after a serious examination, informed the pope that the book contained many unfortunate things and that Lepidi had opposed the vote of the committee.

In the archive there are several additional documents that explain what Cicognani meant.[101] Two of them were written by Lépicier, who was not a consultor of the Congregation. The first of these documents, unsigned, is a four-page report on Maignen's book, *Le père Hecker est-il un Saint?*[102] The second document, five pages long, is a list of propositions culled from *The Life of Father Hecker*, with their respective doctrinal status.[103] On the last page of this document some statements have been added from Ireland's preface and from Keane's article, "America as Seen from Abroad" (*Catholic World*, March 1898). The propositions appear to have been written by a scribe, while doctrinal comments written in the margins are in the hand of Lépicier, who signed the document on July 2, 1898. The calligraphy of the annotations is identical to that of the first document, identifying Lépicier as the author.

It is possible that these two documents represent the work of a committee, which would indicate that the other two committee members had entrusted the preparation of the report to Lépicier, having reached a unanimous conclusion regarding the books of Elliott and Maignen. But the committee's stance is clear, even if the document before us was Lépicier's personal view. Just as Cicognani wrote, the committee's view of *The Life of Father Hecker* was very negative, which is even clearer in view of Lépicier's endorsement of Maignen: "The author Charles Maignen is a good judge, although perhaps he exaggerates a bit."[104]

The following pages are not really a report on the books of either Maignen or Elliott, but a judgment of the life and ideas of Hecker, emphasizing only negative or surprising aspects of them. Lépicier concludes: "This is, in brief, Maignen's book, useful not only for its critique of Elliott's *Life of Hecker*, but also of the doctrine that Hecker's fanatical disciples extract from the seeds that he sowed, new doctrines and, for that reason alone, very dangerous for the Church, inasmuch as in the Catholic Church the canon of Lerins is always preferred: 'Antiquity was preserved and novelty rejected.'"[105]

According to Cicognani's report, the pope did not take this as definitive but submitted the full file, including the book for the scrutiny of Alberto Lepidi, master of the Sacred Palace. Lepidi's report has also been preserved in the archive, and allows us to better understand the master of the Sacred Palace's complex stance on evolution and the reasons that might have been behind Leo XIII's final decision.[106] This is an extensive document in which Lepidi comments on the same propositions from *The Life of Father Hecker* that Lépicier had so harshly criticized. In each case, Lepidi's assessment is that the proposition can be understood according to Catholic doctrine. His comments frequently conclude with phrases like "and in this sense [the proposition] cannot be condemned"; "It is true"; "No danger can come from these words." His final comment is that Hecker's ideas have no great theological precision: they are "aspirations," frequently expressed in emphatic, hyperbolic language. Surely, they display a tendency to exalt individual liberty, which could work to the detriment of authority (thus Hecker insists on the need to let oneself be guided by the Holy Spirit). But for this reason they have been appropriated by those who "do not accept the authority of the Church."

There is yet another report written by Lepidi. Titled "Notes on Americanism," it is an attempt to define its various meanings, distinguishing between political Americanism (the attitude of Catholics in American society and public life), religious (doctrinal ideas attributed to Americanism by its critics), and politico-religious (the attempts to change the internal structure of the Church, especially as regards the role of the hierarchy).[107] Lepidi's meetings with O'Connell must have played an important role in formulating these distinctions. The dates coincide. Lépicier's report is dated July 2, and Lepidi's meetings with O'Connell took place in early July.

The Pope's Letter on Americanism

We have not seen Steinhuber's report, if in fact there was one, but everything seems to indicate that Lepidi's reports played a decisive role in bringing the debates over Hecker to an end. The result of all the discussion was a letter from the pope to Cardinal Gibbons, archbishop of Baltimore, titled *Testem benevolentiae*, dated January 22, 1899. This letter reflects the distinctions made by Lepidi. On the day that Cicognani wrote his Diary note, February 6, 1899, *Testem benevolentiae* had not yet been made public. On February 22, Cicognani marked the publication with another long note in the Diary and the same day the letter was published in the *Osservatore Romano*.[108]

In the United States the letter was made public on February 23, in Gibbons' translation, published in the *Baltimore Sun*. Two days later Cicognani noted in the Diary of the Index Ireland's full acceptance of the papal document, which he made public in a letter to the pope in *Osservatore Romano* on February 24.[109] Copies of other letters in which different members of the American hierarchy or of the Americanist movement also accepted the pontifical document were also filed in the Archive of the Index, along with the material already described.[110]

In *Testem benevolentiae*, a letter whose title is taken from the two first words of the original Latin text, Leo XIII wanted to express his benevolent feelings toward the people of the United States. After praising the American Church, he proposes to put an end to the controversy caused by Americanism:

> We send to you by this letter a renewed expression of that goodwill which we have not failed during the course of our pontificate to manifest frequently to you and to your colleagues in the episcopate and to the whole American people, availing ourselves of every opportunity offered us by the progress of your church or whatever you have done for safeguarding and promoting Catholic interests. Moreover, we have often considered and admired the noble gifts of your nation, which enable the American people to be alive to every good work which promotes the good of humanity and the splendor of civilization. However, this letter is not intended, as preceding ones, to repeat the words of praise so often spoken, but rather to call attention to some things to be avoided and corrected; still because it is conceived in that same spirit of apostolic charity which has inspired all our letters, we shall expect that you will take it as another proof of our love, the more so because it is intended to suppress certain contentions which have arisen lately among you to the detriment of the peace of many souls.
>
> It is known to you, beloved son, that the biography of Isaac Thomas Hecker, especially through the action of those who undertook to translate or interpret it in a foreign language, has excited not a little controversy, on account of certain opinions brought forward concerning the way of leading Christian life. We, therefore, on account of our apostolic office, having to guard the integrity of the faith and the security of the faithful, are desirous of writing to you more at length concerning this whole matter.
>
> The underlying principle of these new opinions is that, in order to more easily attract those who differ from her, the Church should shape her teachings more in accord with the spirit of the age and relax some of her ancient severity and make some concessions to new opinions. Many think that these concessions should be made not only in regard to ways of living, but even in regard to doctrines which belong to the deposit of the faith. They contend that it would be opportune, in order to gain those who differ from us, to omit certain points of her teaching which

are of lesser importance, and to tone down the meaning which the Church has always attached to them. It does not need many words, beloved son, to prove the falsity of these ideas if the nature and origin of the doctrine which the Church proposes are recalled to mind.[111]

Throughout the letter, the pope opposes those who reject the need for an external guide, entrusting all to the judgment of the individual guided by the Holy Spirit. He rejects the downgrading of the passive virtues (humility, obedience, self-control) with respect to the active ones. He calls attention to the disregard for the religious life and the vows that accompany it. He cautions prudence in attempts to attract non-Catholics. Finally, he refers explicitly to Americanism, using the term, and evaluating the different meanings it could have:

> From the foregoing it is manifest, beloved son, that we are not able to give approval to those views which, in their collective sense, are called by some "Americanism." But if by this name are to be understood certain endowments of mind which belong to the American people, just as other characteristics belong to various other nations, and if, moreover, by it is designated your political condition and the laws and customs by which you are governed, there is no reason to take exception to the name. But if this is to be so understood that the doctrines which have been adverted to above are not only indicated but exalted, there can be no manner of doubt that our venerable brethren, the bishops of America, would be the first to repudiate and condemn it as being most injurious to themselves and to their country. For it would give rise to the suspicion that there are among you some who conceive and would have the Church in America to be different from what it is in the rest of the world.[112]

The Americanist leaders wasted no time in announcing their agreement with the pope's letter. The arguments that O'Connell had used in his conversations with Lepidi, enumerating the diverse forms of Americanism and which were acceptable or not, had achieved their objective. At the same time, Archbishop Corrigan of New York and his Roman friends could be well satisfied: the pope had condemned Americanism.

During the same months, appeals for and against Zahm's book were made, in great part by the same cast of characters.

Zahm's Objective: Stop the Decree

The decision taken by the Congregation of the Index about *Evolution and Dogma* took place during the most intense moments of the Americanist question. The Preparatory Congregation had taken place on August 5, but since the Roman Curia's traditional period of summer vacation had arrived, the General Congregation was deferred until September 1. The secretary's audience with the pope took place two days later, on Saturday, September 3. In those days, O'Connell, Brandi, Keane, Ireland, and others involved concentrated all their efforts either to provoke or to avoid the condemnation of Americanism. After his negotiations with Lepidi, O'Connell had left Rome for Fribourg, where he remained during the months of August and September.[113] Nearly a month had passed until the notice of the decision began to be known, and not even Brandi, who was generally up to date on what took place in different Vatican circles, alludes to the condemnation of Zahm's book in his correspondence with Corrigan in the month of September.

This time the news strictly followed the channels prescribed by canonical legislation. We noted that on September 9, Cicognani wrote in the Diary of the Congregation that he had already communicated to the superior general of the Congregation of Holy Cross the decision of the Index. Here we reproduce the entire text because it reveals an important ambiguity: "Concerning the book of Father Zahm, the secretary notified the Father General of the Institute of Holy Cross what the Eminent Fathers decreed in the last General Congregation, with the faculty of communicating the Decree to the same Father Zahm, and to ask him if he wants to submit, etc."[114]

When he refers to communicating the "Decree" to Zahm, Cicognani is not referring to a decree in the strict sense, that is, an official, public document, issued by the Congregation and approved by the pope. In Zahm's case it was decided not to publish the decree until it was known whether Zahm accepted the condemnation, but, in fact, such a decree never reached print. The double sense of the term "decree" has led some authors into manifest confusion, as for example R. Weber, who speaks of an "edict," which certainly did not exist.[115]

The notice of the prohibition reached Zahm only at the end of September and gave rise to a period of tension and great activity, not only by himself but also his friends, who strove to stop what they viewed as part of a more general attack on Americanism, with which they all identified. The state of alarm lasted several months, during which the leaders kept in touch through frequent correspondence, mainly between Zahm and O'Connell, but also between them, Keane, Ireland, and a few others. Through this correspondence it is possible to follow the activities of the Americanists and their opponents with respect to the prohibition of evolutionism almost from day to day.

In the months preceding the condemnation, Zahm had been busy with his duties as provincial. Early in August, he had been reconfirmed in the post during a meeting of the General Chapter of his order in Canada.[116] The chapter also approved the creation of a college in Washington, alongside Catholic University, following a plan that Zahm had been developing for some time. For this reason, when the Chapter meeting was over, he went to Washington to make preliminary arrangements for its construction. These negotiations in Washington busied him for perhaps a month. He was back in Notre Dame by September 24, when the Provincial Council approved the acquisition of the property that Zahm had negotiated in Washington. Around this time he received visits from Ireland and from Keane, who stayed in Notre Dame for a week. When Keane returned to Washington, the news from the Index had not yet arrived.[117]

Français's letter with news of the book's prohibition must have reached Zahm the last week of September. It was a hard blow, as his friends reported. Some weeks later Ireland wrote O'Connell that "Poor Zahm is mightily discouraged."[118] His first letters after hearing the news were not addressed to the general superior nor to the Congregation of the Index—to retract, to seek clarifications, or just to lament—but rather to his friends. These are confidential letters ("Sacredly Confidential," he writes at the top), in which he is looking for advice on devising a strategy to stop the decision against him or at least diminish its importance in the eyes of the public.

Right after he heard the news Zahm wrote to O'Connell, Ireland, and Keane, who was probably the first to be informed. He wrote Zahm from Washington on September 28, lamenting the decision and suggesting a prudent and respectful line of action, but one that does not yield on the issues.[119] The same day Zahm wrote to O'Connell, to marshal his aid and his influence in Rome:

> Sacredly Confidential
> Sept. 28th. 98
>
> My dear "Doc,"
> Only a word to tell you that I have received *official* information that "Evolution & Dogma" is at last on the *Index* & that the decree is to be published soon, unless you & Fr. David, & Serafino can have it prevented; Please take up the matter at once with these do & see what you can do. It will hurt the cause & embarrass me if the decree is published. If there are any propositions in the book which are *contra fidem* I shall of course eliminate them when they are pointed out. See Dave at once, & give him no rest until he has secured a stay in the proceedings.[120]

Zahm's objective was clearly defined: to try to halt the publication of the decree. For this he was counting, above all, on whatever approaches that O'Connell, David Fleming (consultor of the Holy Office), and Cardinal Serafino Vannutelli could achieve. Zahm viewed the prohibition as tied in with the Americanism question. The immediate cause of the condemnation, as Zahm saw it, was the publication of the Italian edition of *Evolution and Dogma* and the fact that it was publicly praised by Geremia Bonomelli, bishop of Cremona, a very controversial figure in Italian public life who had included in the first volume of his work *Seguiamo la ragione* (1898) an appendix praising Zahm's book.[121] Through the intervention of his Roman friends, Zahm hoped to avoid the publication of the decree, but in the event they were unsuccessful, O'Connell should see to it that the news media did not present the condemnation in an overly negative tone.

O'Connell was eager to do everything possible. He reported to Zahm on October 15. He went to see Fleming, but he was in England, so he would try to contact him. Then, he was going to Genazzano, where the Cardinals Vannutelli were, and attempt to persuade Cardinal Serafino, the most influential of the two, to write a letter supporting Zahm. He would also attempt to interest Bonomelli in the matter.[122] A few days later, after exchanging impressions with Cardinal Serafino Vannutelli, O'Connell telegrammed from Genazzano to give Zahm some encouragement in idiosyncratic language, *spero nihil timendum scribo* (which seems to mean, "I hope there is nothing to be feared. I will write").[123]

The general superior, Gilbert Français, must have been more than a little anxious. On September 29, in a letter mainly devoted to business of the congregation, he alludes discreetly to the question, exhorting Zahm once again to submit to Rome's decision.[124] Finally Zahm wrote to Français, declaring his will to submit to the decision of the Congregation of the Index and asking him to transmit his compliance to Rome. As we have seen, Français did so on November 4, the same day he petitioned that the decree not be published.

On October 31 O'Connell relayed to Zahm a conversation he had had with Cardinal Serafino Vannutelli. O'Connell asked him if it would be possible to stop the publication of the decree, and the cardinal told him it would (recall that this cardinal had been prefect of the Congregation of the Index, at the time was a member of the same Congregation, had participated in the meeting in which Zahm's book was discussed, and was an influential figure). The cardinal promised to speak with Prefect Steinhuber and with the pope. He even admitted that, if the decree had not been published, it was probably owing to the presence at the meeting of a voice friendly to Zahm (his own, no doubt). The decree would not in any case be published, he added, before the cardinals of the Index had met again, and he did not think that such a meeting would take place before the following March. Still and all, Vannutelli also expressed the hope that Zahm would forward his submission: he could publish a letter to the effect that he had reflected over the matter and had changed his mind. O'Connell did not agree with this suggestion, because he thought that a public retraction was the equivalent of publishing the decree.[125] That had been Leroy's fate.

The same day, Zahm wrote to O'Connell "Ireland has written a strong note to Rampolla and this, too, will have its effect. If, however, the *réfractaires* want to have a repetition of the Galileo case let them condemn Evolution."[126]

In his note Ireland had protested vigorously and asked Rampolla to read his letter to the pope. He also wrote to Français, asking him to go to Rome to support this action.[127] O'Connell also asked Français to go to Rome, and Zahm did the same, although he observed in a letter to O'Connell that the father general was very shy, and asked O'Connell that he and Keane assist Français when he went to Rome so that he not give in unless it were absolutely unavoidable, and insisted they not fail to do all that was necessary to avoid the publication of the decree.[128]

Still one more surprise occurred before the end of October. On the 26th the newspaper *Lega Lombarda* published a letter from Bishop Bonomelli, written four days before, in which he disowned the appendix to his book, *Seguiamo la ragione*, in which he had taken a position favorable to Zahm and evolution. Bonomelli could no longer be considered a prospective ally in the campaign to halt publication of the decree, and O'Connell ceased mentioning his name in that regard. The Americanists viewed Bonomelli's retraction as a kind of defection, as the Countess Sabina Parravicino (the Italian translator of some of John Ireland's writings) expressed it in a letter to O'Connell on November 10.[129]

Negotiations in Rome

Français left Paris, bound for Rome, on Friday, October 28, the same day he received Zahm's letter and his act of the submission to the Index's decree.[130] He was accompanied by Brother Ernest, a member of his order. Besides Ernest, his principal aide in Rome was the procurator general, Jacques Legrand. Français arrived in Rome Sunday morning, October 30. Following Zahm's request, he met first with Denis O'Connell that afternoon and was briefed on the positions of all the parties concerned. It was of paramount importance for Français that the decision of the Index would not interfere with Zahm's leadership of his province. In presenting Zahm's acceptance of the decree, he hoped to secure a benign response from the Index.

O'Connell, by contrast, saw the issue from the perspective of the "battle" that was unfolding between the Americanists and those they called the "réfractaires." So his main objective was to maintain a certain equilibrium in the position that the Americanists held in the Curia and to prevent, through all available means, the prohibition of Zahm's book from becoming a victory for their opponents. Français's moderation and O'Connell's diplomacy were sufficient to maintain cordial relations between them and a plan of common action, at least in appearances. But their differing attitudes and, more importantly, their underlying motivations were apparent to all, on both sides. Français favored acting clearly and quickly with respect to authorities of the Curia, presenting the anticipated declaration of submission. O'Connell proposed a more "tactical," watchful attitude. Writing to Zahm the day after his first meeting with Français, he expressed the worry that Français, in an effort to resolve the question rapidly, would end up with some public notice of Zahm's retraction, which would be represented by their enemies as a defeat for the Americanists.[131]

Français soon realized that O'Connell's advice concealed something more than simple interest in finding the best way to resolve the problem. In the detailed report that he sent to Zahm several days later, he said that, according to

O'Connell, Zahm should not write to the Congregation before Keane and Fleming returned to Rome, so they could act conjointly. And he added: "This strikes
me as a bit radical and even in a sense dangerous, for I would have been afraid of
hurting the case I was pleading in placing myself behind a group of men who
would have been attacking the Congregation of the Index."

His opinion was reinforced after several days of negotiations in the Curia. At
the end of the same letter, he adds: "Moreover, I have learned that Monsignor
O'Connell is not well regarded by the court of Rome, that they find him rather
imprudent in speech. He has little influence, and can even imperil one's reputation."

Français quickly evaluated the way the wind was blowing in Rome, not only
with respect to evolutionism, but also in regard to the Americanist group, of
which Zahm, by his own choice and in the eyes of all concerned, was a member.
Nor did he avoid expressing his resentment, although in order to do so he appropriated the words of Cardinal Ledochowski, prefect of Propaganda Fide, the congregation of which the United States was a dependency:

> The following day I went to see Cardinal Ledochowski. He knew about the con-
> demnation of your book and he led me to understand that it did not surprise him.
> He added that he had been displeased with you when you were Procurator; that
> instead of taking care of the Congregation of Holy Cross you had made the mis-
> take of involving yourself in general Church affairs; that you had let yourself be
> drawn into intrigues and to be carried away on that score, etc.

Zahm's opinion of his superior general, whom he considered "very shy," is be-
lied by the facts. During his week in Rome, Français took complete charge of the
situation and made the decisions that struck him as necessary to avoid what he
most feared—repercussions that might be harmful to the program of the Congregation of Holy Cross in the United States. At the same time, he managed to
give a clear signal to the Curia of Zahm's open stance with respect to the finding
of the Index, and he obtained some guarantee of what Zahm wanted: that the
decree not be published. Throughout the week, appointment followed appointment. The first was with Ledochowski, who asked that Zahm be warned to take
more care with rash doctrines. The second official visit was with Cardinal
Serafino Vannutelli, without doubt the person who could provide the most direct
assistance. He assured Français that he would speak with the pope in an attempt
to avoid publication of the decree. But Français also realized that this would not
resolve the whole problem. In spite of his friendship with Zahm, Vannutelli did
not hesitate to make clear the difficulties that he had with the evolutionist theory, as Français related to Zahm: "Certainly he [Vannutelli] is your friend, but I
do not believe he shares your ideas on evolution in general or on how you con-
vert Saint Thomas and Saint Augustine into supporters of the theory. He also let

me understand that he would ask you, without doubt, to acknowledge the valor and authority of a few observations that he will share with you."

The most important meeting, naturally, was the one he had next with the prefect of the Congregation of the Index, Cardinal Steinhuber. Français followed O'Connell's advice and did not reveal the existence of a letter from Zahm formally submitting to the decision of the Congregation, but only his full disposition to do so. Steinhuber then asked Français himself to send him a document declaring such a disposition. Français then broached the question of the publication of the decree. The prefect's response was diplomatic:

> At this time I told him how the publication of the decree would be harmful, given the nature of the American people and their current state of mind. He replied that he was not the only person responsible in this matter, that he would appoint a chairman and that each member would vote. I took it upon myself to tell him that I was counting on his personal vote and was aware of his influence. He started to smile and let me understand that he would not forget my visit.

Français met a second time with Cardinal Ledochowski, was unable to see David Fleming, and did not see Keane until Saturday, November 5, the last day of his stay in Rome, when Keane had just arrived from Naples. The former recommended a specific kind of prudence: "He said that the Italian translation had attracted too much attention, and he gave me to understand that in this matter a good dose of prudence is necessary." As for Keane, Français himself asked him to take no action, unless circumstances demanded it, for to apply too much pressure could be dangerous to the cause.

In the letter to Zahm that we are citing, Français sets forth clearly his conclusions, which appear both very objective and appropriate to the reality of the circumstances:

> 1. It is useless to think that this decree of the Index might be revoked. It would be like hitting a stone wall.
>
> 2. There is every hope that the publication of the decree can be prevented.
>
> 3. The most unappealing aspect of your book is that you have interpreted Saint Thomas and Saint Augustine in such a way as to make them evolutionists. The closing lines of your conclusion are also displeasing. After having maintained (or defended) evolution as a hypothesis, you end up presenting it as a thesis, giving us to understand that the truth lies there.
>
> 4. They will not oppose evolution in general, but they will make an exception when the Bible is at stake. They will say, for example, that the doctrine of evolution cannot be taught safely for what concerns the body of man. Otherwise, what explanation could we give for woman's body? We would have to say that it is a myth and anyone can readily see the final danger in such a tendency.

Therefore, all would be softened and greatly modified, inasmuch as they have great consideration for you. But something will be done to stop the Doctrine of Evolution insofar as it concerns the Bible.[132]

In sum, Français urged Zahm to be prudent: that he not speak out against the Congregation of the Index, or write anything, and to be assured that if he condemns the Index publicly, many who seem now to support him will desert him, and others would form a terrible movement against him.

Results of the Negotiations

Before leaving Rome, Français sent Cardinal Steinhuber, as he had agreed, a letter declaring Zahm's will to submit to the decision of the Congregation of the Index, accompanied by another letter in which he asked that the decree not be published. That same day the secretary of the Index, Marcolino Cicognani, noted in the Diary of the Congregation Zahm's submission transmitted by his superior general, which was thus officially accepted.[133]

From Français' perspective, the stay in Rome had been a success, because he had observed some interest in accepting his petition. Denis O'Connell was not so optimistic. He wrote Zahm again from Rome on the activities of the superior general: "Fr General left Sunday rather optimistic, but I said he didn't know Rome. I said I would continue.[134]

In fact, O'Connell was exploring other roads to exercise influence on the pope, besides continuing to count on the help of Keane, Fleming, and others. In the same letter, he said to Zahm that one way was to approach Cardinal Vaughan, archbishop of Westminster in England, taking advantage of the fact that an article by Bishop John Hedley, praising Zahm's book, was quite recent, and that Hedley was very well considered, both by the pope and by Vaughan. The same day O'Connell sent Zahm a telegram with only two words, in Latin: "Nihil timendum" (There is nothing to fear).[135]

Nevertheless, it soon became apparent that any other way would be impractical. On Tuesday, November 8, O'Connell was finally able to see David Fleming, in whom he and Zahm had invested great hope. Fleming was a consultor of the Holy Office and was considered in Rome as the expert theologian on questions related to Anglican affairs, besides having a close relationship with the English Catholic evolutionist St. George Mivart. But the meeting was a disappointment. Fleming himself wrote to Zahm directly: he promised to do everything possible, but made it clear that he did not consider it prudent to try to exercise any pressure on the Congregation of the Index. "I am afraid that it is now too late," he twice repeated. Fleming attributed the problems in part to the attitude of O'Connell in Rome. And he added:

The question to my mind is very obscure as to whether the *immediate* formation of Adam's body by the Creator has been *revealed*. The question of Eve's body *practically* comes in & though no decision has been given in this point, the Index can & does proceed summarily & economically. It is not necessary that there should be anything against faith & morals; it is enough that a thesis may be looked upon as inopportune or premature. It was certainly more unwise *to force it* on the authority here by the Italian translation.[136]

Vannutelli Intervenes

On Monday, November 7, in an audience with the pope, Cardinal Serafino Vannutelli put the question directly before Leo XIII and asked him not to publish the decree. The pope accepted the petition, and so for the moment the matter seemed to have been resolved. News of the pope's decision soon reached Zahm. Keane wrote first, right after Vannutelli came to visit him on Wednesday: "Card. Serafino has just been to see me. Last Monday he asked the Pope that your sentence sh'd not be published. The Pope agreed willingly. So it is settled."[137]

O'Connell did not get the news till several days later. When he heard from Vannutelli, he immediately sent Zahm a telegram in Latin that read: "Iubente Papa decretum supprimitur" (By order of the Pope the decree is suppressed).[138] On November 12 he wrote to Countess Parravicino: "It is true a strong campaign was waged against Zahm. They did not mind him until they saw him in the appendix of Mgr Bonomelli. But now I believe it is all over. So the H. Father decided and last Monday the Prefect of the Index has been so informed."[139]

O'Connell went immediately to see Father Legrand to give him the notice, which the latter transmitted immediately to Français:

Rome, November 11, 1898

Very Rev. Father:
The affair is over with. Monsignor O'Connell just left the house and told me that Cardinal Serafino Vannutelli saw the Pope Monday and spoke to him about the condemnation. The Pope answered: "I withdraw it with great pleasure." Cardinal Steinhuber is already informed of that. Leo XIII has not placed even the least condition to the withdrawal of the decree. Now it remains to be seen whether there will not be an offensive objection from the Cardinal Prefect to the Holy Father. In that case, there will be an interesting dialogue between the two cardinals.

According to Monsignor O'Connell, a Jesuit with a German name, that I do not remember, has been instructed by the Pope to make a commentary on the Bible and in his work this Jesuit says that there is nothing contrary to faith in the simian origin of man. Leo XIII, it is known is very lenient in granting to science men a full liberty in the discussion of points not yet defined.

Well, then, for the time being, we are, I think safe, and you, you can be more at ease.[140]

On the other side of the Atlantic, Zahm received the news more slowly, which made him quite anxious. On December 27, O'Connell sent him a detailed account of the situation:

You know that by order of the Pope the decree will not be published. In compliance with his promise S. [Serafino Vannutelli] spoke to the Pope. The latter said he would be happy to comply if it were not already too late, for just this minute Rampolla has read me a letter from Ireland requesting the same thing. S. replied "I know it is not too late, for I am of the Congregation," *va bene,* tell Steinhuber "not to publish it." So off S. went joyfully to Steinhuber, and returning gave the news to Keane whom he had not yet met since his return, tho' K. had called. So that is the first step and the most important one. And really to me it seems wonderful that it was done so easily in spite of so many enemies and without the aid of some we thought were friends.[141]

Matters still were not fully resolved. The request that Zahm make some kind of public retraction was still unaddressed. Steinhuber had asked Français for at least Zahm's declaration of submission to the decree. Even Vannutelli thought that some retraction was necessary because, even though this was not a matter of faith, he thought that Zahm had gone too far, especially in attributing an evolutionary viewpoint to Saint Augustine and Saint Thomas.[142]

O'Connell certainly wanted to avoid a public retraction, as did Zahm. Zahm's stance was one of prudent silence, as Français had advised, in the hopes of calming the waters.

The end of the year found Zahm a bit less anxious. He had received a letter from Keane in Rome on December 10, assuring him he need worry no more: the matter was resolved. Serafino Vannutelli had said some kind of retraction would be necessary, but O'Connell signaled his disagreement, as did Keane himself (who thought it unnecessary to take any action). Keane added that the authorities wanted him to withdraw the book from sale, which could be accomplished without public fanfare.[143]

Ireland also wrote to Zahm just before Christmas with news that also seemed positive. Rampolla had written him, probably in reply to the letter of protest that Ireland sent him, and assured him the publication of the decree was off the table until such a time as they discuss the matter personally (Ireland was planning a trip to Rome around the end of the year).[144] Zahm fully trusted Ireland, as he indicated to O'Connell a few days later: "I hope the Index business is settled for good, although Rampolla has written Ireland that action is only suspended until his arrival in Rome. I feel, however, that St. Paul [Ireland] will convince the Vatican that it will be better not to take up the matter again."[145]

Was the Case Really Closed?

In the letter in which Legrand notified Français of the success of Cardinal Vannutelli's audience, he wondered whether Steinhuber might attempt a counterattack. This did not happen, at least not openly. But symptoms of a new campaign began to appear early in 1899. In part, at least, it was a conscious, organized campaign. On January 2 Salvatore Brandi wrote to Archbishop Corrigan to bring him up to date on the situation in Rome, mentioning both *Evolution and Dogma* and Americanism. He announces that the encyclical *against* Americanism (the emphasis is Brandi's) is finally ready and would be made public in a few days, in spite of opposition to its publication. He adds: "In the current number of the *Civiltà*, Your Grace will read an article of mine on *Evolution and Dogma*. It preludes, I am pretty sure, to a decree against Dr. Zahm's work."[146]

Brandi says nothing about the reasons for his certainty of the condemnation of Zahm's book. The article mentioned by Brandi appeared in the last issue of *La Civiltà Cattolica* of 1898 and had the same title as Zahm's book, "Evoluzione e domma."[147] Brandi criticized an anonymous article published by one of the principal liberal journals of the times, the *Rassegna Nazionale* of Florence, which echoed the favorable opinion of Zahm's book adopted by John Hedley, bishop of Newport (in the *Dublin Review*, October 1898). After criticizing the *Rassegna* article, Brandi turned to Zahm's book. At the end of the article, Brandi referred to the condemnation of Leroy's book and his public retraction. On January 12, 1899, Français wrote to Zahm:

> Reverend and very dear Father:
> I am informed by a genuine and certain source that the question of your book and of the decree referring to it, and the publication of this decree are back on the table.
> My formal impression at the time I left Rome was that it would not be published. . . . I had for all intents been given full assurance that all would go well. . . .
> Inasmuch as the arrangements are now menaced with change, it must be that you have not sufficiently persisted in the prudence and silence that I counseled.[148]

Around this time an additional worry arose for Français in regard to Zahm's situation before the Roman Curia. The decisions the General Chapter held in August, in which Zahm had been reelected as provincial, had to be ratified by Rome. The papers had been in the hands of Propaganda Fide for some time, but there was no word about their approval. Français feared that the delay was due precisely to the provincial's predicament. In February he wrote to Zahm that he was concerned by the possible reason for the delay, namely Zahm's book and the Americanist movement, both of which had many dedicated enemies in Rome.[149] But this apprehension turned out to have been unjustified. The acts were approved some weeks later, with no objections to the appointments, and only a few small changes in matters having nothing to do with Zahm.[150]

Meanwhile the document on Americanism, in the form of a letter from Leo XIII to Cardinal Gibbons, had been published. The reaction of Ireland, Gibbons, and others implicated in the movement was to immediately endorse the papal document. One front was thus closed, with a result that both sides evaluated distinctly. For the supporters of Americanism, the document did not deal with their proposals; what it principally condemned was a doctrinal movement located more in Europe, especially in France, than in America. For its opponents, the document had materialized in spite of all the efforts of Ireland and his allies, who now supposed that an explicit condemnation of evolution was to follow. Salvatore Brandi wrote to Corrigan at the end of March:

> There is no doubt about the decree of the Holy Office condemning Fr. Zahm's work "Dogma and Evolution." But it has not been published because of the intercession of his Superiors in France, who being notified of the fact, have promised to get Fr. Zahm to disown and repudiate that work, submitting himself to the same retraction to which submitted Fr. Leroy O.P. in 1895. I have been told by a competent authority, that such retraction is to be signed by Fr. Zahm in Rome before the Holy Office.[151]

What was the origin of Brandi's assertion, which matches the fears that Français had expressed at the start of the year? Information on this point is incomplete, but one of the basic documents survives in the Archive of the Index. On February 3, 1899, the cardinal prefect of the Index had an audience with Leo XIII. Among other matters discussed was Zahm's case. The pope's decision is registered in the Diary of the Congregation: "In the same audience the Pope ordered the suspension of the publication of the Decree of September 1, 1898, by which Father Zahm's book, *Evolution and Dogma* was prohibited, although the author submitted in a praiseworthy manner and repudiated the book, until Father Zahm, who soon will come to Rome from America, is heard."[152]

As had been suggested to Français, some officials expected a public retraction from Zahm, perhaps in a newspaper, just as Leroy had done. It is possible that, having seen no notice of any such retraction, Steinhuber might have reopened the question in his meeting with the pope, or that external pressure might have been brought to bear (Brandi's article being a case in point).

In terms of the procedures of the Index, the pope's decision was unusual. Authors were not summoned to justify the substance of their works. That would have been a gesture of deference to Zahm. For him, in any event, who hoped that the question be closed with no further conditions, the pope's action was not a victory, because the question was left hanging with no final outcome. Another surprising fact is that this decision was not embodied in any kind of notification, not even a private one to Zahm or Français. Therefore it could not be considered an official summons for Zahm to appear in Rome. It looks like a simple expedient to leave the question open. Nor was it purely circumstantial that a few days before his audience with Steinhuber on January 22, the pope had signed the letter *Testem benevolentiae* on Americanism. Inasmuch as Zahm was considered one its principal exponents, the pope's decision could have been a way to signal that for the time being no further action should be taken.

On March 30, Zahm wrote to O'Connell. He first comments on a letter that Ireland had written to *l'Osservatore Romano* a few days before about his acceptance of *Testem benevolentiae*:

> What a magnificent *coup* Ireland's letter was! It saved the situation completely. That is the general verdict there. It was a Quixotic tilt against wind-mills that the *réfractaires* got the Pope into. They themselves feel that & are saying little about the condemnation of "Americanism." One of the German papers says Ireland has been condemned three times already, but, adds sorrowfully, "he won't stay condemned."

Zahm goes on to rehearse the current status of his book:

> The evening press announces that "Evolution & Dogma" will soon be on the Index. Corrigan stated before all his seminarians & a number of priests, that he had positive information that Dr. Zahm would soon be condemned. They had quoted me approvingly in some papers that they had written, & in the interest of orthodoxy he felt called upon to denounce me in advance of Rome. What about the matter, anyhow? I felt sure from your letters & telegrams that the matter was definitely settled, & that the decree would never be published. Abp. Ireland said he would take up the matter with the Pope, & have the decree suppressed for ever. I feel that he did so, but I will be grateful if you will reassure me about the matter. The constantly repeated reports that I am to be on the Index convinces me that the enemy is not sleeping, but is determined to have revenge for the *tour* Ireland played them. Keep your eyes open, & don't let them catch you napping. Yourself, Serafino, & Abp. Ireland—I depend on all of you—can more than checkmate their moves & nullify their fondest schemes.[153]

The next day Zahm also wrote to Ireland, who had been in Italy from the beginning of the year, posing the same question as he had asked O'Connell.[154]

These were difficult days for Zahm, anguished by a new threat, but ready to continue the battle, on his own, to prevent the case from being reopened. Meanwhile, news of the condemnation of *Evolution and Dogma* was spreading. One of his close collaborators at Notre Dame, James A. Burns, left a reflection in his diary of the news reaching Notre Dame. It was said in Rome that Zahm was about to be summoned and that he would be deposed as provincial. The morning of April 5, Burns had spent a few hours with Zahm:

> He admits the gravity of the situation. He attributes it all to the Jesuits and Cardinal Satolli, who is one of the bitterest enemies of evolution. He showed me a telegram from Arch. Ireland rec'd a few days ago from London and reading: "No surrender," but thinks it has not reference to him, but to "Americanism." Of course, he is working tooth and nail to offset his enemies' efforts in Rome. A petition to Rome from the Superiors of the Province may be necessary. Nothing that is necessary will be left undone. I believe Fr. Zahm will win.[155]

On April 12, O'Connell wrote to Zahm, bringing him up to date on the case, which was still in limbo. It is an interesting document because it describes the situation in Rome and suggests some reasons for it:

> I am now of the impression that we can never count on anything here as certain. The Pope did, on two occasions, order that the decree should not be published, but they seem determined to publish it in spite of that. Then on New Year's Day, when Cardinal Serafino told me to send you that cable *"nolite timere"* a young friend at the table and Consultor of the Index gave me a look and after the dinner told me the Cardinal did not know all that was going on and that all danger was not past, that they were working like demoniacs against the Americani. Then came the article of the *Civiltà* with the publication of the letter of Leroy and Serafino said that looks as if they wanted a similar letter from Zahm. Then Ireland spoke to the Pope again. He said Zahm ought come to Rome and make an explanation. Ireland suggested the distance, your many occupations, &c. Then came the broad hint in the correspondence of the *Freeman's* [*Journal*], and Ireland spoke to Rampolla. He too said "Zahm ought come to Rome." Again a few days ago in our meetings, Van Ortroy said Zahm is coming to Rome.
>
> Now it seems to me the object is to bring you here and exact a personal retraction from you which they will publish afterwards as they did with Leroy. In that case it seems to me that it is no favor to you at all to withhold the publication of the decree on such terms, and that for you the lesser evil would be to abandon that favor and submit to the decree when it appears as hundreds have done before you.
>
> I hear they have some difficulties in publishing the decree fearing to seem in conflict with the theory of evolution, and to meet particular difficulties in England and Germany, and that their end would be better attained, as far as you are concerned, by getting a personal letter of retraction from you repudiating the theory as false, and publishing it, leave the responsibility with you.

A perfect spirit of madness prevails here at present and we can do nothing. Father David [Fleming] went back on us completely and became a most hurtful enemy. The Cardinals are kept in the dark about much that is going on. Here Ireland could do absolutely nothing and was made to feel that his influence was ended. There is no hope at present.

I wish I could have a long talk with you. Letters are so unsatisfactory and I hated to write you such news. But I think it is true, and not my pessimism. All our friends, without exception, all the savants we know, have absolutely the same views. They even advise abandoning the future Catholic Congress.

I would love to meet you, but I would hate to see you called to Rome. The inspired press would make everything of it.[156]

Who was the consultor of the Index who advised O'Connell of the plans afoot against Americanism and Zahm? O'Connell mentions a "young friend" who was consultor of the Index and had been at the New Year's dinner with Cardinal Serafino Vannutelli. It must have been someone quite a bit younger than O'Connell, who was nearing his fiftieth birthday. Most of the consultors of the Index were of similar age. So, the most likely suspect is Rafael Merry del Val, only thirty-three years old. Moreover, Merry del Val's presence at Vannutelli's dinner is completely understandable. Cardinal Vannutelli had been apostolic nuncio in Belgium between 1875 and 1880, and there he established a close relationship with the Merry del Vals. Rafael's father was Spanish ambassador to Belgium. He studied at the College of Our Lady of Peace, in Namur and, later, from 1878 on, in that of Saint Michel, in Brussels. The friendship continued later on in Rome.[157] In 1898 Merry del Val had been in Canada for a few months as apostolic delegate, which also explains his relationship with O'Connell and other American clerics.

Born in London the son of a diplomat and educated in England and Belgium, Merry del Val had been secretary of the committee to review the Anglican orders, where he must have worked in close quarters with David Fleming. In spite of his youth, Merry del Val was connected with the principal circles of the Roman Curia. At the time he lived in the Vatican and was in close contact with the Secretariat of State. He had been appointed consultor of the Index in 1897. His information must have been quite trustworthy. Later, under Pope Pius X, he was cardinal secretary of state (1903) and, still later, prefect of the Holy Office (1914).

It seems clear that the idea was to stop Zahm and obtain a personal retraction that would become known, without resorting to a public condemnation. Zahm confided his dilemma in his closest colleagues, as James Burns attests in many entries in his diary, including this note a few weeks later:

Our Superior General is very much alarmed about the fact of Father Zahm's book being on the Index. Father Zahm thinks that what restrains the publication of the decree is the fear that the condemnation of the book may be taken by the world as a condemnation of Evolution, and its case be linked to that of Galileo. It is for this reason that they want him to go to Rome himself and condemn the book or rather retract. He has not yet been summoned, however, and may not have to go at all.[158]

The French Edition

In the summer of 1896, just after his arrival in Rome, Zahm had closely followed the preparation of the Italian and French translations of *Evolution and Dogma*. The Italian edition was published quickly, and after the denunciation and condemnation of the book he was asked to withdraw it from sale. So the French translation, prepared by the priest J. Flageolet, appeared in 1897.[159] But the Congregation of the Index knew nothing about it until March 1899, when a review of the book appeared in *Annales de Philosophie Chrétienne*. Only then did the Congregation of the Index decide to warn Zahm again, which it did in an official letter to Father Gilbert Français:

Rome, April 25, 1899

Most Reverend Father:
Owing to the duty of my office and in the name of this Holy Congregation, I write with urgency to call your attention to an article in *Annales de Philosophie Chrétienne*, March 1899. In the "Varietés critiques" section there is a review of the French translation by the priest Flageolet of Father Zahm's book, *Evolution and Dogma.*

I was greatly surprised by such a review, completely favorable to the book and aimed at publicizing it, after Zahm's book had already been prohibited in all translations, in accord with number 45, title II, chapter IV of the General Decrees. When the Decree of condemnation was communicated to you, only the Italian translation was mentioned, because the Congregation was unaware of the French translation. Had it known, it would have ordered Father Zahm to stop its distribution also, which could not have been done without his consent.

Neither will I hide from you that comments have made the rounds that lead me to suspect that this French translation will be given the widest possible distribution, including distributing it among Seminary students.

You are advised of all this so that you might order Father Zahm to stop the distribution of the French translation of his work, as was done with the Italian translation.

I renew on this occasion the expression of my esteem, with which I am
Of your Most Reverend Father

Devotedly
Fr. Marcolino Cicognani, O. P.
Secretary[160]

As was customary, the dispatch of the letter was noted in the Diary of the Index.[161] The letter must have caused Français more than a little anxiety. His reaction was rapid. He sent the Congregation's letter to Zahm, urging him to act, and he also wrote to the Index, begging its pardon and assuring the Congregation that Zahm would expend all the effort necessary to withdraw the French translation.[162]

Zahm also acted quickly. On May 16, probably soon after receiving the letter from Français, he wrote two letters. The first was addressed to his French translator, asking him to do everything he could to withdraw the work. The wording is carefully studied so as not to explicitly mention the prohibition or give the slightest hint of a retraction:

> My dear Abbé Flageolet,
> I learn from unquestionable authority that the Holy See is opposed to the further distribution of your translation of my work "Evolution & Dogma," & I, therefore, beg of you to use all your influence to have the work withdrawn from sale. Knowing as I do your devotion to the Holy See I am sure you will join me in executing its wishes without delay. By so doing you will put under special obligations
>
> Very sincerely yours
J. A. Zahm, C. S. C.[163]

Zahm wrote directly to the secretary of the Congregation of the Index in Italian, alleging that he was unaware of the facts because, since he had given his permission to proceed with a French translation three years before, he heard nothing since. He also enclosed a copy of the letter he had just written to Flageolet.[164]

This rapid response achieved its objective and, a few days later, the publisher of the French translation, Lethielleux, withdrew the book from sale, although understandably he was dissatisfied from a financial perspective and sent Français a very stiff bill. Français was unable to meet such a demand and tried to negotiate with Lethielleux. Faced with the publisher's inflexible reply,[165] Français opted to forward correspondence directly to Zahm on May 27, advising him of the importance of complying with all of Rome's demands. He mentioned that he was still held to blame in Rome for not having examined Zahm's work before it was published, and that Lethielleux's complaint to that effect had some basis in fact: the publisher complained because he had taken it as given that the work had complied with all the ecclesiastical requisites. For the first time in his letters to Zahm, Français is impatient:

> The English edition of your book and its broad distribution in America should have yielded some financial resources. Therefore, use them to liquidate your debt with M. Lethielleux. The French translation is important, and if it continues in

circulation, you will be denounced again to Rome, and what we have been able to stop till now will again be used against you and against us all. Write now to M. Lethielleux and make him an offer. I cannot hide from you that I want this business ended quickly. If I had money of my own at my disposition, I assure you I would use it to end this matter immediately, because it is so painful for me and so sensitive for the interests of our Congregation with respect to the Holy See.[166]

The problem persisted for a while. At the beginning of June, Français, who had just sent Zahm's letter to the Index, again insisted on the importance of resolving the question as soon a possible:

I have sent your letter to Father Cicognani. If M. Lethielleux is not compensated, since nobody suspects he knows about the condemnation of your book, he will continue to sell and advertise it. Then Rome could be irritated with you and me and decide to publish the decree.

You know the situation as well as I. So see what arrangement you can work out directly with M. Lethielleux.

Father Flageolet has nothing to do with this matter and can do absolutely nothing. M. Lethielleux paid him 600 francs for his translation. Therefore it is with Lethielleux that you should negotiate.[167]

In mid-July, when the issue with Rome finally seemed resolved, Français reminded Zahm of the need to settle with Lethielleux.[168] But three months later he received another bill from Lethielleux, for almost 2,000 francs: "I have not replied to him. This is a matter between you and him. He reproaches me for not having approved your book. He says he had trusted the Congregation, etc. etc. His trust has been misplaced."[169]

Only toward the end of November was Français finally informed that the issues with Lethielleux had been definitively resolved.[170]

A Toast and a Letter

May and June were disquieting because news of Zahm's condemnation and retraction continued to spread. Alarming new news arrived from Rome at the beginning of June. The procurator general of the Congregation of Holy Cross in Rome, Legrand, had recently been replaced by Dr. Linnebarn, who,[171] perhaps not being as circumspect as Legrand, sent some news that must have worried Zahm. James Burns wrote in his diary for June 2:

It is now becoming known that Dr. Zahm's book on Evolution is on the *Index*. Dr. Linneborn—our new Procurator Gen. in Rome—told Brother Boniface so in a letter recently received, and said that the ratification of the decrees of the General Chapter were delayed so long through the unwillingness of the Propaganda to ratify the election of Dr. Zahm. The Cardinal—Ledochowsksi—asked the Doctor how and why we "came to elect such a man to such a high office," and he says that Father General had to go to Rome and guarantee the withdrawal of the book from circulation before they would ratify the election.[172]

Meanwhile matters were quieting down, although the reasons for it are not at all clear. On June 6 O'Connell, who had recovered his old confidence, sent Zahm a short telegram with the epigrammatic message, "Ogni pericolo passato" (All danger has passed).[173] Zahm had no clue as to why O'Connell had reached such a conclusion. He was on a trip to Washington and New York when the telegram arrived. After he finally read it, he wrote to O'Connell: "I am grateful to you, as you know, for the good news, but am curious to know what new element has entered into the case, & how the happy result was brought about."[174]

In reality, there is every indication that it was Zahm himself who secured the "happy result," thanks to two events: a meeting with the apostolic delegate in the United States, Sebastiano Martinelli,[175] and a letter written to the Italian translator of *Evolution and Dogma*, Alfonso Maria Galea.

Archbishop Martinelli, accompanied by the secretary of the Apostolic Delegation, Frederick Z. Rooker,[176] arrived at Notre Dame on May 10 for a one-week stay. A few days later there occurred an event that, in the opinion of Zahm's biographer, Weber, was the beginning of the positive outcome: "At a banquet in the Bishop's honor on Monday, May 15, Zahm arose and proposed a toast to the Archbishop, a toast filled with words of respect for Papal authority."[177]

The following day, Zahm wrote to Galea, asking him to do whatever possible to slow the distribution of the Italian edition of his book, because he knew the Holy See opposed it. Weber thinks that perhaps he wrote to Galea on Martinelli's suggestion.[178] This letter is related to the Congregation's warning to Français and Zahm on account of the French edition. Around this time Zahm must have received Français's letter of April 29, with a copy of the warning from the Index, because on May 16 Zahm wrote a set of related letters, one to Cicognani, another to his French translator, Flageolet, and a third to his Italian translator.

Until this moment (insofar as we know) Zahm had not fulfilled the stipulation of the Index that he "communicate to Alfonso Maria Galea, who published the Italian version of this book, the prohibition of the work."[179] The relationship of this letter with the one he had just written to Flageolet is evident, because the first few lines are practically identical. Zahm informs Galea of the Holy See's opposition to any further distribution of the book and asks that he exercise his influence to withdraw it from sale:

Notre Dame, Indiana, May 16, 1899

I have learned from an irreproachable source that the Holy See is opposed to any further distribution of *Evolution and Dogma* and, therefore, I beg you to use all of your influence to withdraw the book from the market. You may well have foreseen this outcome and so it won't surprise you. . . . Nevertheless, we can both thank God because we have worked solely for His honor and glory in bringing this book to the public. As for me, I will not suffer seeing the fruit of so much work consigned to oblivion. God rewards intent and our intentions were good.

Most sincerely yours
J. A. Zahm, C. S. C.[180]

What Zahm had not foreseen was that his letter, several days later, would be made public. Perhaps Galea himself decided that a quick way to comply with the wishes of the Holy See was to publish Zahm's letter. Perhaps someone intervened. What is certain is that on May 31, the newspaper *Gazzetta di Malta* published Zahm's letter along with a statement from Galea, associating himself with Zahm's position:

I too join with the illustrious Dr. J. A. Zahm, as translator of his *Evolution and Dogma*, and ask my sincere friends not to read nor give any further publicity to my miserable translation of the work mentioned, as a courtesy to the wishes of the Holy See, always ready to change my mind if I am asked.

Bétharram Sliema, May 31, 1899
Alf. M. Galea[181]

On June 24, when he was still not aware that his letter had already been published, together with Galea's statement, in *La Civiltà Cattolica*, Zahm wrote to O'-Connell:

Some time ago I wrote to Sig. Galea, of Malta, my Italian translator, asking him to use his influence in having his translation of *Evolution & Dogma* withdrawn from circulation. Imagine my surprise on finding my letter published in the *Gazzetta di Malta*. I hope it will not appear in the Italian press as I do not wish the public to know I wrote such a letter. I am not ashamed of it, but I do not wish to give the enemy the satisfaction of knowing that I should have reason to write as I did.[182]

But not only did "the enemy"—in this case, Brandi—find out, but he reproduced the letter in *La Civiltà Cattolica* and ran Galea's statement with it. *La Civiltà* published these documents with minimal commentary. It only recalled that in two previous issues, it had warmed of the "many errors" in Zahm's book. There was no embroidery, just a simple factual note. But it was evident that for Brandi this was a victory. In spite of Zahm's not having indicated that the book was disapproved, the general impression conveyed by the documents was that of a retraction. There was no mention of prohibition, but it was stated that the Holy See opposed the distribution of the book. This was enough to suit Brandi. Indeed, all this material, just as *La Civiltà* published it, has been used for more than a century as an authoritative source for the assertion that the Holy See opposed Zahm's book. Nothing more was required, nor is it strange that immediately thereafter the case was considered closed, even though Zahm neither went to Rome nor offered a retraction

Somewhat later the letter was made public in the United States. On July 2 an article titled "Father Zahm Submits to Rome" appeared in the *New York Tribune.* It presented Zahm's case as if it was still under examination by a committee of cardinals that was about to deliver a verdict. Therefore, Zahm had anticipated the worst and had withdrawn the book from sale, ending the controversy. The article had also picked up the story of how Zahm had declared his loyalty to the pope during Martinelli's visit.[183]

The news spread rapidly. William Seton, a well-known writer recently interested in scientific and biological questions, wrote to Zahm from New York on July 3, asking for information on the exact contents of the condemnation.[184] He took it as given that the book had been condemned by Rome and asked him if what was condemned was what used to be called "Mivart's theory" on the evolutionary origin of the human body, or evolution in general.[185]

Spirits were calmed. Français, satisfied by Zahm's act, immediately wrote him:

> I have received and read with much pleasure the two newspapers you sent me and where you speak so loyally about your book, with total submission to the Holy See.
>
> With this you have performed an act of courage in favor of America, for which I congratulate you and from which you will receive good results. In Italy there was complete dissatisfaction. Nothing remained except the small, black point of the French edition, but I hope that you too will see it disappear.
>
> This matter has been both a test and an obstacle for you. Now you can in complete peace realize your ideas, so creative and practical, for the development of your province and of Notre Dame.[186]

At the end of November the matter of the French edition had been resolved, and in October everything else was settled. On October 6, Zahm wrote to O'Connell:

> I do not think we shall hear anything more about "l'affaire." The toast to Martinelli was spontaneous, & intended as a blow to "our fathers." No one knew anything about it, not even Martinelli, until it was proposed. It had the desired effect. My letter to Galea was intended to be confidential, & I cannot imagine why he published it. I think however it was not after all a bad thing, as it spiked the Dago's guns completely.[187]

Perhaps because of his zeal to put a definitive end to the business, Zahm did not note that the "Dago's" guns did not remain silent but, to the contrary, now began to fire. Zahm had given the "Dago" just what he needed: a written declaration that he accepted the Holy See's censure of his book. Nothing more need have been said. This became the standard line on Zahm for a full century.

In the Archive of the Province of the Congregation of Holy Cross, there is one last document relevant to the question: a copy of a letter dated June 23, written by Cavaliere L. Bufalini, Zahm's Italian publisher, to Alfonso Galea, notifying him that all remaining copies of the Italian edition had been withdrawn from sale: "I am most thankful for your generosity in sending me a check for fifteen pounds sterling to donate Zahm's book to the libraries of Italian seminaries. On Monday I will order the shipments to begin and will add 'sold out' to the catalog [description of the book]. That should handle any problem and satisfy the wishes of all."[188]

The withdrawal of the Italian edition thus was simpler to accomplish than that of the French. Galea and Bufalini's solution, however, is ironic because placing the remaining copies in Italian seminaries was a very effective way of distributing the book.

Hedley and Brandi on Zahm

The public polemic over Zahm's book continued apace. David Fleming, consultor of the Holy Office had, as we already know, written a very favorable review for the *Dublin Review* in 1896. In the same journal's issue for July–October 1898 (precisely when Zahm's book was under examination in the Congregation of the Index), the bishop of Newport, Wales, John Hedley, followed suit with a generally favorable article on Zahm and his work.[189]

Hedley was not easily pleased. In 1887 he published a tough critique of Mivart in the *Dublin Review,* in response to an article by Mivart on the Church and biblical criticism. In 1888 Mivart replied in the same journal, and Hedley added a brief rebuttal.[190] So Hedley's benevolent stance with respect to Zahm was especially significant.

The format of Hedley's article was a review of four of Zahm's recently published books: *Evolution and Dogma; Bible, Science, and Faith; Evolution and Teleology;* and *Science and the Church.* Hedley states that the majority of enlightened Catholics has accepted evolution, if cautiously because the theory is frequently accompanied by unacceptable ideological baggage:

> To argue, therefore, in favour of the evolution hypothesis, and at the same time to dissociate oneself from all the supporters of that hypothesis who in any way contravene the Catholic faith, is a task which requires both a clear head and great independence of judgment. This is what an eminent American Catholic writer and lecturer, the Rev. Dr. J. A. Zahm, has done in the works named at the head of this

article, and chiefly in the series of lectures which he has reprinted under the name of "Evolution and Dogma."[191]

Hedley's commentary points out some problems of Zahm's ideology, but in general he is very positive and includes an important endorsement, namely, "there is nothing in the theory of evolution, or proved by fact, that is really antagonistic to Catholic faith."[192]

Soon enough, *La Rassegna Nazionale* brought Hedley's position to Italy in an anonymous article published November 16, 1898, on Hedley's review of Zahm. At the same time, Salvatore Brandi was awaiting the publication of the condemnation of Zahm's book and on January 7, 1899, published his article, "Evolution and Dogma," in *La Civiltà Cattolica*.[193] In that article, Brandi says that the anonymous author from *La Rassegna* had not read Hedley's article and had only copied an article from the *Tablet* of London, adding ten words at the beginning and a dozen lines in the conclusion.

Brandi's article was a frontal attack on Zahm's book. He takes Hedley's reservations and enlarges on them. On Saint Augustine and Saint Thomas he references articles by Salis Seewis in *La Civiltà* and cites Satolli's *De habitibus*. He criticizes Zahm for lack of coherence and characterizes evolutionism (using the words of Salis Seewis) as a tissue of vulgar analogies and arbitrary suppositions unsupported by the facts. He asserts that the first impediment to accepting evolutionism does not come from fear of opposing the Bible but from the scientific inconsistency of the system: "In this situation, it seems to us someone who, contrary to the traditional judgment of the Church Fathers, obstinately supports the theory which holds that the human body is derived from the monkey or from whatever other animal, can certainly be called rash."[194]

At the end of his article, Brandi mentions that Bonomelli, Hedley, and Zahm all cited Leroy in support of their ideas unaware that Leroy had retracted, and to clear up the confusion he reproduced Leroy's letter of retraction, as published in *Le Monde.*

Three years later, when the Zahm case had long since concluded, the differences between Hedley and Brandi surfaced again. On this occasion, Brandi alludes to Zahm's letter to his Italian translator, of May 16, 1899, in which he said he had learned from a credible source that the Holy See was opposed to the further distribution of his book and for that reason would withdraw it from circulation. He comments that Zahm had made this public declaration, just like Leroy, to avoid Rome's censure.[195] Some of this was true, but Zahm's letter was private, and Brandi converted it into a kind of public retraction.

Zahm Cited by Arintero

In 1898, in the introduction to his work *La evolución y la filosofía cristiana*, González de Arintero cites Zahm repeatedly, relying on the Italian translation—for example, when Zahm asserts that evolution is confirmed by Genesis as well as by the Church fathers; when he discusses the concept of species; when he says that evolution will have the same fate as the ideas of Galileo; when he presents Albertus Magnus as the first transformist; and when he discusses the probability of evolution.[196] It is clear that Arintero was very familiar with Zahm's work.

Arintero finds it strange that Zahm does not accept Cardinal Zeferino González's reservations regarding the origin of the human body while, at the same time, he accepts the position of Mivart, an author with works on the Index. He adds that he does not know why these works are on the Index, but he thinks (wrongly) that his position on evolution must have been involved.[197] Arintero thinks that both Leroy and Zahm overstep the limit when they favor the evolutionary origin of the human body, which, he says, has provoked criticisms that he considers just.[198]

Barry Brundell on Zahm

In his article on evolutionism based on documents from the Archives of the Congregation of the Faith, Barry Brundell's treatment of Zahm is more accurate than his discussion of Leroy.[199] His basic argument, that is, the central responsibility of the Jesuits of *La Civiltà Cattolica* for the onslaught against evolution, is supported only by data external to the Congregation, concretely in the articles in *La Civiltà* itself. Evidently, this thesis is more defensible in Zahm's case than it is for Leroy, because *La Civiltà*'s critiques of Zahm were very explicit and harsh indeed. But Brundell confuses the issue too when writing of Zahm. So, he says that the denunciation of Zahm's book is "only" based on the articles in *La Civiltà* and on Mazzella's book. This is not true because Zardetti's denunciation adduces more arguments. Zardetti says that Zahm cites Mivart in his favor after Mivart had been condemned; he cites Satolli's *De habitibus*; he says that Zahm cites Leroy in his favor, despite Leroy's retraction.

According to Brundell, one internal proof, based on the documents of the archive, is that *La Civiltà* was familiar with Buonpensiere's recommendation, because in an article written afterward, Brandi used the word *disinvoltura* to describe Zahm's attitude, and this word, which appears in the report, is unusual. The coincidence is clear, but it does not in itself constitute proof that someone was leaking information to Brandi from within the Congregation.

Brundell cites two passages of Buonpensiere's report stating the consultor's proposal, adding that the proposal was not accepted, and that a more private procedure was authorized, to send a decree of personal condemnation to Zahm via the superior general of his congregation. This is inexact because, as we have seen, the matter was quite a bit more complex. Initially the Index wanted to publish the decree, but publication was only avoided after months of long, complex negotiations that passed through a number of stages and in which the pope was personally involved.

Afterthoughts

A central aspect of Zahm's case was the social projection of his life and work. Author of a number of books, he was a famous proponent of the harmony between science and faith. He played an important role in the International Catholic Congresses that were held in Europe in his time and was closely connected with the representatives of the Americanist movement. No wonder, then, that ecclesiastical politics were involved in his case, with influential personalities both for and against him.

In a certain sense Zahm represented the cultivated scientific wing of the Americanists. They could be considered as progressive, but surely they were faithful to the doctrine and the authority of the Church. In an epoch of political convulsions in Europe, which included the isolation of the pope within the new Italian nation, the Americanists considered separation between Church and political power in the United States as a very good solution to the traditional conflicts that existed in the Old World. They also considered the typical American virtues as a paradigm for the modern Christian. Zahm's character and his openness to modern science were congenial with all this. Probably the connection between evolutionism and Americanism extended only up to this point. But this was enough for Zahm to be considered a danger by those who, at the same time, were both afraid of evolutionism and opposed to Americanism.

Evolutionism was only one of the pieces of the puzzle. The process that led the authorities of the Index to decide on the prohibition of Zahm's book was very straightforward. The antecedent of Leroy's case shortly before had paved the way. The complications began once that decision had been taken. Zahm and his Americanist friends were ready to fight to the end in order to prevent the publication of the decree, while their opponents also worked for its publication with untiring constancy. Much of this was already known through the correspondence of the personalities involved. The Vatican archives provide new documents, but even now some mystery still remains. Apparently Zahm had to go to Rome, presumably to produce a written recantation that could be published, but this trip never took place. Instead, a private letter from Zahm to the Italian translator of his book was enough. Zahm asked the translator to prevent the further diffusion of the book because he had known that the Holy See opposed it. When the letter was published, without the knowledge or consent of Zahm, this was enough, as the letter appeared as a testimony both to the Holy See's attitude and to Zahm's acceptance.

Surely, the case of Zahm can be considered as an exemplar of the kinds of complexities involved in the relationship between science and religion. Also, it shows that the authorities of the Church were satisfied if they could obstruct the spreading of evolutionism without taking any strong public decision.

Condemned for Evolutionism?

Geremia Bonomelli

For more than a century, Bishop Bonomelli has been cited as one of the authors who retracted his defense of evolutionism because of pressure from the Holy Office. The case is significant because Bonomelli was an important figure in Italian public life. The Archive of the Holy Office furnishes new data, although Bonomelli's correspondence had already shown that his retraction was not the result of any official action of the Holy See, but rather a strategic retreat by the controversial bishop in the face of a possibility of a new condemnation: he already had another work on the *Index of Prohibited Books*, which had nothing to do with science nor with evolution, and he was not disposed to suffer another listing on account of evolutionism.

The Vatican and the New Italy

Geremia Bonomelli was born in Nigoline, Brescia, on September 22, 1831. His father was a small rural proprietor. He entered the Seminary of Brescia in 1851, was ordained in 1855, and was then sent by his bishop to Rome to continue his studies in the Pontifical Gregorian University. After receiving his doctorate in theology in 1858, he became professor of philosophy of religion and hermeneutics in the Seminary of Brescia. At the same time, he gave Lenten sermons and directed spiritual exercises for the clergy and "popular missions," within his diocese and outside.[1] In 1859 he was posted to the parish of Adro (also in Brescia) and in 1866 was named parish priest of Lovere. In this period he wrote the three volumes of his book *The Young Student Instructed in Christian Doctrine*, published between 1871 and 1874.[2]

In 1871 he was named bishop of Cremona, likewise in the north of Italy, where he remained until his death in 1914. Throughout his forty-three years as bishop of Cremona he was involved in numerous controversies that had public resonance. When he died, an obituary in *La Civiltà Cattolica* included the cryptic comment: "How and why he almost suddenly changed the nature of his public conduct, at times even becoming a critic of the Church, of the clergy, and of Catholic laity, history will say. . . . But let us leave judgment to the Lord. The blessing of Pius X comforted him in his last hours, and we beseech peace for his soul."[3]

Bonomelli was not an eccentric nor did he entertain heretical doctrines. He was especially known for his open support for the new Italian state, which was at odds with the Vatican's opposition to it. For many centuries Italy had been divided into multiple territories: Milan, the Republic of Venice, the Grand Duchy of Tuscany, and the Kingdom of Naples were among the principal ones, together with the central part of the peninsula, occupied by the Papal States, whose sovereign, spiritual as well as political, was the pope. In the nineteenth century there arose the idea of unifying the whole territory in a single nation and this undertaking concluded when the troops of the new Italian state occupied Rome in 1870. The pope was dispossessed of his temporal power and was confined to the Vatican. The official posture of the Vatican was to refuse to accept the new situation, giving rise to severe tension with the new Italian authorities. The problem, commonly called "the Roman question," was not resolved until 1929, when Italy and the Vatican reached an accord that resulted in the current status.

Bonomelli proposed a solution similar to the one that was finally adopted: that the Vatican recognize the new state of Italy and that, to guarantee the independence of the pope with respect to events of Italian policy, the pope would remain as sovereign of a minuscule territory in the Vatican. Bonomelli was well connected to figures in Italian public life, including the royal family (Italy was then a monarchy), and he favored resolving the Roman question as soon as possible. In 1889 Bonomelli published a pamphlet in which he set forth his solution, but his ideas clashed head on with the official stance of the Vatican, and the pamphlet was placed on the *Index of Prohibited Books* that same year.[4]

A Controversial Bishop

In the Holy Office archives there is a thick file on Bonomelli, which reflects the controversies in which he was involved owing to ideas and behavior that, without being heretical or overly strident, were the object of strong opposition from part of the Catholic press and from priests from his own diocese. He was denounced to the Vatican and was on the brink of losing his bishopric. These controversies were well documented before the opening of the Holy Office archives,[5] although that archive does provide some new information.

For example, in the Holy Office archives there is a letter with a dateline "Monaco," May 11, 1892, signed by Antonio, archbishop of Cesarea, saying that he had completed his task and stating that Bonomelli is disposed to renounce the

Geremia Bonomelli. Archivo Storico Diocesano de Cremona.

bishopric of Cremona.[6] The letter is signed by Antonio Agliardi, born in 1832 in a village in the diocese of Bergamo, ordained in 1855, bishop from 1884, who had been apostolic nuncio in Bavaria since 1889 (the letter, therefore, was written from Munich, Monaco in Italian). Agliardi was named cardinal in 1896 and was a close friend of Bonomelli;[7] therefore, he had all the qualifications required to ask Bonomelli, in the name of the Holy See, to renounce his bishopric, something that, nevertheless, did not come to pass, because finally the Holy See did not make such a decision.[8]

One of the most important controversies began with the publication in Italian of the sermons preached at Notre Dame by Father Monsabré, a Dominican. As we have seen, Jacques Marie Louis Monsabré (1827–1907) was a very well-known orator in France. He preached in the principal cities of France and Belgium, even in London. His fame was owing principally to the sermons he preached at Notre Dame de Paris over a twenty-year period starting in 1869 (with the interruption explained by the political events of the period), laying out in an ordered fashion a complete account of Catholic dogma, with a great audience in attendance. Bonomelli translated and published these sermons, adding his comments in footnotes, in two works that filled 22 volumes.[9]

But Bonomelli's notes soon raised their own comments, and the matter reached the Holy Office. On March 27–28, 1890, the newspaper *L'Osservatore cattolico* published on its front page a letter from Monsabré (in the original French, with an Italian translation) in which Monsabré disapproved of the Italian edition of his work, giving as a reason Bonomelli's comments, especially with reference to the relationship between politics and religion.

A short time later (April 12, 1890), *Il Messaggero di Cremona*, on its front page, reported that on March 13 Monsabré had written to Bonomelli explaining that he had received protests from Italian prelates and clerics owing to Bonomelli's notes, and thus he refused all responsibility. This was the same report that had been published in *L'Osservatore cattolico*, but *Il Messaggero di Cremona* added that Bonomelli responded to Monsabré expressing his surprise, inasmuch as he had received many congratulations, and the work had the requisite ecclesiastical permission. Bonomelli admitted that perhaps there were some inexact phrases and makes clear his willingness to change any errors that are pointed out to him. Monsabré replied on March 19 in a friendly tone, saying that he did not know Italian, and that he would transmit the matter to the father general (of the Dominicans). On March 28 the father general replied to Monsabré begging his pardon, saying that nothing had happened, and that he had been aware that "a Monsignor who held a high post had attributed to him faculties he did not have." The newspaper lamented the whole affair, but Bonomelli was looked upon favorably, in an article signed by the vicar general of the diocese of Cremona. The newspaper commented that once again one had to admire the virtue and wisdom of Bonomelli, his generosity and forgiveness, and his heroic submission to the Holy See.

An Evolutionist Intellectual

Bonomelli was friendly with the "Americanists," in whose circle was found John Zahm, the defender of evolutionism. In a letter to Countess Sabina Parravicino, Bonomelli asserted that Americanism and evolutionism were "two related things."[10]

Reconciliation between science and faith was one of the outstanding features of Bonomelli's thought, and one of its elements was the reconciliation between evolutionism and Christianity. In this he concurred with the writer Antonio Fogazzaro (1842–1911), a close friend of Bonomelli, a senator of the Kingdom of Italy.

In 1893 Fogazzaro published a book favoring evolutionism, which was harshly criticized in *La Civiltà Cattolica*.[11] The book was based on a lecture delivered by Fogazzaro on March 2, 1893, in the presence of the Queen of Italy, on the origin of man and religious feeling. Bonomelli knew Fogazzaro well and enjoyed his full confidence. On March 17 he advised Fogazzaro by letter that, if he still had not sent his lecture to the printer, he should have it read by some authority that could defend it, bearing in mind the attacks on him in the Catholic press.[12] On March 31 he told him that he had read his piece several times and had even given it to two professors in Rome, admirers of his, and would now say what the three of them opined: although there was nothing objectionable in the first part of the lecture, the second part, in which he discusses the human soul, was open to many objections, because he was unclear about the origin of the soul, specifically whether it was created directly by God, and moreover it might appear that he accepted the false, condemned doctrine of Rosmini according to which God is a natural object of the human mind.[13] Fogazzaro took the advice and included an explicit reference to the divine origin of the soul, and on July 9 Bonomelli expressed his satisfaction.[14] In spite of their efforts, the book received a severe critique by the Jesuit Francesco Salis Seewis in a long review published in *La Civiltà Cattolica* in October of the same year.[15]

Salis Seewis began his review by expressing his respect for Fogazzaro, a well-known literary figure who proclaimed himself a Catholic. But then he recounted, one after another, objections to evolution, scientific as well as philosophical and religious. Fogazzaro was not a scientist, and his exposition was filled with literary and poetic allusions—for example, "A clear inner voice told me that the question of the origin of man, in spite of its great scientific and philosophical problems, is in great part a question of feeling and of taste." Salis Seewis sought to persuade his readers of the lack of scientific arguments in favor of evolutionism and of the theological weakness of the arguments of Fogazzaro, who certainly provided ample material for such critiques. Toward the end of the review,

Salis Seewis includes a few comments that can be taken as representative of the kind of criticisms made at the time against the attempt to reconcile evolution and Christianity:

> When, through the work of unbelievers around forty years ago this phantom of evolution arose to break and harm Revelation, Catholic apologists displayed unanimity and took the position that the subject required, staying within the compass of science and from there relentlessly unmasking the emptiness of that hypothesis and its ostensible demonstrations. The result was as it should have been and as Fogazzaro himself confesses: the total defeat of the system, which now cannot anymore be sustained because it is demonstrated, but only because someone *wants* to retain it. In this situation, our author proposes the truly poetic and novelesque strategy of abandoning our victorious position and joining the enemy, perhaps for the pleasure of hoisting him on our shoulders, during the proclamation of a Christian evolutionism. I hope the author might excuse us, for we cannot follow him. Science having declared that evolutionism, whether baptized or not, is a myth, has only us believers to sustain it against the crude dogmatism of the unbelievers. We cannot be traitors to her, even if only for the honor and independence of our reason. Then, religious feeling, at least in the way we and most believers experience it, rebels against the idea of denying both science and the doctrines taught by the Church in order to base the concept of Creation upon the dream of minds agitated by incredulity. That would be like erecting some pagan idol on our altars to more worthily represent to us the objects of our sublime Christian cult. To conclude, in this as in any struggle, the false science of unbelief must itself recognize its error. In fact, it is so doing. It certainly is not our role to back the cause like unexpected patrons and participate in its shame.[16]

Bonomelli and Evolution

In his letters to Fogazzaro, Bonomelli expresses himself freely and reveals a certain acceptance of evolutionism: except when he works miracles, God always acts through secondary causes, whence is explained the law of evolution:

> A stupendous, universal law which relates all the creatures one to another, which develops its forces, respects all liberties . . . that from maximum unity tend to maximum variety, which creates order and beauty. Creation, as it was understood in the past, by blows, by lightning bolts, by jumps, from day to night, etc. etc. is poetry: now we have science. Science agrees with faith in the original creation out of nothing: but science requires evolution, and faith should not exempt anything. Insofar as concerns the human soul, insist *strongly* on its creation *from nothing*; after, the way is clear.[17]

Nevertheless, Bonomelli was not at that time (1893) a convinced evolutionist. He appears to feel the power of the scientific arguments in favor of evolution and encourages Fogazzaro, who himself was not a scientist, in his enthusiasm for evolution. But he also perceives the difficulties and expresses them in his letters to Fogazzaro:

> Above all, I confess that my knowledge of the great law of evolution and transformation is too slight for me to dare pronounce. . . . Galileo launched the old heaven into the void. Geology is turning the Earth upside-down. History together with paleontology and with the help of modern science reverses a mountain of prejudices. It is natural that the world today divides into two groups, the group which, alert, grips the past to the letter and the group, drunken with the new horizons appearing, intoning hymns of joy, set forth on the great road of progress and shout: forward! forward! I am with the latter, but always with the fear that we might break our necks by running so fast.[18]

In the following months of 1893, Bonomelli read Fogazzaro's piece several times more and was increasingly persuaded of the truth of evolutionism, as he wrote him on July 9:

> Rereading [your lecture] for the third time . . . God creates: but he lets things develop in accord with the forces He has placed in them, and here evolution begins. What good work you would do if you developed these two powerful ideas: 1. God always works through secondary causes; 2. And therefore everything is made little by little through evolution, in the three kingdoms, from the atom to man. . . . The whole universe is an evolution or transubstantiation. . . . Galileo turned the heavens around: geology is turning the Earth around. . . . I am like a child who believes he can touch the Moon by climbing a hill. Science magnifies everything, overvalues everything, and faith receives the reverberation that sublimates it.[19]

In 1896 Fogazzaro was named senator of the Kingdom of Italy, although he did not actually begin his tenure until four years later. The correspondence between the two friends was continuous, as always. On October 26, 1897, Fogazzaro wrote:

> A great consolation for me. I am reading a book titled *Evolution and Dogma*. The author is Father Zahm, professor of Philosophy in a Catholic university in America. I have the Italian version published in Siena with the *imprimatur* of its Curia. It is on sale at an ecclesiastic bookshop in Turin. The book seems like an expansion of my lecture *For the Beauty of an Idea*. One reads things like this: "It is wrongly asserted that simian origin makes man baser; such an origin ennobles the simians." I declare in sum: what a long road has been traveled! And *Civiltà Cattolica*, still recently, spoke of my ideas on evolution as the ideas of a novelist![20]

Bonomelli replied on November 1, conveying a seemingly insignificant fact, but one that was to have far-reaching consequences: "I have ordered the book you mention: I will read it with pleasure."[21] Zahm's book would mark an important moment in his life and, indeed, in the history of the relationship between evolutionism and Catholicism.

The Appendix to 'Let Us Follow Reason'

John Zahm published his book *Evolution and Dogma* in the United States in 1896, and it was published in Italian that same year, as we noted. As a result of the elegy to Zahm's book that Fogazzaro wrote in his letter of October 26, 1897, Bonomelli bought and read it, and it made a strong impression on him. Bonomelli was inclined toward evolutionism while maintaining some doubts and reservations, and he found in Zahm a kindred spirit, who argued for evolution with clarity, interpreted Catholic tradition in its favor, and held that evolution is not only compatible with Catholic doctrine but helps it and gives it new wings.

Just at that time Bonomelli was winding up the publication of an apologetic book that he titled *Let Us Follow Reason,* in which he attempted to explain in the best way possible, utilizing reason, the principle truths of Catholic faith. He divided the work into three thematic volumes: the first on God the Creator, the second on Jesus, and the third on the Church. The three volumes were published between 1898 and 1900. The first appeared a short time after Bonomelli had read Zahm's book, in 1898. He was so impressed by his reading that he decided to add an appendix to this first volume, summarizing and praising the book. It is titled, simply "Important Appendix," and it occupies thirteen pages.[22]

The first part of this appendix (pp. 201–3) is titled "On Evolution" and is a very flattering presentation of Zahm's book for the Italian reading public—flattering to the point where Bonomelli states that book had changed his own ideas. The purpose of the appendix was to rectify what Bonomelli says in the eighth chapter of the same book as well as to dispose the reader to accept evolutionary ideas.

In the eighth chapter, Bonomelli discusses evolutionism in relation to the origin of man. He distinguishes between evolutionism as a scientific theory and its ideological interpretations of a materialist type, and asserts that evolutionism as it once was presented is no longer accepted, but what was true in it still stands:

> How did the first man and woman appear? Were they possibly the producers and authors of themselves? That is a patent absurdity. . . . Were they the result of change? Chance is nothing, and nothing cannot produce what exists. . . . Were they perhaps an effect of chemical combinations? . . .

Here *transformism* is presented with all its scientific apparatus and it promises to explain to us that man did not arise from man but from other organisms, without any need to recur to an external cause, person, intelligence, by way of some kind of *autogenesis* or *spontaneous generation.* In truth, this system, which caused so much noise around twenty years ago, now is in decline, like so many others, leaving behind what is true about it. *Transformism,* formulated with much ingenuity by the Englishman Darwin, is a system in which many errors are mixed with many truths, as generally happens in all scientific systems. At first, surprised by its novelty and by the truth that the system contains, and by the detailed, painstaking, and rigorous observations of the author, it all seemed clear and evident, and so what was false was accepted along with what was true, and perhaps more than that. But time has worked, and is working, its justice. . . . Let us examine this system quickly, separating the good from the bad and rejecting the *materialism* that it conceals. (pp. 117–19)

In his book, Bonomelli had displayed an attitude that combined a certain acceptance of evolutionism with scientific reservations, to which he added some more of an ideological nature, but he wanted to separate out materialist ideology from the scientific theory. But Zahm's book convinced him to come out decidedly in favor of the reconciliation between evolutionism and Christianity, and so he added the appendix in which he modified the ideas expressed in the same book. He begins the appendix with these words:

Whoever is not estranged from modern scientific questions knows very well the importance the numerous and so diverse theories about evolution and transformism have acquired, especially after Darwin, and what abuse has been made, and still is made, of them. One could fill a large library with the books written on this subject. At first, some Catholics, afraid of the consequences they foresaw from that quarter and seeing dogma so fiercely attacked, they perhaps launched headlong into their defenses, rejecting that which could be conceded without risk as well; they did not see evolution and transformism except as a formidable machine against the principles of the faith. Now things are better clarified, and it has been recognized that many concessions can be made, while the confines of dogma remain safe.

I myself should modify that which I have written about man and the transformation of species, above, in chapter 8. Supported by the authority of great naturalists, including those now living, I could have left what I have written as it was; but after having thought things through better and after I'd read Professor Zahm's book, *Evolution and Dogma,* it strikes me as useful to add this Appendix, where you will find a summary of the book. We must ease the road for those who are outside the Church, that they might enter, in such a way that no one will find reproachable; on this point I too intended to show that certain difficulties can be eliminated or at least diminished. Professor Zahm is a Catholic and enjoys great fame as a learned man. His book, translated by Alfonso Maria Galea, has been printed in Siena, with the approval of the Archiepiscopal Curia. (pp. 201–2)

Bonomelli goes on to say that he makes no claim to subscribe to all of Zahm's opinions, some of which strike him as disputable or weakly based. But he makes clear he is sympathetic to Zahm's attitude, which was his own: namely, do not fear science and point out that evolution, well understood, can lead to a still greater idea of God:

> Do I subscribe to all the opinions of that fine professor [Zahm] and make them mine? Not at all. There are things that don't seem sufficiently demonstrated, others which, in my opinion, are not sufficiently clear, and others that perhaps can be rejected. In any case, unless I am mistaken, this book, published in Italy, is a symptom of the new direction that these studies can and should take, and it is good that clergy are not reticent to embark on the new road which, while maintaining the inviolability of dogma, shows that they welcome open research and all truly scientific progress.
>
> Here I present a precise summary of Professor Zahm's book, particularly the great question of evolution, which opens immense new horizons and which far from diminishing the idea of Creation, marvelously broadens it. Just like Galileo's discovery enormously enlarged the idea of God in the celestial world, so too the law of evolution, well understood and well applied, will enlarge it in the terrestrial world. (p. 202)

This is an attitude characteristic of Bonomelli. As he himself recognizes, he is not the one to evaluate the scientific value of evolutionism. Caverni, Leroy, and Zahm, all of them priests, have devoted themselves to the serious study of scientific questions, and Zahm, indeed, was professor of sciences at Notre Dame. Having found a person like Zahm, who was well acquainted with science and who promoted an attitude open to progress, Bonomelli associated himself with Zahm in his advocacy of the reconciliation of evolution and Christianity, which is an aspect of the harmonization of science and faith:

> My cry is this: No fear of science in whatever form it may present itself. If it is true science, it can only come from God and lead us toward God. To fear that science might destroy faith is to err with respect to faith itself, to doubt the divine fundamentals on which it is based. We Catholics should applaud all the new conquests of science, which are conquests of truth, of God himself which is its source: we should compete with the same enemies of the faith to cultivate all of science, and if possible, vanquish them and use their arms to throw a light ever more vital on the truths of Catholicism. Standing firmly on the base of the dogmas proposed by the Church, let us take care not to confuse them with the opinions or hypotheses

more or less probable of some theologians and certain narrow and flimsy interpretations of the holy books that expose us to the ridicule of the learned. Let us
travel spiritedly along the broad trails that Saint Thomas and Saint Augustine
blazed for us, especially the latter, who fifteen centuries ago laid out the laws of
evolution in their Christian sense in broad strokes—read Zahm's book *Bible, Science and Faith* (Siena, 1895), especially Chapter IV. (pp. 202–3)

After this enthusiastic presentation, there follows a discussion of "Evolution
and Dogma," the second part of his appendix (pp. 203–13), which is a full and
clear summary of Zahm's book, subject by subject: the arguments in favor of
evolution, difficulties of the theory and how to resolve them, the compatibility of
evolutionism with Christianity. It is a simple summary, without comment, a
faithful and convincing résumé of Zahm's ideas. Although Bonomelli had said
that he is not convinced by everything Zahm says, this summary leaves the
reader with the impression that everything stated here, which includes all of
Zahm's principal points, is true.

A Spontaneous Retraction

The appendix was published at an unfortunate moment. Shortly before, on November 5, 1897, Zahm's book had been denounced to the Congregation of the
Index and was being examined at the time. On April 15, 1898, the Dominican
Enrico Buonpensiere, consultor of the Index, submitted his report on the book,
which was so completely negative that he even proposed that evolutionism
applied to the origin of the human body should be clearly condemned. The
paeans of praise that Bonomelli showered on Zahm's book put him in a precari-
ous position.

Bonomelli could not escape the fallout from attacks on Zahm's book. Monsignor Denis O'Connell stated that Bonomelli's appendix had been the stimulus
of the Congregation of the Index's interest in Zahm: "It is true that a strong campaign has been mounted against Zahm. They paid no attention to him until they
read Monsignor Bonomelli's Appendix."[23]

But this could not be true, because Bonomelli's book was published in 1898,
when Zahm's book had already been denounced (on November 5, 1897). Still, it
is clear enough that the existence of the appendix was known immediately and
that it landed Bonomelli in an embarrassing situation, as he himself explained to
his friend Fogazzaro, on August 4, 1898:

> You probably are aware of the bothersome problems I have had and perhaps will
> continue to have. You need to arm me with patience, a lot of it: I wrote a sum-
> mary of Zahm's book on evolution and published it, with a few reservations, as
> an Appendix to my volume *Let us Follow Reason*. Immediately a firestorm from the

usual journals raged against that Appendix. I was denounced to Rome: I was told nothing from there. Because beforehand they would have to condemn Zahm and others more advanced than I. I know that great efforts are being made by the usual people to obtain some decisions against *Americanism* [and] *evolution* from the Holy Father. I hope that, learning from the past, they go slow.[24]

We do not know on what basis Bonomelli says he was denounced to Rome: only Bonomelli himself mentions the denunciation. But he then contacted his close friend, Cardinal Agliardi, who sent him his comments and advice. In a letter dated September 26, 1898, Agliardi says that evolution, especially when applied to the human body, is not proven, and that seemed, "at the least, dangerous" if one wished to salve the biblical text of Genesis. In another letter written October 20, he adds that the appendix on Zahm had caused a storm in Rome, and that he had sounded out Cardinal Steinhuber, prefect of the Congregation of the Index, who advised him that Bonomelli should publish a rectification in a Catholic journal.[25]

Bonomelli acted accordingly. Two days later he wrote a letter of retraction, dated October 22, 1898, which was published on the first page of *Lega Lombarda* of Milan a few days after, with a very brief introduction ("We have received the following and hasten to print it"):

Nigoline, October 22, 1898

In the first volume of my new work *Seguiamo la ragione,* published by Cogliati in Milan early in the present year, speaking of man and his origin, I explicitly demonstrated that evolution is repugnant to reason and theology, and I supported the common doctrine professed by Catholic philosophers and theologians. While that volume was in press, there fell into my hands Professor Zahm's book *Evolution and Dogma,* published in Siena with ecclesiastical approval, which impressed me greatly. I immediately made a summary of it and added it to my book as an appendix. In truth, I did not publish this appendix as a *thesis,* but only as a *hypothesis;* because, as I said in the preface, I did not understand many things that Zahm said, and some others did not seem to be demonstrated. I published that appendix with the intent to call attention to possible surprises that modern science holds for us, to motivate both clergy and lay believers to better study the matter, and above all, just as Father Leroy has done in France, as an effort to extend our hands to some intellectuals who vacillate between faith and error.

But, right after the publication of my summary and ever since, many kind friends who are reliable both for their knowledge and their authority, have told me in conversation and in writing that this doctrine, even as a simple hypothesis, cannot be made to agree with the common interpretation given by the Church. So that it may never happen that I might propound—even though only as a *hypothesis,* a doctrine which is not in perfect agreement with the teachings of the

Church, and, in this way, to eliminate any doubt or twisted interpretation, freely following the advice of very dear friends, I deem it appropriate and necessary to ask those readers who take as my own the ideas expressed in the main part of the book and which I have always professed in my other works, and not those which, as a *hypothesis*, I presented in the appendix, based on Professor Zahm's authority. In a while a new edition of my book will be prepared, and then the appendix will be eliminated. A few months ago, I made a statement similar to this one (also in *Lega Lombarda*—in number 164, June 22–23); but it seems useful to restate it more explicitly. Most grateful for your publishing [this letter], I profess my great esteem, your most devoted,

Geremia, Bishop of Cremona[26]

It is not clear the appendix was denounced to the Congregation of the Index, which at this moment was occupied with Zahm's book, and Bonomelli is not mentioned in the archival documentation for the period.[27] But, as Cardinal Agliardi had observed, Bonomelli's appendix, so tightly tied to Zahm's book, had caused a stir. As Bonomelli himself says, he had received no indication of any kind, much less an order, from Rome. But, the advice of his friend Agliardi, who had spoken with the cardinal prefect of the Index about the matter, was enough to provoke his immediate public retraction. In a certain sense, it was a spontaneous retraction, made to head off further problems, and it had the desired effect. Cardinal Agliardi wrote to Bonomelli on October 28, 1898, that the appearance of the letter had put out the fire and Bonomelli need not worry about the matter further.[28]

Justifying the Retraction

The rapidity with which Bonomelli made his retraction shows that he didn't have a moment's doubt. This is completely understandable because, although he was a supporter of evolutionism, he had not studied it seriously and he recognized that it presented problems and uncertainties. Bonomelli was a bishop who was very active in public life, who already had a publication on the *Index* (since 1889), and who must not have been pleased by the prospect of adding another title. He continued to demonstrate that, when he was convinced of something, he exposed himself to serious risks. Indeed, some years later, in 1906, Bonomelli was dressed down personally by the pope (Pius X, by then), but this had to do with the relationship between the Church and the Italian state, which had always preoccupied him and was, in fact, the one that had caused the prohibition of 1889.

Evolutionism did not occupy so important a place in his hierarchy of values. Soon after his retraction, he wrote to Fogazzaro to explain what had just happened:

Cremona, November 1, 1898

Most esteemed and dear Friend:
Important and not very pretty things. I am writing this to you only: you are the first to know this and perhaps there will be no others.

I had a sure notice from America that Zahm's book (*Evolution and Dogma*) has been listed on the *Index;* perhaps the Decree will not be published. Zahm said he was disposed to explain himself and to rectify what he has written. Naturally, along with the book my summary of it has also been pounded. I wrote to Cardinal A[gliardi] immediately. . . .

He suggested I make a declaration, which you will have seen in *Lega Lombarda* (October 19 or 20). That pleased the higher-ups and was sufficient. But more, this declaration of mine might impede the publication of the Decree against Zahm because it was my *summary* that really alarmed Rome, because it appeared that my authority lent credit to this theory in Italy. In any event, I came out of it safely and perhaps because of me Zahm will also be saved.[29]

Besides the inexactitude of the date (the letter was published on October 25–26), the letter might give the impression that the Congregation of the Index made some decision about his appendix, which was not the case. Nor is it certain that it was this appendix that provoked the action against Zahm's book, which had already been denounced before, although it may well have alarmed the authorities, just as Bonomelli says.

Once again it is clear that a retraction in the form of a letter was sufficient for the Vatican authorities, because in this way they achieved their objective, which was to stop the promotion of evolutionary ideas. The same thing had happened with Leroy's retraction letter and Zahm's letter to his Italian translator, which wasn't even a retraction, although it made clear the Holy See's opposition to the distribution of his book. Under these conditions, it was easy to avoid a personal condemnation. Bonomelli had done so, and he was at peace with himself.

Not everyone saw things this way, however. On November 10, 1898, Countess Sabina Parravicino wrote to Monsignor Denis O'Connell:

Bonomelli's letter has left me *desolate.* He had made up his mind that Zahm was headed for the *Index* and had no wish to follow him with his famous *Appendix.* It looks like someone told him that his book *Seguiamo la ragione* had been denounced to the Index on account of the famous appendix, and then he wanted to preempt any action. When I wrote to him, I was not able stop from expressing my shock at his act.[30]

Nevertheless, Bonomelli justified his actions to his friends. Fogazzaro hadn't been pleased by the retraction either, and on November 6 Bonomelli wrote him:

I must explain: if the theory of evolution were demonstrated and were not just a magnificent *hypothesis,* but a truth, I too, just like you, would say: even though I might lose my honor and my life, I do not yield, and what I say stands and I affirm it. But as of now it is not a truth, it is a *hypothesis,* which presents all the characteristics of a future thesis. In this case one can submit to the competent authority. It is known that the opposite of the evolutionist *hypothesis* is not heresy or even error. Surely condemnation is not infallible, inasmuch as the authority is a respectable Congregation, that is playing an inside game. The time could arrive when the *hypothesis* becomes a certain *thesis,* like Galileo's theory: but in the Church, authority must be respected and I respect it: if not, goodbye to discipline and order! All this is reasonable.

The whole cosmic order and plan of Providence, not only in the economy of nature but also in the supernatural, require evolutionary theory: I am with you: the spectacle of the universe proclaims it: but, for now, let us stop here. . . . So let's not worry a lot or a little: truth conquers all and it is necessary to fight so that it does triumph.[31]

Bonomelli and the English Bishop

Bonomelli was not the only bishop favorable to evolution. As we have seen, John Hedley, bishop of Newport, Wales, had written a favorable review of *Evolution and Dogma* and other books by John Zahm in the *Dublin Review* for October 1898.[32] Hedley's article found a favorable echo in a long comment published in the *Tablet,* in London,[33] and also in Italy, where *La Rassegna Nazionale* reproduced the comment from the *Tablet,* adding a few paragraphs at the end.[34] Countess Parravicino, who was in the Americanist circle, provides some interesting details identifying her as the author of this article in the same letter in which she told O'Connell that Bonomelli's retraction had left her desolate:

> Do you know what I did after reading *The Tablet?* I made a nearly complete extract of Monsignor Hedley's book review and sent it to *La Rassegna* for publication, having added contextual material appropriate to its readers. So as not to lose any time I did it without telling him. I did not sign the article, so that people won't imagine it had been inspired by you, under a pseudonym.[35]

Salvatore Brandi was quite right when a bit later, in the article "Evolution and Dogma" that he published in *La Civiltà Cattolica,* he said that the anonymous author in *La Rassegna* had copied a text taken from the *Tablet,* adding ten words at the beginning and a dozen lines at the end.[36]

At the end of November 1898, Bonomelli wrote again to Fogazzaro, expressing his agreement with the evolutionist vision that the two friends contemplated in a mode more poetic than scientific. The letter contains some news about the case, based on what Hedley had published:

> The Bishop of Newport [John C. Hedley], England [Wales] came to the defense of Zahm's evolutionism in *The Tablet* and the *Dublin Review*. What will happen after Zahm's condemnation, which still has not been published? I am curious to know. Perhaps my letter impeded the publication of Zahm's condemnation, as a Cardinal wrote me: but now that another bishop makes Zahm's theory his own, with no restrictions, what will happen?[37]

The problem was more complex, however. Bonomelli did not provoke Zahm's problem, nor did his letter of retraction play a role in the decision not to publish the decree. Bonomelli had acted quickly to avoid a problem for himself, but he continued to follow Hedley's activities with keen interest. In another letter to Fogazzaro on November 28, he again asserted his support of evolution and his conviction that in the end it would be accepted.[38] On January 13, 1899, he invited Fogazzaro to lecture in Cremona and noted that "the anger and jibes against evolutionism and Americanism continue."[39] On March 4 he gave Fogazzaro his version of the pope's intervention in the question of Americanism.[40]

Those who sought a condemnation of evolutionism made good use of Bonomelli's retraction letter, published in *Lega Lombarda*, October 25–26, 1898. This letter was immediately reproduced, with a short introduction, in the "Contemporary Chronicle" section of *La Civiltà Cattolica*, in its issue of November 5, 1898.[41] *La Civiltà* showed its satisfaction with the bishop of Cremona's retraction. In his letter, Bonomelli does not mention the Holy See or any of its organisms or authorities. The persons who were competent by reason of their "authority" and had convinced him to reverse his views might have been bishops, but even in the event that they were Vatican officials, Bonomelli does not say they had acted in an official role: they might have done so personally (which is what appears to have happened). Still, everything pointed to some action by the Holy See, and the important bishop was quick to distance himself from the evolution controversy: the result, from any perspective, redounded to the favor the position upheld by *La Civiltà Cattolica*.

Italian Anticipation

In fact, the only intervention of the Vatican in this case was a rumor coming from Rome on the occasion of Bonomelli's praise of Zahm's book. The rumor was transmitted to Bonomelli by a friend and included a comment by the cardinal prefect of the Index, who apparently desired a public rectification by Bonomelli. The rectification was published immediately. Bonomelli had too many controversies around him and did not desire to get involved in a new one, taking into account that the reason for the new conflict remained quite far away from his main interests. He also justified his attitude by highlighting that evolutionism remained a hypothesis that had not been proved yet.

Famous and controversial among the Italian bishops, Bonomelli repeatedly attracted the attention of the Vatican, in whose archives he occupies quite a significant place. Evolutionism, however, was a minor subject among Bonomelli's interests and activities. This is why his case is interesting, as it shows how an incident without major relevance could be transformed in an argument against evolutionism that would be repeated generation after generation in textbooks and histories of theology.

"The Erroneous Information of an Englishman"

John C. Hedley

Among the evolutionists who retracted, John Hedley, bishop of Newport, Wales, is usually listed, alongside Bishop Bonomelli. Gruber presents him as "Mivart's first champion among the hierarchy" and attributes to him a retraction that can be summed up in the following line: "the 'Mivartian' theory . . . can no longer be sustained."[1] However, in Hedley's original text, this phrase is preceded by two limiting conditions. Hedley says that "according to *La Civiltà*" Mivart knew that his work had been examined in Rome by the competent authority, and adds: "'The authority' here referred to must, I presume, be that of the Holy Office."[2] In other words, what Hedley says is that if the information in *La Civiltà Cattolica* is true, and if the Holy Office had condemned Mivart's theory (on the origin of the human body by evolution), then this theory cannot be sustained. But the documents in the archive show that the Holy Office had not condemned this theory. Thus, the retraction attributed to Hedley never occurred and, as we will see, Hedley himself said so a short time after the events in question.

In an article titled "The Erroneous Information of an Englishman,"[3] *La Civiltà* accused Hedley of being mistaken due to lack of information. But the article in *La Civiltà* itself contained gross mistakes that were responsible not only for the error that led to Hedley's false retraction but also for the story's repetition over a period of decades as part of the evolutionism debate in the Catholic Church.

An English Bishop

In the period covered here, the Catholic Church in Great Britain was a dependency of the Congregation for the Propagation of the Faith (Propaganda Fide), like other "mission" countries such as the United States. In 1850 the Catholic administrative hierarchy was reinstated many centuries after the last bishop recognized by Rome had died, in the reign of Elizabeth I. Dioceses were created and bishops designated for the whole country. On September 29, 1850, the new diocese of Newport and Menevia was established in Wales (in 1895 this diocese was divided, leaving Newport as a single unit). Its first bishop was the Benedictine monk Thomas Joseph Brown, who directed the diocese until his death in 1880. He was succeeded on February 18, 1881, by John Cuthbert Hedley, also a Benedictine. The only male religious community then existing in the diocese was the Benedictine abbey of Belmont, Hereford, and by custom the canons of the Cathedral were likewise Benedictines.

John Cuthbert Hedley was born in Morpeth, Northumberland, on April 15, 1837. He received the Benedictine habit in 1854 and was ordained a priest on October 19, 1862. He was professor of philosophy in the Benedictine Novitiate at Belmont Abbey. From 1873 he was auxiliary bishop of the diocese of Newport, which he led as bishop from 1881 until his death on November 11, 1915. Hedley

John Cuthbert Hedley. Reproduced from J. Anselm Wilson, *The Life of Bishop Hedley* (London: Burns, Oates and Wahsbourne, 1930).

was very well known for his theological works, and his strong personality informed all of his writings. Between 1879 and 1884 he was editor of the *Dublin Review*.

When Hedley died, Wilfrid Ward wrote an article about him in the *Dublin Review* in which he stressed, along with his strength of character, his mental openness, his broad culture (not only literary but also musical), his good taste in art, and his fine literary style.[4]

Hedley and Brandi

Hedley's "retraction" occurred in the context of a controversy with Salvatore Brandi, which took place in two phases: the first in 1898, the second in 1902. In 1898 Hedley published an article in the *Dublin Review* that praised, albeit with some reservations, the intent to reconcile evolution with Christianity that Zahm had outlined in his book, *Evolution and Dogma*. In London, the *Tablet* published a favorable comment on Hedley's article,[5] and this commentary was reproduced in its totality in Italy, with some brief additions, in *La Rassegna Nazionale*.[6] In reply, Salvatore Brandi published an article in *La Civiltà Cattolica* in which he criticized Zahm's book and which included, in its last section, the letter of retraction published by the Dominican Leroy several years before. Hedley replied immediately. He published a letter in the *Tablet* in which he said that if the Holy Office publicly weighed in against Mivart's theory about the evolutionary origin of the human body, this theory could not be sustained. Thus Hedley's statement had an important qualifier: the theory could not be sustained only if the Holy Office had condemned Mivart's theory, as it had Leroy's, as *La Civiltà Cattolica* had claimed.

The second phase of the controversy took place in 1902, when a letter was made public in which Hedley cast doubt on the veracity of *La Civiltà*'s report on the Holy Office's presumed condemnation. Brandi again replied in *La Civiltà* saying that Hedley was not well informed, and so he published his version of what had really happened. Some of Brandi's misinterpretations, however, have plagued this episode practically from the beginning.

Hedley on Science and Religion

The controversy was originally provoked by Bishop Hedley's article published in the *Dublin Review* in 1898.[7] The article was a review of four works by Zahm, focusing on his book *Evolution and Dogma*. Hedley analyzed the book, particularly Zahm's argument for the compatibility of evolutionism and Christianity. Although Hedley praised Zahm, he made some carefully considered distinctions. On the one hand, he saw no difficulty in squaring evolution with the existence of a creating God, in the providence with which God governs the world, and with the existence of a purpose in nature reflecting the plans of divine rule. But, on the other side, he warned that the matter was not so simple when one considered more specific aspects of Christian revelation, especially biblical statements of matters that appeared to conflict with certain scientific results. Most of his review of Zahm focused on the first set of questions.

Hedley began with a long preamble on the relationship of science and religion. His general outlook was very cautious, so much so that his views might seem overly negative. He declared that a Catholic would always find it hard to adopt new hypotheses proposed by scientists who rejected Christian revelation. Although it is obvious that physical theories should be evaluated by scientific methods, it is also clear that scientists are not simply machines that accumulate data. A non-Christian scientist (Hedley says "infidel") is human and can easily succumb to the temptation to press conclusions beyond what the data permit, for example, in tailoring a point for a specific audience, or out of a desire to present complete explanations or aesthetically pleasing ("pretty") results. In this way a scientist might present a theory that, although solidly based, might also carry excess baggage (particularly of an ideological nature) owing to the personality of an individual scientist. This kind of thing happens frequently, and when it does, Catholics—especially if scientists themselves—do not remain indifferent. It is evident that Hedley did not have any simplistic view, nor did he minimize difficulties. "There are innumerable ways," he stated, "in which physical theory may affect the study of dogma, of scripture, and even of morals" (p. 242).

Hedley did not limit his comments to physics. When he spoke of "physical theory," he meant what today we would call "scientific theory" in general. Indeed, the article on Zahm dealt with one such "physical" theory in particular, the theory of evolution, which of course is biological and anthropological in nature. In his elegant, thoughtful style, Hedley proceeds methodically, considering the theory's difficult points while avoiding overly facile simplifications or generalizations. His point of departure is that Catholics are accustomed to receive scientific theories mixed with philosophical or religious errors ascribable to scientists' lack of background in these subjects. Thus it is unsurprising, states Hedley, that Catholics frequently reject new scientific theories, even though they might be partly true:

> It is for reasons of this kind that Catholic feeling seems so often to have at first set itself against novelties in science. Whether it was the helio-centric theory, or the stratification of the earth, or the critical analysis of the Bible, or the system of evolution as propounded in our own days, the men who were identified with them have too often been men who lost no opportunity of dealing a blow at revelation. The Church, therefore, as represented by her pastors, has been obliged to assert her doctrines and her belief, and her children have followed her; and if the

result has been to involve some physical truths in the same condemnation as religious error, however much such a result is to be deplored, it is really not the Catholic body which should be chiefly blamed, but rather the intemperate utterances of men who have gone beyond their province, and ought to have been better advised. (pp. 242–43)

Scientific truth, Hedley observes, has a sacred side, as does every truth, because it is the manifestation of a law formulated by God himself. But, at the same time, it occupies a secondary place when compared with moral, spiritual, and supernatural truth. And this applies even more so in the case of theories that are not demonstrated or, in the present state of our knowledge, can only be presented with many limiting factors and cautions. This is the context in which Hedley begins to talk about evolutionism.

Hedley on Evolution and Christianity

It is plain to see that Hedley was not a person who let himself be carried away by facile enthusiasms for science when religious values were in play. Thus it is significant that, when people began to talk about evolution, Hedley recalled (looking back from 1898) that in those times the majority of educated Catholics had accepted the evolutionary hypothesis in a general way. Then he immediately adds that they had done so cautiously and carefully. Whatever one might say about Darwin's religiosity, says Hedley, his followers, in England as on the continent, had given his theory a latitude that made it untenable and made it impossible to speak of God, of Creation, of spirituality, and of responsibility. As a result, the Catholics who accepted evolution were obliged to propose numerous distinctions, which necessarily brought them into contention with evolutionists like Herbert Spencer and Ernst Haeckel, who are openly antireligious.

Only after presenting these strictures did Hedley begin to discuss Zahm's work. He presents Zahm as an eminent Catholic writer and lecturer who, owing to his independence of judgment and his mental clarity, is capable of dissociating what is acceptable in evolutionary theory from what is unacceptable. Zahm argues enthusiastically in favor of evolution, although he recognizes that there are many obscure points, especially with respect to the mechanisms of evolution. Moreover, according to Zahm, the entire material world had come into being through evolution, with the exception, of course, of the human soul. What Zahm promotes is evolution as different from the "special creation" of the diverse species of living beings through special acts of God. Hedley repeats that the majority of educated Catholics have abandoned the idea of special creations in consequence of geological, paleontological, and embryological arguments, and adds: "still there are amongst us not a few who consider that the theory of evolution leads to atheism and to materialism—that it is equivalent to a denial of Divine Providence—that it leaves God without witness in the world—that it is against Holy Scripture—and that it cannot be reconciled with Catholic philosophy" (p. 246).

With Zahm's help, Hedley proposes to clarify these doubts. But he does not restrict himself to an exposition of Zahm: he comments on his ideas and partially corrects them. He states that the theory of evolution, as it is usually expressed, is frequently conjoined with atheistic or materialist perspectives, and it strikes him that Zahm is overly optimistic when he declares that evolution postulates Creation and divine action in the world. Hedley is more cautious, stating that, insofar as the theory of evolution confines itself to facts, it is completely reconcilable with Christian faith: it has absolutely nothing to say about God and about divine action. It leaves the believer ample margin to affirm divine action as the basis of evolution, and to affirm, as he must, special divine intervention, at least in the case of the human soul, and also, as seems more probable, when life began and in the formation of the body of the first man. Hedley concludes that with these reservations, which do not affect science because science can say nothing about such questions, the believer can reasonably adopt the theory of evolution, without compromising his faith (pp. 248–49).

With respect to materialism, Hedley states that almost all the scientific leaders of evolutionism are materialists, but that evolution does not postulate materialism. The motives are similar to those just reviewed with respect to atheism. And something similar attains with divine providence and the finality of nature. Hedley analyzes each one of these problems and, in the same line as Zahm, concludes: "Thus, it may fairly be said that, whatever the assertions and generalizations of unbelieving philosophers, there is nothing in the theory of evolution, or proved by fact, that is really antagonistic to Catholic faith" (p. 253).

Hedley then devotes a few pages to the examination of the current status of the evolutionist ideas most diffused in England: those of Darwin and Spencer. He shows that only with difficulty can either be regarded as a complete theory and again concludes with an affirmation of the compatibility between evolutionism and Christianity:

> It appears to me that the Catholic student who carefully studies the pages of Dr. Zahm will have no hesitation in dismissing all fear that to accept organic evolution is in any degree to endanger the faith. What is more, he will probably conclude that he would be shutting his eyes to scientific truth if he did not admit evolution as a useful and probable explanation and co-ordination of facts. As to the metaphysical deductions, made by unbelieving scientists, in the regions of theology and mental science, he can be a good evolutionist without giving up one iota of his faith or his Catholic philosophy. (p. 258)

Hedley's Reservations

In all the issues mentioned till now, Hedley lavishes praise on Zahm and stresses his erudition. But then he examines a whole other order of problem, concretely those offered by the interpretation of Scripture. This was an especially vexed issue in those times and had merited the attention of Leo XIII, in his encyclical *Providentissimus Deus*, of 1893, which Hedley mentions. Although here he also praises Zahm, he still makes clear his reservations:

> But on the subject of Scripture it can hardly be said that Dr. Zahm, however useful his studies may be to the ordinary Catholic, and even to the preacher or controversialist, really gets to the heart of certain questions which are at the present awaiting definite statement, if not solution. It is not enough to utter strong expressions about "liberty and license," and to assert in general terms that "revealed truth and dogma are compatible with the most perfect intellectual freedom." Neither is it sufficient, at the present moment, to show that no scientific facts can be quoted to disprove the Mosaic cosmogony, the narrative of the Deluge or the origin of man, as these things are presented in Holy Scripture reasonably interpreted. It is certainly not literally true that "Catholics . . . will not admit that they are in any way hampered in the pursuit of science by the exigencies of dogma." [Zahm, *Bible, Science and Faith*, p. 40.] On the contrary, there are some matters so clearly revealed as to be out of the field of question or investigation. There is, for example, the point of the unity of the human race, as Dr. Zahm himself admits. But there are also many questions, especially those relating to the primeval man, to the human soul, to language, and, I may add, to the constitution of material things, in which it would be not only a mistake, but also an offence against religious faith, not to start with a firm hold of what is taught by the Church—taught, that is to say, indirectly, and implied in theological dogma. (pp. 258–59)

It is clear that Hedley offered Zahm no across-the-board praise. Rather he endorses concrete aspects of Zahm's theory, especially those that refer to the relationship between God the Creator and nature. But he makes clear his reservations with respect to problems of biblical interpretation. Most Protestants, Hedley states, have adopted one solution: exclude human facts from revelation and limit the inspiration of Scripture to a vague presence of God. Catholics, guided by the encyclical *Providentissimus Deus*, propose interesting explanations to save the inerrancy of Scripture. According to Hedley, Zahm should have more explicit about these questions.

In which of Zahm's two spheres, according to Hedley, is the evolutionary origin of the human body found? As we have seen, he puts it in the first sphere, the least problematic. But it seems to him more probable that Catholics should admit a special intervention by God in this case, which does not agree with what Zahm says. According to Zahm, to preserve Catholic doctrine it is enough to admit that God creates the human soul and it is not necessary to admit, as Cardinal González does, a special divine act to prepare the body of a lower animal so that it is ready to receive a soul.

On this point, Zahm and Mivart agree, but Hedley dissents. Thus, Gruber is wide of the mark when he presents Hedley as "Mivart's first champion among the hierarchy." The praise that Hedley heaps upon Zahm overshadows the reservations he displays on questions as important as the origin of man.

'La Civiltà' against 'La Rassegna'

Bonomelli's retraction was published near the end of October 1898 (the letter is dated October 22 and it was published in *Lega Lombarda* for October 25–26). This was an authentic retraction, even though it was not in response to any official act of the Holy See but due only to personal advice. This retraction desolated the Countess Sabina Parravicino,[8] one of the most active participants in what has been called "Italian Americanism," which, the same as the properly American brand of Americanism, defended both Americanism and evolutionism. But the Countess soon found an outlet for her ideas. When she received notice of Hedley's article, she began to use it to propagandize in favor of Zahm and evolutionism.

Hedley's ideas did not reach her in their original version, as published in the *Dublin Review*, but rather via a comment that appeared in the *Tablet* of London. She herself explained this to Monsignor Denis O'Connell in a letter dated November 10, 1898.[9] She said she had made a nearly complete extract from the *Tablet*, added some context to it, and had sent it to *La Rassegna* to publish signed with the pseudonym "Theologus."

Salvatore Brandi quickly seized on the occasion by publishing in *La Civiltà Cattolica* one of the two articles that became the source for future writers on the Catholic Church's relationship with evolution. Under the same title as Zahm's book, "Evolution and Dogma,"[10] he got right to the point:

> The self-styled "Theologian" of the liberal *Rassegna Nazionale* de Florence (the article is anonymous, signed only by Theologus), wounded to the quick of the heart by the noble declaration of the Bishop of Cremona against Dr. Zahm's evolutionism, has quickly found a balsam for his pain in an article friendly to that professor's theory, published by the bishop of Newport in the latest number of the *Dublin Review*. (p. 34)

It looks as though Brandi had read Parravicino's letter, since he adds that the Theologian had not read Hedley's article in the *Dublin Review* because, if he had read it, he could not have omitted Hedley's doubts and reservations. In his zeal for pitting one bishop against another, Brandi remarks, the Theologian had trusted blindly in a review in the *Tablet*, which *La Rassegna* reproduces exactly, having added only ten words at the beginning and a dozen lines at the end. The Theologian attributes to Hedley the notion that believers can accept evolution so long as they accept divine intervention in the creation of the human soul and perhaps the origin of life. Brandi comments, again correctly, that Hedley states that a special divine intervention in the formation of the body of the first man should probably be accepted also. He adds, this time with some exaggeration, that Hedley asserts the need to accept divine intervention for the origin of life (what Hedley really says is that, at the beginning of Creation, God endowed matter with the power to produce life). In any case, Brandi states that Hedley's article surely favors evolution of the type promoted by Zahm and, therefore, offers evolution's supporters a reason for satisfaction.

How is it possible, Brandi asks rhetorically, that evolution is palatable once one adds all the reservations that are repugnant to the basic principles of evolution, as defended by Spencer, Haeckel, Darwin, Wallace, Mivart, and all the eminent professors? Brandi's reply is that Catholic writers, wanting to present themselves with the halo of science, do not fear being illogical and incoherent; in this case as in others, under the pretext of saving the confused, they themselves become confused.

Brandi then launches into a ten-page critique of Zahm and concludes that the theory of evolution "is a fantastic edifice and there is no better way to describe it than as a tissue of vulgar analogies and arbitrary suppositions which are not supported by the facts" (p. 45).

Condemnations of Evolution

More important for Hedley's case was the end of the article, where Brandi examines the attempts to reconcile evolution and Christianity from the perspective of theology. In Catholic theology, four types of argument are used: biblical interpretation, an appeal to Catholic tradition, the teachings (Magisterium) of the Church, and arguments based on reason. The first two refer to the sources of revelation and are complementary, since tradition, expressed by the consensus of the Church fathers and other evidence such as liturgy, is used to interpret Scripture in a Catholic sense. Both of these provide the basis of the Church's Magisterium. In his article, Brandi himself uses the four types of argument informally. He claims to have demonstrated rationally the scientific weakness of evolution, and on this basis he concludes that it would be foolhardy to abandon the natural and traditional interpretation of what the Bible says about the origin of man:

> The first impediment to accepting evolution for educated Catholics comes not from the fear of contradicting the Bible, but from the *scientific insufficiency* of that

system, that is, the absolute lack of evidence that confirms it, whether as a theory or as a hypothesis.

In this situation, it seems to me that whoever stubbornly defends the theory of the human body's descent from a monkey or any other animal, against the traditional views of the Church Fathers, can with good reason be called rash. Besides the respect due to the Bible, it is certainly required that the words of eternal Truth not be interpreted and warped on the basis of gratuitous hypotheses, to make them say today in obedience to one theory, what will be said tomorrow in obedience to another. (pp. 45–46)

There still remained arguments based on the Magisterium of the Church. Brandi recalls that, according to Zahm, the Church had never condemned evolution. Brandi comments that just because the Church has not condemned an opinion is not reason enough for a Catholic to support it, inasmuch as a Catholic, in addition to being an obedient believer, should also exercise reason. Moreover, Brandi adds, alluding to an apostolic letter of Pius IX:

A Catholic ought to reject not only the opinions formally condemned by the Church and those that are opposed to the doctrines it has defined or taught through its ordinary Magisterium; he should likewise repudiate ideas that he knows are contrary to the opinions that the common and consistent consensus of Catholics holds as theological truths and conclusions, as certain as the opinions contrary to them, though they cannot be called heretical, still merit theological censure. (pp. 46–47)

Brandi goes on to mention the retraction of Monsignor Bonomelli, bishop of Cremona, summarizing it and referring to the text published in *La Civiltà Cattolica*. The bishop, Brandi says, knew that the theory of Mivart and Zahm had not been condemned up to now in a public act. Nevertheless, "advised, as he assures us, by some friends, who were *very competent by reason of their knowledge and authority*, that the doctrine, even as a simple *hypothesis*, can not be reconciled with the common interpretation of the Church," he had no qualms about repudiating it publicly, and asked that the opinion that he reproduced, purely as a hypothesis, as an appendix of his book, supported by Zahm's authority, should not be considered as his own (p. 47).

Clearly the common opinion among Catholics was against evolutionism. Bonomelli was attacked in the Catholic press for his praise of Zahm. And we also know that he was aware of opinion in Rome through his friend Cardinal Agliardi who, in a letter of October 20, 1898, told him that his appendix on Zahm had provoked a storm there, adding that he had sounded out Cardinal Steinhuber, prefect of the Congregation of the Index, who advised Bonomelli to publish a rectification in a Catholic periodical.[11]

Of what value was such a rectification? It was not a response to any official action of the Holy See, but the prefect of the Index had advised it. Bonomelli said he had acted under the advice of *very competent* persons, among them Cardinal Agliardi and the cardinal prefect of the Congregation of the Index. Although his reference to these persons was only a generic one, Brandi presented it as an authoritative argument. The peculiar closeness of *La Civiltà Cattolica* to the Holy See could make up for the lack of more information.

Brandi left the main course for the end. In his retraction, Bishop Bonomelli said that with his praise of Zahm's stance he thought to hold out his hand to educated people who vacillated between faith and error, "just as Father Leroy had done in France." Brandi says that Hedley had done the same (although it doesn't look like Hedley had cited Leroy). Certainly, Zahm had appealed to Leroy's authority, mentioning that his book had been published with the requisite ecclesiastical permissions. Brandi then introduces something that everyone seemed not to know: that Leroy had been called to Rome and that he had retracted publicly when he found out that his work, examined in Rome by the *competent authority*, had been judged unsustainable, above all for what it said about the human body, because it was incompatible with biblical texts and the principles of sound philosophy. So that no doubt remain, Brandi included the original French text of Leroy's letter, just as it was published in 1895 in *Le Monde*.

If until that moment many had been unaware of the existence of Leroy's retraction, from then on, thanks to Brandi, everyone knew. This article damned the four clerics associated with the evolutionist cause: Leroy, Zahm, Bonomelli, and Hedley. It made it clear that Leroy had retracted, it criticized Zahm's posture fully, it alluded to Bonomelli's retraction, and Hedley's stance was given in evidence. Only one significant name was missing: Mivart. He is mentioned a few times in Brandi's article, but it would be Bishop Hedley who, pressed by this article, was going to complete the quintet adding, most conspicuously, the name of Mivart.

Hedley's "Retraction"

Hedley received *La Civiltà*'s challenge and reacted immediately. Within a few days, on January 14, 1899, the *Tablet* of London published a letter from Hedley, under the title "Physical Science and Faith," which is usually considered his retraction:

To the Editor of The Tablet
Sir:—My attention has been called to an article in the *Civiltà Cattolica* of January 7, entitled "Evoluzione e domma," in which the writer speaks unfavourably of the Rev. Dr. Zahn [*sic*], and of the paper contributed by me to *The Dublin Review* of last October, under the name of "Physical Science and Faith." There is nothing that

calls for any special remark in the greater part of the article. It is not very clear whether the writer altogether rejects Evolution in the physical world, or not. But the last paragraph is important, as we find there a letter from the Dominican writer, Père Leroy, from which it appears we must conclude that the Holy See has spoken on the subject of the formation of the body of Adam. It is well known that Dr. Mivart's view is, that it is perfectly open to a Catholic to hold that the body of Adam was not formed directly from the dust, but might have arisen from a non-human animal, the rational soul being subsequently infused. Père Leroy had defended this view in a work published in 1891. In February, 1895, however, according to the *Civiltà*, Père Leroy was summoned to Rome *ad audiendum verbum,* and the result appears from the following sentences of a letter dated February 26 of that year, which is printed at the end of the article referred to. I translate from the French: "I have to-day learnt that my thesis, after examination at Rome, by competent authority, has been judged untenable, especially in what relates to the body of man—being incompatible with the text of Holy Scripture and with the principles of sound philosophy." The "authority" here referred to must, I presume, be that of the Holy Office. The "Mivartian" theory, therefore, can no longer be sustained. As to my own paper, it will be observed that I have carefully abstained from saying a word in favour of his theory. More than thirty years ago, in an article in *The Dublin* on "Evolution and Faith," I felt obliged to consider the theory at least "rash." I am aware, however, that views which may at one time be theologically speaking "rash," need not always be rash; for the note of "rashness" is given to those propositions which are either contrary to the common doctrine of theologians or are put forward without any reasonable grounds. Still, I need not say that if the "competent authority" has decided in the sense in which it appears to have done, the view that the body of Adam was "evolved" must still be pronounced "rash"—and something more.

+ J. C. Hedley, O.S.B.
January 11, 1899[12]

The operative sentence of this letter is: "The 'Mivartian' theory therefore, can no longer be sustained." In his biography of Mivart, Gruber takes this as Hedley's retraction. Nevertheless, this sentence is set in the context of three conditions that must be fulfilled to consider the letter a retraction. Hedley says that:

—From Leroy's letter, "it appears we must conclude that the Holy See has spoken on the subject of the formation of the body of Adam." In fact, the matter was not entirely clear. Leroy does not mention the Holy See or any of its dependencies in his letter. He speaks only of the "competent authority" which had examined his work in Rome and had judged it unsustainable, and that he retracted his work as "a docile son of the Church." He might be referring to the authority of the Dominicans in Rome. Moreover, it is not stated that the Holy See had spoken publicly. Indeed, it had not. Later, Hedley would return to this important point. Now, a century later, we know which authority was meant, thanks to the archival documents.

—In February 1895, according to *La Civiltà*, Leroy was summoned to Rome *ad audiendum verbum* (to communicate something to him), and he retracted via the letter mentioned. But Leroy never said he was officially summoned to Rome; *La Civiltà* so reported. How did it know that? Who were its sources? Which authority summoned him?

—Hedley supposes that the "authority" mentioned must refer to the Holy Office. The archival documents show that this was not so. We do not know whether Brandi knew this. Later on he intimated that the actions were taken by the Holy Office. This is a significant point.

These three points can be summarized as follows: Leroy retracted because the Holy Office summoned him to Rome and asked him to retract. This demonstrates that the consequence—"The 'Mivartian' theory, therefore, can no longer be sustained"—was also conditional. "Therefore" means "If the previous conditions are fulfilled." But now we know that the conditions were not fulfilled, and at the time Hedley himself was not sure they would be fulfilled: for this reason, he expresses himself conditionally. If to this we add that Hedley did not accept Mivart's theory on the evolutionary origin of the human body, we must conclude that Hedley's letter cannot be considered a retraction in the strict sense. In his letter Hedley says: "As to my own paper, it will be observed that I have carefully abstained from saying a word in favour of [Mivart's] theory."

Without doubt, recalling his position favorable to Zahm's evolutionism, Hedley's letter can be considered as a kind of retreat, but only relatively so. It is, above all, the manifestation of an attitude of obedience to the authority of the Church: if the Holy See had spoken, he is prepared to accept what may have been established. As for Mivart's theory, Hedley had already shown his reservations, and he would have converted them into an outright rejection if the Holy See had so decided.

Did the Holy See Intervene?

Hedley's doubts did not dissipate. On the contrary, new information led him to conclude that neither the Holy Office nor any other Roman congregation had intervened in the case of Leroy, which is the same as saying that *La Civiltà*'s information was not trustworthy. Hedley expressed just this conclusion in a letter to a Protestant friend, the Reverend Spencer Jones, and he gave him permission to reproduce it in Jones's 1902 book, *England and the Holy See*. Hedley told Jones:

No article or portion of any article of mine has ever been censured by the Holy See. The facts are these. I wrote the article you refer to a year or two ago. In that article I spoke of the question of the body of the first man—whether it was created instantaneously for the purpose of the infusion of the soul, or whether for that purpose a pre-existing animal was taken. I said I had always considered the latter opinion "rash," to say the least, but that now, as Dr. Mivart had published the opinion and it had not been condemned, it might perhaps be "rash" no longer. On this point the *Civiltà Cattolica* rejoined that the opinion had been condemned in Rome. I then wrote to the *Tablet* saying that if this was the case, I should withdraw my statement as to its not being "rash." I added in effect that the *Civiltà* quoted no decision of any Roman Congregation, but only spoke vaguely of "authority." I have since been informed that the condemnation in question, if it ever was pronounced, emanated merely from the Dominican Superior, and not from the Holy See at all. I need not say that you are perfectly at liberty to publish this or anything contained in my article. But there has been no action or intervention on the part of the Holy See, or of any tribunal of the Holy See.[13]

In reality, *La Civiltà* had not stated that the Holy See had intervened. Brandi's article did not mention the Holy See. But he insinuated as much when he said that "the theory of Mivart and Zahm had not been condemned until now (*in a public act*)" (Brandi's emphasis through the use of italics seems to indicate that there had been some kind of unpublished condemnation); when he said, again using italics, that the persons who had advised Bonomelli were "*competent enough in terms of knowledge and authority*"; when he said that Leroy had been called to Rome *ad audiendum verbum* (the italicized Latin in the original insinuates an official act); and when he reproduced the letter in which Leroy speaks of the *competent authority* (Brandi again emphasizes the phrase by using italics).[14]

If it was certain, as Hedley said, that there was no action by the Holy See, then the insinuations of *La Civiltà* were unfounded, and it even seemed that *La Civiltà* had spread an erroneous conclusion. *La Civiltà*'s reaction came quickly.

Unpublished Decrees

This time, *La Civiltà* did not publish a long article. It limited itself to a three-page comment title "Evolution and Dogma. The Erroneous Information of an Englishman."[15] This new article by Brandi became another source of data for historians.

Brandi had no easy task. He knew that decisions were made in the Holy See that were not published. Leroy's retraction and Zahm's letter to his Italian translator had been provoked by these decisions. But the Holy See had no intention of making its actions in these cases public. Moreover, Brandi did not know the facts in detail (we will see that he published mistaken information). How could he clarify events with which he was not well acquainted, in a very well-known journal, using only material in the public domain? This was Brandi's dilemma

The only reference to the Holy See in this whole matter was contained in Zahm's letter to his Italian translator of May 16, 1899. Brandi seized upon it and wrote:

> That our judgment of Zahm's work was not exaggerated is made clear by the declaration [*dichiarazione*] that [Zahm] himself made public [*fece di pubblica ragione*], four months later. In this document he claims to have known "from a trustworthy source that the *Holy See* was opposed to the further diffusion of his book *Evolution and Dogma*," and for that reason he wished that "the work be retired from circulation." (pp. 75–76)

The information was based on real facts, but Brandi distorts them. Zahm did not intend to make any "declaration," let alone a "public" one. He wrote a private letter to Galea, and certainly he did not want anyone to find out what he had written in this letter. We do not know how it came to be published by the *Gazzetta di Malta*, whence *La Civiltà* obtained it. Leaving speculation aside, the fact is that Brandi made a private letter intended as confidential into a public declaration. With the information provided by Brandi, the reader had to conclude that Zahm, warned by the Holy See, had made a public declaration, when the truth is that Zahm used all means to avoid doing this and, indeed, he did not do it. Brandi converted Zahm's letter into a kind of retraction. The Holy See certainly had made a decision against Zahm's book and seemed to await his public retraction. But it is certain that Zahm never did so, and that Brandi used the letter to Galea as a substitute.

Brandi goes on to clearly insinuate that Zahm had made this declaration at the instance of the Congregation of the Holy Office or of the Index:

> Whoever knows the wise norms prescribed by Benedict XIV and observed in all cases by the Congregations of the Holy Office and the Index; moreover, whoever also is familiar with the procedure that these Congregations follow in particular cases, that is, when it is a matter of works written by Catholics who enjoy a certain fame, will have no difficulty in understanding all the force and the genuine theoretical and practical significance of the mentioned declaration [by Zahm]. Concretely, Zahm's work met the same fate that, four years before, had befallen another work on the same subject written by the Dominican father Leroy. He also had advocated the theory of the origin of the human body from a brute animal; his work was likewise denounced to the Holy Office, and he too, in order to avoid a public censure, made a public declaration according to which he "disallowed, retracted, and condemned the said theory" and "wished to retire from circulation, insofar as was possible, the copies of his book." In both cases, the *competent*

authority that examined and judged the works, and whose orders Leroy and Zahm obeyed, was that of the Supreme Tribunal of the Holy See. (p. 76)

Our archival documents show that Brandi's assertions are inexact. To claim that Zahm made no public declaration, Brandi was forced to use Leroy's words, adding that Zahm had done something similar. Furthermore, neither Leroy's book nor Zahm's was denounced to the Holy Office, but rather to the Congregation of the Index. And the authority that decided those cases was equally the Index, not the Holy Office (which was the Supreme Tribunal of the Holy See). These inexactitudes are important, because the action of the Index was generally limited to placing publications on the list of prohibited books. And its decrees never mentioned the motives: it may be that the publication contained heresies or errors, maybe only "reckless" doctrines, or perhaps only because it was viewed as not opportune or as potentially scandalous at a some particular point in time. For example, in Zahm's case, as we have seen, one aspect that weighed negatively when the book was judged in the Index was his insistence on presenting Saint Augustine and Saint Thomas as supporters of evolutionism. On the other hand, the Holy Office also published censures that referred directly to doctrines or matters related to faith or morals.

Finally, Brandi refers to Hedley's main assertion: *La Civiltà* cites no concrete action of the Holy See because in fact none existed. Brandi responds that Hedley is misinformed. All he need do, as a bishop, is to ask the competent authority in the Holy See and surely he would receive more detailed information, perhaps even confidentially. And *La Civiltà* cited no concrete action of the Holy See because, as Brandi pointed out in his previous article, "the Holy See, for the best of reasons, has not believed it opportune until now to condemn *by a public act* that theory [the evolutionary origin of Adam's body]" (p. 77).

We now know, in fact, that both in Leroy's case as in Zahm's, the Congregation of the Index decided to proscribe both books, but it did not want to publish the corresponding decrees. Rather it asked Leroy to retract, which he did, and hoped for the same from Zahm, which never arrived. Zahm's letter to his Italian translator was a poor substitute. This manner of proceeding had the advantage of avoiding any harm to the authors and their religious institutions and, moreover limited somewhat the spread of evolutionism applied to man without officially compromising the Holy See. But that was not the equivalent of a public act. And it also had the drawback of lacking clarity. Thus it was that the only public sources of information were Brandi's articles in *La Civiltà Cattolica*, which then became the necessary reference for future theologians and historians.

Happiness in Hell

St. George J. Mivart

St. George Mivart is the author most cited whenever the difficulties of reconciling evolution with Christianity are mentioned. This English scientist published a book in 1871 in which he accepted evolution, said it was compatible with Christianity, and, at the same time, criticized the importance, exaggerated in his view, that Darwin attributed to natural selection as a mechanism that explained evolution. His objections were important, to the extent that Darwin gave close attention to them in a new chapter that he added to the sixth edition of *The Origin of Species.*[1]

Mivart believed that evolution was at work up to the formation of the human body—the most controversial point from the religious perspective. Nevertheless, his book was never censured by the authorities of the Church

Mivart's relations with Church authorities were excellent during almost all of his life. But they deteriorated in his last years. First, he published three articles in which he proposed that the punishments of hell were not eternal; these articles were listed in the *Index of Prohibited Books.* Mivart accepted the condemnation. Later, however, ever greater conflicts arose. He published some articles that the English Church authorities deemed unacceptable, and he was asked by them to sign a profession of Catholic faith. When Mivart refused, his bishop, Cardinal Vaughan, prohibited him from receiving the sacraments. Mivart died soon thereafter. His family asked that he be given a Catholic burial, claiming that his last conflicts were the result of a serious deterioration of his health.

It has always been known the cause of Mivart's problems was not his advocacy of evolutionism. The archival documents throw new light on the way the events unfolded.

The Genesis of Species

St. George Jackson Mivart was born in London on November 30, 1827.[2] His parents were evangelical Protestants, not Catholics. He converted in 1844. Although he was initially interested in law, his love of science induced him to study biology. Mivart first studied biology privately with George Waterhouse, entomologist and correspondent of Darwin, and with the morphologist Richard Owen. In the early 1860s, while still a Darwinian, he studied with Thomas H. Huxley, the great champion of Darwinism, whose demonstrator he was in lectures at the School of Mines. Huxley and Owen supported his candidacy for a lectureship in comparative anatomy at St. Mary's Hospital medical school, where he taught until his retirement in 1884. He was elected a fellow of the Royal Society in 1867.

Around 1868 Mivart's research on the taxonomy of primates led him to believe that morphologically similar classes and orders appeared to have originated independently of each other. Around the same time, for religious and ideological reasons, he rejected natural selection and felt obliged to tell his mentor so:

> After many painful days and much meditation and discussion my mind was made up, and I felt it my duty first of all to go straight to Professor Huxley and tell him all my thoughts, feelings, and intentions in the matter without the slightest reserve, including what it seemed to me I must do as regarded the theological aspect of the question. Never before or since have I had a more painful experience than fell to my lot in his room at the School of Mines on that 15th of June, 1869. As soon as I had made my meaning clear, his countenance became transformed as I had never seen it. Yet he looked more sad and surprised than anything else. He was kind and gentle as he said regretfully, but most firmly, that nothing so united or severed men as questions such as those I had spoken of.[3]

Mivart had up to that time had a more than cordial relationship with Darwin, expressing his respect and admiration for him in a number of letters. After his change of heart, however, his demeanor changed sharply, as indicated in a review of Darwin's *Descent of Man,* published in the *Quarterly Review* in July 1871. The crux of Mivart's criticism of Darwin was that variations, which Darwin presented as small and continuous, were, in fact, large and discontinuous. Mivart's critique of natural selection had two related faults: first, he supposed that Darwin's variations were of rare occurrence, calibrated individually, rather than being always present in the population. He had difficulty, that is, thinking in terms of populations. Second, variation seemed random to him and therefore (using a term apparently drawn from Aristotelian philosophy) "accidental." But, as Chauncey Wright pointed out at the time in a detailed, philosophically acute critique, "This class of variations, that is, 'individual differences,' constant and

St. George Jackson Mivart. Reproduced from Jacob Gruber, *Conscience in Conflict: The Life of St. George Jackson Mivart* (New York: Temple University Publications by Columbia University Press, 1960; reprint, Westport, Conn.: Greenwood Press, 1980), frontispiece.

normal in a race, but having different ranges in different races, or in the same race under different circumstances, may be regarded as in no proper sense accidentally related to the advantages that come from them."[4]

Darwin was shocked by Mivart's animosity after having first received a series of respectful letters: "yet in the Q. *Review* he shows the greatest scorn and animosity towards me, and with uncommon cleverness says all that is most disagreeable. He makes me the most arrogant, odious beast that ever lived. I cannot understand him; I suppose that accursed religious bigotry is at the root of it. Of course he is quite at liberty to scorn and hate me, but why take such trouble to express something more than friendship. It has mortified me a good deal."[5]

Mivart tried to extract anything he could from Darwin's writing that appeared to support a notion of the limit of variation within any group. For example, "The proposition that species have, under ordinary circumstances, a definite limit to their variability, is largely supported by facts brought forward by the zealous industry of Mr. Darwin himself."[6] Mivart's strategy was always to quote Darwin against himself, to show internal inconsistencies in his arguments. However, he also could not resist from giving his own "spin" to Darwin's data. As Vorzimmer observes, "The tendency to variation itself . . . was seen as a manifestation of an internal force, as part of the original Divine plan of creation. Equally he felt that the tendency for many organisms to vary in the same direction was evidence of the operation of such a force. Even the *continued* tendency toward variation in the same direction—also espoused by Darwin—lent itself to such an interpretation."[7] In his *Quarterly Review* essay on *The Descent of Man* (1871), Mivart (who was one of those responsible for pressuring Darwin to modify his views on evolutionary mechanisms) trumpets that "The assigning of 'natural selection' to a subordinate position is virtually an abandonment of Darwinian theory; for the one distinct feature of that theory was the all-sufficiency of 'natural selection.'" This review became a centerpiece of religious anti-Darwinism, but it was an exaggeration. Natural selection remained Darwin's primary mechanism.[8]

In 1871 Mivart published a book titled *On the Genesis of Species* in which he accepted evolutionism but opposed the central role that Darwin assigned to natural selection as the motor of evolution.[9] This separated him from Darwin and Huxley, but his reputation as a biologist was not diminished thereby, and Darwin himself was obliged to reply to Mivart's objections, which he did at some length in a new chapter added to the sixth edition of *The Origin of Species*. In a countertheory that he called "Specific Genesis," Mivart proposed that the life histories of organisms were played out within a limited range of variation, something like oscillations in a stable equilibrium (an extension from Galton's statistical results on hereditary genius to the organic world generally).

Mivart's book has twelve chapters. In the first, Mivart asserts that the great problem of the origin of the different classes of organisms seems about to be resolved. But he points to a number of difficulties. The first is that almost nothing is known about the development of individual embryos and that, with an even greater motive, little can be known about the origin of species. This was a realistic observation that took into account the inexistence of the field genetics at that time. It also refers to critiques of Darwinism, at times quite inaccurate, on the part of some theologians, and he attributes them in part to the logical reaction of believers to the attacks that quite a few evolutionists had directed against theology. He also observed that there is no opposition between Creation and evolution and that many notable Christian thinkers accept both.[10]

The remaining chapters are devoted to scientific issues, with the exception of chapter 9, which addresses evolution and ethics, where Mivart shows that natural selection is insufficient to explain morality, and chapter 12, titled "Theology and Evolution," in which Mivart responds to objections raised by Herbert Spencer against the existence of God. He goes on to distinguish between Creation in a primary or absolute sense (the absolute origin of something out of nothing) and derived (something is formed from the potentialities that God himself has put into Creation), and he asserts that the ostensible opposition between creation and evolution arises from a misunderstanding of these senses. In reality, the creation in a derived sense requires creation in an absolute sense, against which science can say absolutely nothing.[11]

These clarifications lead Mivart to assert that Christians are absolutely free to accept the theory of evolution. In spite of its being a recent theory, Mivart continues, one can assert that ancient authors have affirmed the compatibility of evolution with theology. Mivart illustrates his point with quotations from Saint Augustine, Saint Thomas Aquinas, and Suárez. Thus he seeks to show that the oldest and most venerable theological authorities advocated derivative creation, which can be harmonized with all the requisites of modern science.[12] This is a rationale that later on those, like John Zahm, who followed Mivart's line would further develop.

With respect to finalism, Mivart quotes Huxley, who recognizes that there is a teleology that is not affected by evolution: mechanism and finalism are not opposed, and the more one asserts that determinate mechanisms exist in nature, the more one can also think that these mechanisms respond to a prior plan.[13] In spite of his philosophical concerns, Mivart was well within the evolutionary consensus. As David Hull has observed, "Mivart could easily have become a Darwinian. His views about evolution differed in no important respect from several key Darwinians. Like Huxley, he thought evolution was more saltative than Darwin did. Like Gray, he thought it was directed. And like so many Darwinians, he did not think natural selection could do all that Darwin claimed of it."[14]

The Origin of Man: Soul and Body

Near the end of the book, Mivart broaches the problem of the origin of man and resolves it by distinguishing between physical and spiritual levels. The natural sciences cannot affirm or deny the existence of the spiritual soul. There is no problem in believing that the body of the first man was made from other organisms while asserting that, at the same time, his soul was created immediately and directly by God, which is something like what happens in the life of each new human being.

> In this way we find a perfect harmony in the double nature of man, his rationality making use of and subsuming his animality; his soul arising from direct and immediate creation, and his body being formed at first (as now in each separate

individual) by derivative or secondary creation, through natural laws. . . . That Divine action has concurred and concurs in these laws we know by deductions from our primary intuitions; and physical science, if unable to demonstrate such action, is at least as impotent to disprove it. . . . We have thus a true reconciliation of science and religion, in which each gains and neither loses, one being complementary to the other.[15]

Mivart observes that, in fact, evolutionary ideas are very old, and Aristotle's opponents even expounded the idea of natural selection, as Aristotle himself says. He ends the book reaffirming that there is no opposition between science and religion and specifying the book's objective with regard to the origin of species:

The aim has been to support the doctrine that these species have evolved by ordinary *natural laws* (for the most part unknown) controlled by the *subordinate* action of "Natural Selection," and at the same time to remind some that there is and can be absolutely nothing in physical science which forbids them to regard those natural laws as acting with the Divine concurrence and in obedience to a creative fiat originally imposed in the primeval Cosmos, "in the beginning," by its creator, its Upholder, and its Lord.[16]

At the same time, Mivart notes the differences separating man and the higher animals and presents the arguments of Alfred Russel Wallace in favor of a special creation of the human body.

Mivart presented and extended his ideas on human nature and the peculiar origin of humanity in his *Lessons of Nature* (1876), and also in *On Truth* (1889), a systematic treatise of aspects of Christian philosophy presented as the fruit of a lifetime of reflection.[17] In the last chapter of *On Truth*, he discusses human evolution and explains its mechanism:

The origin of the human species must, however, belong to a different category, since, as we have seen, in spite of the exceedingly close resemblance of the human frame to the structure of apes, the soul of man possesses powers so utterly distinct in kind from those possessed by any other known existence in the material universe, that it merits to be distinguished by a radically distinct denomination—that of "spirit."[18]

Clear and Cloudy Skies

In recognition of his merits, Mivart was named doctor of philosophy by Pope Pius IX in 1876 and awarded an honorary doctorate of medicine by the University of Louvain in 1884. He was a prolific scientist, and his articles were published in leading British and American journals. But he also provoked confrontations and debates. In the realm of science, his criticisms of Darwin and the way in which they were interpreted provoked a break with Darwin and his circle. Mivart believed that Darwin, in *The Descent of Man* (1871), presented in the name of science an unacceptable image of human beings that would lead to their degradation, and he published a strong critique that set him apart definitively from Darwin and Huxley in the same year. After examining this process in detail, Gruber concludes: "The wounds opened by Mivart's attack upon Darwin and through him upon his friends and disciples were never healed; they continued to fester throughout the lifetimes of the participants. Until the day of his death Mivart was haunted by the hostility, latent and overt, of the small circle which had surrounded Darwin."[19]

In the theological arena, Mivart's *Genesis of Species* provoked a warm response by a reviewer in the *Dublin Review*.[20] The anonymous author congratulated him for the positive reception that the book had found in the greater part of the periodical press, and among those who, like the staff of the *Dublin Review*, sincerely wanted to defend revelation, while doing justice to science at the same time. Thus Mivart was warmly praised for this rigor. By pointing out that scientists cannot with their own method demonstrate or refute anything in the supernatural or hyperphysical area, Mivart had made a worthwhile contribution to science. The journal left for its next issue a review of Darwin's *Descent of Man*, which had been published the same year, but after Mivart's book.

The review shows that evolutionism was accepted without much difficulty in British Catholic circles. After identifying Darwin as one of the most important scientists of the epoch, it laments that Darwin is opposed to revelation and also that Darwin, absurdly, thinks that religious people pose a threat to him. The reviewer asserts that most of what Darwin wants to prove has nothing to do with atheism and could even be an aid to revelation. Mivart accepts evolution, but insists that natural selection should be complemented by the action of other natural laws that have yet to be discovered, and he shows that objections that aim at extracting an antireligious message from evolution are unfounded.

Darwin acknowledged that he had perhaps exaggerated the importance of natural selection and said outright that natural selection was not the only mechanism of evolution. Mivart suggested that some internal force or tendency could be a significant agent of evolution, even the principal one. The author of the review provides an ample exposition of Mivart's objections to natural selection. Finally he says that there is nothing objectionable in Mivart's reflections on evolution and theology, and he promises to treat the question in greater detail in another article.

In effect, the review of Mivart's book appeared in the January–July 1871 number of the *Dublin Review*, and in the next issue (July–October) there appeared a very long article, also anonymous, on evolution and theology, in the form of an essay-review of four books: *The Descent of Man* by Darwin (1871), *Contributions to the Theory of Natural Selection* by A. R. Wallace (1871), *Instinct; Its Relations with Life and Intelligence* by Henry Joly (1870), and Mivart's *The Genesis of Species* (1871).[21] In the new article the author proclaims the absence of contradiction between science and Christian faith, in spite of declarations by those who use evolution to pose objections to faith. Some aspects are treated in greater detail, and of those, the key to the discussion of evolutionism in that period was the origin of the body of the first man. To this subject the author turns in the second half of the article.

The anonymous author argues that one must choose between two alternatives: either God created the body of the first man instantaneously and infused it with a soul, or else he infused a human soul into a previously prepared animal. The tradition of the Church fathers favors the first alternative, but it should be recognized that what the fathers emphasized is the special dignity of the human being. The article concludes that it should be accepted as Catholic doctrine that Adam began his existence as an adult. Moreover, this corresponds with the obvious, literal sense of Genesis and with what believers generally hold (*sensus fidelium*). To deny this would be at least rash and dangerous. The reviewer points out in a footnote that Catholic theologians consider a proposition to be "rash" if it accepts assertions without sufficient evidence, or if it is contrary to the common doctrine of the fathers, or if it contradicts the common opinion of theologians, without a substantive reason or basis in authority.

Mivart inclines toward evolution but, as with the author of the *Dublin Review,* he clearly asserts the spirituality of the human soul, created and infused by God. This is the main point of the discussion, the *Dublin Review* observes,[22] and the longest part of the article is devoted to the defense of this spirituality. In this context, the author completely agrees with Mivart. The author concludes by encouraging Mivart to continue in the same line with a critical analysis of Darwin's recent book on the origin of mankind.[23]

In the next number of the *Dublin Review* (January–June 1872) the same subjects turn up again in another anonymous article in the book review section, on the occasion of Huxley's recent attack on Mivart and the latter's response.[24] The journal criticizes Huxley while lavishing praise on Mivart. But it also takes the opportunity to return to the origin of Adam's body, summarizing the position taken in the last article: again it asserts that an evolutionary origin is unfounded, based not so much on Scripture, as on theological consensus and tradition, and advising readers that this is a matter of faith because of the supernatural destiny of human beings. Similar considerations apply to the peculiar formation of Eve from Adam. Mivart, the author continues, ought to have no difficulty in agreeing to all this, inasmuch as he accepts the possibility of miracles, and the immediate formation of Adam's body by God requires less of a miracle than the infusion of a human soul in a pure animal. Both the arguments marshaled and the language in which they are expressed make clear, however, that this is still not a defined doctrine of faith.

The Polemic with Bishop Hedley

Mivart had good reason to be self-assured in his incursions into theology. He was viewed as an authentic benefactor of the Catholic cause and had been awarded a doctorate in 1876 by the pope personally.

In his biography of Mivart, Jacob Gruber presents his final falling out with the ecclesiastical authorities as the culmination of a long process, as if it were impossible to reconcile the scientific spirit with religious faith, and this tension, which had always been present in Mivart, had finally reached the boiling point —a logical progression, not an aberration owing to the illness that Mivart endured in his last days.[25] Still, as Gruber himself explains in great detail, Mivart devoted a great part of his life to establishing the inexistence of the presumed opposition between science and Christianity. This he did consistently and with good result.

Between 1885 and 1887, Mivart found himself in a double polemic, with the Reverend Jeremiah Murphy and the Benedictine John Cuthbert Hedley, bishop of Newport. Murphy had stated that evolution applied to human beings could not be considered an orthodox opinion. In 1885 and 1887, Mivart published two articles in the journal *Nineteenth Century* in which he spoke of the freedom that Catholics had with respect to questions of science that appear in the Bible. Bishop Hedley then published in the *Dublin Review* a long article criticizing Mivart and inviting him to correct himself.[26] According to Mivart, the Galileo episode shows that God has assigned the clarification of scientific matters, whether mentioned in Scripture or not, to scientists and not to theologians or ecclesiastical tribunals or the Roman congregations. Bishop Hedley thought that such an assertion was dangerous because it was partially true. Hedley focused on the lessons that the Church could extract from the Galileo case and what attitude Catholics should adopt when considering those scientific points that are mentioned in the Bible. He reproached Mivart who, instead of acting with more sensitivity, made a hyperbolic presentation of an idea that, taken literally and in its most general sense, was disconcerting and incorrect.

Mivart replied in an eight-page letter addressed to the editor of the journal.[27] First, he introduces an article by Jeremiah Murphy that had also been cited by Hedley. According to Murphy, Catholicism and evolution are incompatible, contrary to Mivart's well-known position. Mivart used the occasion to lament the harm that priestly ignorance of science causes Catholics; he appealed to his own experience and work in this area and characterized Murphy's position as "intolerably pernicious." Murphy said that it is not necessary to abandon the literal meaning of Scripture when speaking of the creation of man, unless the evolutionists show there is sufficient reason to do so, thus discouraging any attempt to prove that evolution is right. Mivart replied that while it is not necessary to abandon literalism, it does not follow that others should be *prohibited* from abandoning it. "According to this view a theologian may *insist* upon a proposition being believed while all the time he holds it may possibly turn out to be untrue; providing himself, as it were, with a little back door to slip out at, should an inconvenient evolutionist one day show him 'sufficient reason' so to effect his escape. This to me is a revolting position, at once skeptical and shocking from the indifference to truth it makes manifest."[28]

Returning to his critique of Hedley, Mivart says that what moved him to formulate his idea so openly was the danger faced by Christians worried by evolutionism when they encounter priests like Murphy. It is possible, he adds, that from the point of view of the bishop his advocacy may sound bad, but when confronted by Murphy's assertions, he felt obligated to speak out. With this type of person, he continues, both the "prudent silence" and the "respectful protest" that Hedley had urged upon him are useless.

For the rest, if Mivart replied to the various criticisms that Hedley had made, he did so respectfully, from a Catholic stance, and always with arguments. Nothing in this reply would lead one to think it would produce a break with Church authorities. Mivart tried to show, precisely, that his ideas fit perfectly the Catholic perspective.

In the same issue of the *Dublin Review*, right after Mivart's letter, came Hedley's two-page reply.[29] The bishop defended Murphy and briefly specified what, in his opinion, constituted the heart of the matter, again summoning Mivart to peace and prudence. It is a serene response, aimed at clarifying the issues, and ends with a long paragraph in which, leaving a few differences aside, he praises Mivart's work and concludes that Mivart merits the thanks of English-speaking Catholics for his writings against materialism and in defense of spiritualism.

Unanimity of Theologians

Mivart's attacks now put Jeremiah Murphy on the defensive. He replied with an article in the same number of the *Dublin Review*.[30] The article is interesting, because it clearly lays out the reasoning that in those days, and at least till the end of the nineteenth century, Catholic theologians opposed to the evolutionary origin of the human body used.

First, Murphy asks whether a Christian can admit that the human body derives from evolution, without admitting an *immediate* act of God apart from the creation of matter and his cooperation in the evolution of potentials that God himself endowed matter upon creating it. His response is negative. In the first place, that God created man is an article of faith, revealed in Genesis and taught by the Church. As such, it is the responsibility of the Church to explain, with certainty and without error, the meaning of this truth, and the theological position on the subject should have as much weight and value as it has with respect to any other revealed doctrine. To determine this meaning, he had appealed to the sources from which Catholic theologians draw their arguments: Scripture and tradition.

But, as Cardinal Mazzella (whom Murphy cites) states, theologians, taking into account the authority of Scripture, as the Church fathers understood it, respond unanimously that the body of the first man was made by a direct and immediate act of God, distinct not only from the first creation of matter but also from God's cooperation with the action of secondary causes. If this assertion is correct, says Murphy, to apply evolution to man is contrary to the faith:

> The revealed account of man's creation is precise and clear, and, if it must be taken literally, there is no room left for evolution. . . . The Fathers and theologians, who are the best qualified to interpret such texts, tell us, with the most extraordinary unanimity, that the literal sense is the true one . . . the case against evolution, which rests, not on Scripture text alone, but on the meaning put on that text by a constant, uncontradicted tradition. . . . And what is the ground for contradicting it here? The evolution theory—a theory, however, confessedly not *proved*, not even *provable*, but pointed out as "scientifically probable by "analogy"—"a misleading guide."[31]

Murphy adds that, keeping all this in mind, it is not necessary to abandon the assertions of Scripture on the creation of man. Mivart thinks that this expression is ridiculous; so Murphy says he will express it in another way: "We have very strong reasons—conclusive reasons—for assenting to the literal meaning of Gen. ii. 7; but have no reason, or at best a "misleading" one, for dissenting from it; then I say, we *cannot reasonably abandon that meaning*: to abandon it would be against reason, as well as against faith."[32]

Murphy refers at some length to the Church fathers and to Catholic theologians to demonstrate that they are unanimous in their assent to the thesis expounded and then concludes:

> Fathers and theologians alike teach this doctrine as part of the Divine deposit of faith. They appeal to Scripture to prove it, and thus record their belief that the doctrine is contained in the sacred text. Thus, then, the doctrine comes to us as a Catholic tradition, the interpretation of a revealed truth, uncontradicted for 1800 years. Such a tradition is evidence that the doctrine so taught is part of God's

Word, of Divine faith, infallibly true. Fathers and theologians, as far as I have quoted them hitherto, knew nothing of evolution. Granted: but their teaching is conclusive, though indirect, against it [evolution]. For they teach that man's creation was a special act, distinct from the creation of matter and from the operation of natural laws. Evolution denies this. And therefore, if Fathers and theologians teach truly—*and they do*—evolution as applied to man must be false. But, now that the evolution theory is known, is has been, as Dr. Elam says, "weighed in the balance, and found wanting." It has been carefully studied and examined by our best modern theologians, and it has been formally and explicitly condemned by them as irreconcilable with faith. . . . I admit that some recent writers have been quoted as at least tolerating evolution. Well, I think I do these authorities no wrong, and I certainly mean them no offence, when I say that they do not disturb the balance of Catholic teaching as against the doctrine.[33]

The conclusion of Murphy's article is unequivocal and could not be more categorical: evolution applied to man is contrary to the faith. Equally unequivocal is the argument that leads him to this conclusion:

Thus, then, we have on the one hand, a consensus of Fathers and theologians teaching the immediate formation by God of the body of the first man, and this by an act distinct from the creation of matter and from His co-operation with Nature's laws—a doctrine that is incompatible with evolution applied to man—and we have, on the other hand, absolutely certain authority for holding that such a consensus is a conclusive proof of the truth of the doctrine so taught. Such a consensus is the voice of the ordinary *Magisterium* of the Church, and the doctrine so taught is part of the Divine deposit of faith, infallibly true. And as the evolution theory applied to man contradicts this doctrine that theory is false, and against faith. This is the difficulty that confronts Mr. Mivart.[34]

Murphy's analysis is clearly stated and lineal. He places Mivart in confrontation with the Magisterium of the Church and against the faith. But not everyone saw things that way, not even in the Catholic world. The editor of the Catholic *Dublin Review* felt himself obliged to add a discrepant editor's note at the end of Murphy's article:

It is true, as Father Murphy has well pointed out, that theologians generally teach the immediate creation of the human body by God. What, however, has never yet been made sufficiently clear is, whether theologians taught this as the current and unquestioned belief of their time and as their own opinion, or whether they taught it as a truth revealed by God and binding upon the conscience of all the faithful. The question is not whether they considered their interpretation of Genesis as true, but whether they taught that it was an article of faith. This, we venture to think, has not yet been placed beyond doubt.

So we see that Murphy's rationale was not viewed as definitive not even by the editor of the *Dublin Review*. Eight years later (1896), when John Zahm published his book *Evolution and Dogma*, which followed Mivart's line, the *Dublin Review* published, as we have seen, a favorable review by the Franciscan David Fleming, who worked in the Vatican, at the Holy Office. When Zahm's book was examined two years later by the Congregation of the Index, it was clear that there was no established doctrine on the problem. Indeed, one of the consultors asked (in writing) that, to clarify the matter, the Holy Office be queried prior to the Congregation's deliberations.[35] Moreover, the consultor who wrote the report on Zahm's book collected all the arguments available, including those of Murphy, to show that Zahm's opinion could not be sustained; he thought that it was Catholic doctrine to assert that God had made Adam immediately and directly from the mud of the earth. He also thought that the unanimous assent of the fathers and scholastic theologians, in matters of faith and customary practices, provided certain witness to Catholic dogma. Yet, he was aware that he could not marshal against the evolutionary origin of the human body more than a few decisions of provincial councils, and so he proposed that the Congregation of the Index condemn the proposition that God made man from the body of a brute anthropoid prepared by evolution out of base matter.[36] But no such condemnation resulted.

Mivart's position on the evolutionary origin of the human body met opposition, but as of 1888 it did not lead to any official condemnation by a Church authority. A few years later Mivart had to face the condemnation of one of his writings, not on evolution, but on hell.

Evolution and Hell

Biological evolution did not extend into the inferno, and Mivart never proposed it did. What he suggested was that hell is compatible with some kind of happiness, and for this his article "Happiness in Hell" was condemned. It had nothing to do with science or with evolution. Yet this condemnation of 1893 was a significant moment in Mivart's life.

"Happiness in Hell" really refers to three articles that Mivart published in 1892 and 1893 in the journal *Nineteenth Century*. The first article bears this title and appeared in December 1892.[37] Mivart aimed to explain hell while avoiding the obstacles that make this doctrine so hard to accept, enabling those scandalized by it not to become estranged from the Church. But Mivart's ideas on the subject could not be accommodated by the traditional teachings of the Church, motivating the bishop of Nottingham, Edward G. Bagshawe (an Oratorian, born in London), to publish in the same month a fifteen-page pastoral letter warning of Mivart's ideas.[38] The same journal immediately published a very critical response by Richard F. Clarke, a Jesuit. Mivart replied with another long article on the same subject in February 1893, and yet a third in April.[39]

In the first article, Mivart tries to harmonize the Christian doctrine of hell with modern sensibility, so adverse to cruelty and disproportionate punishment. His conclusion is that most human beings are incapable of committing a mortal sin and, after death, they reach a state of natural happiness, in accord with their possibilities, influenced by knowledge and natural love of God. Other groups are tested on earth, and they reach a future state proportional to their merits or demerits, which can be equal to the natural happiness just mentioned, or do not reach it. (That is, souls in hell would enjoy happiness in direct relation to the degree of their culpability on earth.) Finally, God has equipped a certain number with faculties that make them capable of supernatural union with Him, a privilege that goes together with the risk of failing; but even those who fail will prefer existing to not existing, and it is permitted to think that there exists an eternal upward progression, although without ever reaching the supernatural state that is repugnant to them: they are left to themselves in the inferior conditions that they have freely chosen. Thus, in the future life no one will suffer the loss of the happiness that he might imagine or desire or that is commensurate with his nature and faculties, except for his conscious and deliberate choice. Therefore, it is possible to speak of happiness *in hell.* Mivart thinks his ideas conform to the teaching of the Church and fall within the scope of legitimate interpretation within Catholic doctrine, and he laments that in past epochs hell has been presented in another way.[40]

In the second article Mivart intends to better explain his ideas in response to Richard F. Clarke's critique. Mivart says he is not trying to establish anything new; everyone believes there is some happiness in hell; he only intends specify what kind. His central idea is that it is not a matter of faith that the aversion to God that the condemned experience be eternal; therefore it could be held that a progressive diminution of this aversion occurs until it becomes a very positive attraction, which not only mitigates his punishments but also betters his moral condition and brings with it a certain grade of happiness. With brilliant style, Mivart states that the opinions common among Catholics of a given period do not necessarily represent the dogmatic teaching of the Church. His intent is to help many who are estranged from the Church and honestly seek the truth, and others who are in the Church but suffer uselessly when they think that what their good sense tells them is opposed to the Catholic faith. According to what an experienced American priest told him, a great obstacle for conversions in America is the moral rejection of the doctrine of hell, as habitually taught. Mivart presents his aim as a third step in his apologetic work: the first step was to have shown, in his *Genesis of Species*, his *Lessons on Nature*, and his article on Galileo, that there is no contradiction between Christianity and science; the second was to show that there is no contradiction between the Church and biblical criticism; now, with the third step, he will show there is no contradiction between ethics and the Church's teachings on hell.

In the third of the series, Mivart refers to the pastoral letter of the bishop of Nottingham. He says the bishop is impulsive and refers to an occasion on which he had to swallow a pastoral letter because so many Catholics complained to Rome and Rome agreed with them. Mivart responds to the bishop's criticisms and energetically advocates freedom for Catholics in everything that has not been expressly defined as a matter of faith. By the time the third article was published in April 1893, the two first had already been denounced to the Holy Office of Rome.

The Holy Office Dossier

In the Archive of the Holy Office, the corresponding documentation is found in the section of Censorship of Books (Censura Librorum), year 1893, folder *Nottingham* (the diocese where the denunciation originated), section *on the work "Happiness in Hell" by George Mivart* (*sullo scritto "Felicità nell'Inferno" di George Mivart*).[41]

Mivart's two first articles were denounced by the bishop of Nottingham to the Congregation for the Propaganda of the Faith (on which the English Catholics depended). This congregation, in turn, sent it to the Holy Office.[42] The denunciation was accompanied by a list of suspicious propositions culled from Mivart's articles.

In the dossier of the Holy Office, we find that Father Louis Hickey was assigned to prepare a report. His first report was printed with the date of March 18, 1893.[43] At the end of the report some propositions of Mivart considered suspicious were collected in a summary. The consultors of the Holy Office, in their meeting on April 14, 1893, asked that another, more complete report be worked up, in which, again, the suspicious propositions, both in English and Italian would be included in a summary. In his second report, dated May 23, 1893,[44] Hickey advised that, through error, the principal suspicious propositions are not found in the first report. Hickey mentions in his first report that the denunciation was made by the bishop of Nottingham along with the Jesuit Richard F. Clarke.[45] Hickey threw into relief the great repercussion of Mivart's writings, and at the end of the report he presents a letter from Clarke to Cardinal Vaughan, on May 12, 1893, where Clarke says that Mivart has written more on hell (the third article); he adds that he plans to respond to him and asks for an accelerated decision because, if it is delayed, it will be too late to avoid the harm that it will do to the English people.[46]

The closing summary of Hickey's second report rehearses statements extracted from Mivart's first two articles that were declared worthy of censure in the original denunciation. Those statements were divided into two groups, of which the first contained the five most serious issues:

1. One can believe that in hell some evolution, that is, gradual progress and betterment in the condition of the condemned, can take place. Still, the condemned will never be elevated to supernatural happiness, inasmuch as the inhabitants of hell will forever remain there (article 1, p. 916).

2. In hell the worst individuals hold that it is better to exist than not to exist, and they do not wish to end their existence, just like many who are miserable in this life. Therefore, one can assume that those individuals suffer less in hell than those who endure the miseries of this life (article 1, p. 915).

3. As for the many who are sentenced to hell, it may be that they are not only conscious of their condition and console themselves with the hope of a better future, but also, for all we know, they might feel at home there inasmuch as by reason of their behavior on earth they are prepared for what they find there. It can happen, therefore, that such individuals might acquire friends and thus might happily embrace their chains, and practice those behaviors and lower desires which they chose for themselves in this life and in which they found consolation (article 1, p. 912).

4. It is not a matter of faith that condemned persons' aversion to God is eternal (article 2, p. 325).

5. Along with the evolution and improvement that I [Mivart] have praised, the condemned's aversion to God might also cease little by little, and end in a positive attraction to Him. Such an attraction would not only soften the punishment, but also, more importantly, would incline their moral condition admirably towards the good. To oppose what is good implies some pain; in the same way, to be attracted to what is good implies some degree of happiness.[47]

The secondary propositions are related to other matters, for example: that ancient pagan cults were, in certain respects, good; that there is no reason to lament the late arrival of Christianity or the defection of many Christians; that many sins which in themselves are mortal, are in practice venal owing to attenuating circumstances; that most people are incapable of mortal sin and will naturally attain happiness, while another group will reach a more or less similar end in function of their merits or demerits. Mivart was also supposed to have said "If the supreme authority [of the Church] declares something false, that physical science shows to be absolutely true, then our faith is in vain," and said something similar with respect to biblical criticism and ethics.[48]

In his reports, Hickey comments on the points mentioned, emphasizing quite a few theological inaccuracies. At the end of the first report he proposes that

Mivart's writings be listed in the *Index of Prohibited Books* and that suspicious propositions be subjected to a new, more profound examination.[49] At the end of the second report he observes that Mivart has, through his writings, provided authentic benefits to the Church, but this cannot be said of his theological work, in spite of his good intentions: he has serious faults both in doctrine and in his way of expressing himself, and a sincere friend would advise him to leave theology to the theologians.[50]

Hickey also reproduces an important letter from Cardinal Vaughan to the Holy Office at the end of his second report:

May 2, 1893

Most Eminent Cardinal:
I have spoken with Mr. Mivart, who says he is disposed to publish a summary of his doctrine on hell, just as I suggested to him. He states that he has a real desire to serve the Church against the rationalists. I write this to Your Eminence first, so that you might know that the case is not hopeless, and second, because I know that some have written urgently seeking a condemnation. A condemnation, if the Church does not require it, would be lamentable in the event that he confirms his will to keep (profess) the true faith. If the matter is to be taken up in the Holy Office or the Index, I beg for a delay of two or three weeks, to see what might be done along the lines you have expressed to me.

P.S. I add that the Mivart case was discussed at the meeting of bishops two weeks after Easter. Except for two who dissented, we were in agreement: 1. in not seeking Rome's condemnation. 2. for now, not to act as one sole body, but to let the bishops act, each one in his diocese, and to instruct the clergy about what should be done in this case.[51]

The Condemnation and
Mivart's Reaction

The Holy Office was not in a hurry. Hickey's second report was dated May 23, 1893. By this date Mivart had already published his third article, an annotated copy of which is in the Holy Office dossier of the case, indicating that it was examined. But the Mivart case itself was not discussed in meetings of the Holy Office until July.

It was first taken up in a meeting of consultors that took place on Monday, July 3, 1893. The assessor of the Holy Office, Francesco Segna (later to become cardinal prefect of the Index), noted that the participants decided that the articles in question, including the third (published by Mivart after the first two had been denounced), should be placed on the *Index of Prohibited Books* by a "Wednesday decree" (after their usual Wednesday meeting), and also that they should write to Mivart's bishop instructing him to communicate with the author and ask whether he wants to accept the decree so that the public retraction would appear in the published decree (an outcome they viewed positively).[52]

That decision, in turn, had to be ratified in a meeting of the cardinals, who took up Mivart's case on Wednesday, July 19, 1893, in the building of the Holy Office. Five cardinal members of the Congregation were present: Lucido Maria Parocchi, Gaetano Aloisi-Masella, Teodolfo Mertel, Isidoro Verga, and Camillo Mazzella (who, in addition, was still cardinal prefect of the Congregation of the Index). Also attending were the assessor and two other officials of the Holy Office. All these participated in the so-called secret part of the meeting. In the public part of the meeting the same persons were present, together with the notary. In the book of official acts of the Holy Office, the date and place of the meeting, the participants, and the decisions were recorded. It was decided to place the writings of Mivart on the *Index of Prohibited Books* by the decree dated July 19, 1893, and to communicate the decision to the Congregation of the Index to execute the contents of the decree.[53]

On July 21, the assessor of the Holy Office, Francesco Segna, wrote to the cardinal prefect of the Index (the Jesuit Mazzella, who had participated in the meeting of the Holy Office), communicating the decision of the cardinals and adding that this decision had been approved by the pope the day before.[54] The letter begins: "In response to Your Eminence's note of this morning," which seems to indicate that Mazzella, who was aware of the decision, was reminding the assessor to communicate it officially to the Congregation of the Index, that is, to himself (moreover, Segna wrote that the decision had been made "in yesterday's Congregation," when in reality he should have said "the day before yesterday"). Segna also provided the Index with Mivart's articles and the bishop of Nottingham's pastoral letter.[55]

The Index had just (on July 14) held a General Congregation of cardinals, in which it was decided to prohibit various books. In the corresponding decree Mivart's three articles were added, with their titles and dates, with a note that they had been prohibited by the Holy Office in a decree dated July 19, 1893. The decree, signed by Prefect Mazzella on July 14 (before the decree of the Holy Office), was published on July 24.[56]

Mivart was immediately informed from the Vatican, because on August 6 he wrote to his friend Edmund Bishop giving him the news and asking his advice. Mivart said that he had received the following authentic notice, that he himself enclosed in quotation marks:

I have just received the following private but most authentic piece of news: "The *doctrine* of your articles has not been condemned. You may hold those views, but it is not opportune to disseminate them in existing circumstances, on account of the misapprehensions to which they have given rise. I do not remember a case in which there has been such a widespread misapprehension. Practically the meaning of the condemnation is to prevent you from publishing those articles apart in book form. . . ."[57]

Of course, the fact that a writing was on the *Index* was not the equivalent of a condemnation of any concrete proposition: a work might be listed simply because its publication was deemed in some way untimely. The notice that was received mostly likely came from Cardinal Vaughan, who had been delegated to ask if he would submit. Bishop advised him on the way in which he might make his submission public, and Mivart thanked him on August 11: "I thank you warmly for your admirable letter which treats the subject (to me very important) in such an exhaustive manner. I read it twice and then sat down and wrote my submission and sent it to the Cardinal who, by the way, has been personally very kind in his way of writing to me about the matter. I am sure you are right and I have acted rightly."[58]

The letter that Mivart wrote to Cardinal Vaughan on August 10, a day before his letter to Bishop, is preserved in the Archive of the Congregation of the Index:

Hurstcote, Chilworth, Surrey
10 August 1893

My Lord Cardinal,
In response to Your Eminence's letter, I beg leave to request Your Eminence to be so kind as to transmit to Rome, for me, my complete and *ex animo* submission to the recent decree of the Sacred Congregation of the Index in my regard.

I have the honour to be, My Lord Cardinal, Your Eminence's attached and faithful servant

St George Mivart
To His Eminence Cardinal Vaughan, Archbishop of Westminster[59]

On August 15, Cardinal Vaughan sent Mivart's letter to Cardinal Mazzella, prefect of the Index, accompanied by the following letter:

London, 15 Aug. 1893

Most Eminent Lord,
I beg to enclose a letter which I have received from Mr. Mivart, which I hope will give satisfaction. He has been sincerely distressed to think that he should have incurred censure. I believe him to be truly desirous to render service to the cause of religion, though he has a most unfortunate way of exhibiting his desire. I think he will not again venture upon matters of doctrine upon which he is so little able to speak with knowledge and accuracy.

I have the honour to be, my dear lord Cardinal, your most humble and obedient servant

Herbert Card. Vaughan[60]

As happened in such cases, when the author accepted the decree, the Congregation of the Index showed its acknowledgment by explicitly mentioning the submission in its next decree, using the customary wording. This is what happened in Mivart's case. In the decree of Friday, June 8, 1894 (published on June 12), signed by the new cardinal prefect of the Index, Serafino Vannutelli, and by the secretary, Marcolino Cicognani, Mivart and his three articles are mentioned, followed by: "prohibited by decree of the Holy Office, July 12, 1894. The author submitted in a praiseworthy manner and repudiated the articles."[61]

For the time being, the matter was over, but with the suspicion that it had been handled poorly. With his goodwill, Cardinal Vaughan smoothed over the proceedings and made everything seem, both to Mivart and the Vatican, easier than what in fact they were, and Mivart acted the same way, following the advice of his friend Bishop. Mivart had been anxious, but Vaughan's explanations and general attitude had put him at ease. It seemed that nothing had happened; it was a matter of bad timing. The matter could have ended there. Mivart had rendered great service to the Church in England and he was recognized for it. It was simply preferable that he not intrude so much in matters that were directly theological. In the same period, Bishop Bonomelli had a work of his placed on the *Index*—he accepted it and continued for many years after as bishop of Cremona, until his death. No further measure against Mivart was contemplated.

Stirring the Waters

It was Mivart himself who provoked the final conflict. According to Gruber, events took their course logically. According to Mivart's family, abnormal behavior provoked by the disease from which he died was at the root of the conflict. A cold assessment gives the impression that Mivart never came to terms with the listing of his articles in the *Index:* it had looked like he had won the battle to reconcile science and the Bible, but now he was faced, for the first time, with an official condemnation, even though he had been assured, correctly, that this condemnation had a very limited reach. Whatever led up to it, the fact is that in 1899, when he was already gravely ill, Mivart took, on his own initiative and with no external provocation, certain actions that finally led Cardinal Vaughan to withhold the sacraments from him unless he retracted.

On August 29, 1899, Mivart wrote to the cardinal prefect of the Congregation of the Index, the Jesuit Andreas Steinhuber, recapitulating the condemnation of his articles in 1893. He was disturbed to find that his articles still appeared in a new edition of the *Index*. In fact, this was normal: when an author submitted, the fact was made known, but the work continued on the *Index*. Mivart lamented that the matter of the *Index* had influenced his removal as professor at the University of Louvain, where he had been professor since 1890. He asserted that the *Index* had no force in England and, in sharp tones, asked to be apprised of the circumstances of the condemnation, threatening to withdraw his act of submission:

Augt 29 1899
77 Inverness Terrace W
London

To His eminence the Cardinal Steinhuber: Prefect of the Congregation
of the Index
My Lord Cardinal
It may be known to Your Eminence that I published (in 1892 and 1893) some articles on hell.

My object in so doing (clearly stated in the first article) would, I thought, command them to the approval of Authority. In fact, however, they were quickly (in 1893) placed upon the *Index*—and this without any communication being made to me!

My great veneration for His Holiness Leo XIII and the fact that no statement made by me was condemned, induced me (as stated in my last article) to submit to the decree.

Although (as Your Eminence knows) the Index has no force or authority in England, it has occasioned me inconvenience in Belgium—leading me to resign my Professorship at the Catholic University of Louvain.

I have also lately heard that these articles have been freshly inserted in a new edition of the *Index*.

It is this which now moves me to address Your Eminence.

I need hardly point out to you, My Lord Cardinal, that the mode in which my condemnation (like that of others) was effected, is a mode profoundly abhorrent to us Englishmen, to Americans also and, indeed, to all English-speaking people.

I therefore, as an Englishman, who feels he has a grievance, venture, with a most profound respect to ask Your Eminence—in your benevolent kindness—to answer the following questions:

(1) Who was it accused me to the Sacred Congregation?
(2) What were the faults imputed to my writings?
(3) Were they reported on, and, if so, by whom and what did he report?
(4) What matters in my articles decided the Congregation to place them on the Index?
(5) What reasons led the Congregation to intervene on that matter and to condemn the articles?

(6) Can my articles be now removed from the Index?

On these points I (and other Catholics also) think information should be afforded me and if I do not receive answers I shall feel impelled to withdraw my submission.

I have the honour to be of

My Lord Cardinal

Your Eminence's most humble and most obedient servant

St George Mivart, F.R.S.

We have already presented the archival evidence that, after more than a century, makes it possible to answer Mivart's questions. But then he had to be content with generalities. The Archive of the Index conserve a draft of its reply to Mivart, with some deletions, in Latin. It is a respectful and courteous reply:

St. George Mivart F.R.S.
77 Inverness Terrace
W. London

Distinguished Sir,
It is fair that you not be denied a reply to the letter you recently sent (the past August 29) to the Prefect of the Congregation of the Index.

First I should tell you that your articles on hell were not proscribed by the Holy Congregation of the Index, but by the Congregation of the Holy Office, even though, as is customary, the proscription was promulgated by the Congregation of the Index. From this you will easily deduce that the reply to the questions you formulated in your letter do not pertain to the Index but to the Holy Office. I can respond in general, however, that the Holy Congregations, when examining and prohibiting books, proceed with maximum maturity, counsel, and prudence, about which Benedict XIV wisely prescribed many things in the Constitution *Sollicita ac provida*. For which reason it can not happen that books by Catholics be prohibited recklessly.

You should also know that the Holy Congregations do not condemn authors, but only aim that books, the reading of which could provoke harm to the integrity of faith or customs, be taken from the hands of the faithful. Indeed, no one doubts that this right and duty, surely a serious one, was imposed on the Apostolic See by Christ himself.

With respect to what you say about the Index lacking the force of law in England, perhaps at some moment it could have been said without lacking a basis, but nowadays, after the Constitution *Officiorum ac Munerum* of Leo XIII, it is completely false, inasmuch as in it obedience to new decrees is enjoined "for the Catholics of the whole world."

As for the reasons for the prohibition of your comments, it would be very useful for you to consult about the doctrine contained in your writings a serious and learned theologian, who will demonstrate that the opinions that you enunciate on theological issues cannot be reconciled with what the Church, the column and firmament of the truth, has also taught and still teaches about the punishments of hell.

With my full concern
Most Illustrious Sir[62]

All of this reflects the reality of the situation. But it could hardly satisfy Mivart, who presented himself as a victim demanding justice.

In these circumstances, Mivart took various steps that eventuated in his separation from the Church. On October 17, the *Times* newspaper published a long letter from Mivart, on the Dreyfus case, which had roiled public opinion throughout the world, in which he expressed very harsh criticisms against the French Catholics (especially the bishops), the Roman congregations, and the pope. Whether the judgments were shared by other Catholics or not, the tone was that of a quite disagreeable critique. Furthermore, in January 1900 he published two articles, in two different journals, which contained a whole set of assertions over what the Church should be and of criticisms on what it is, that practically sealed Mivart's estrangement from Catholicism.[63] According to his own confession, Mivart gave them a strongly critical tone to provoke a clear reaction, either favorable or contrary.

As might have been expected, the reaction was contrary. On January 6, the Catholic journal, the *Tablet*, published an unsigned article, strongly critical of Mivart, setting off a spirited and rapid interchange of letters between Mivart and Cardinal Vaughan, which began and ended in January 1900.[64]

Dialogue and Condemnation

On January 6 Mivart responded personally to Cardinal Vaughan, as the authority (real or supposed) responsible for the *Tablet*, protesting, as an English gentleman, that he had been called a coward and a liar. On January 9 Vaughan responded saying that, since in his articles Mivart had impugned the most sacred doctrines of the faith, he needed to know whether he still considered himself a Catholic. To ascertain this, he enclosed a long and explicit profession of Catholic faith, asking Mivart to return it signed. The cord was now stretched out to where it was becoming easy to prognosticate a break.

On January 11 Mivart replied seeking reparation for the offense that had been inflicted on him in the *Tablet*: he enumerated three ways in which he would accept this reparation. Now the cord was stretched even tighter. On January 12 Vaughan replied to Mivart that, with respect to the *Tablet*, he should direct himself to its editor, and that he continued to see himself obliged to ask him to sign the formula of faith that he sent him.

On January 14 Mivart replied. He insisted that the cardinal, as owner of the *Tablet*, and as a matter that fell within the area of ordinary ethics, should order the journal to rectify the offense it had done him, and that before this were done, he would make no step, given that the organ of the cardinal (the *Tablet*) and hence the cardinal himself considered him a coward and liar. On January 16 the cardinal wrote Mivart a very short letter, saying expressly that, as his bishop, he gave him a third and last opportunity to send in signed his profession of faith. Vaughan repeated that if Mivart had anything against the *Tablet* he should direct himself to the editor, and that he write to him exclusively as his bishop and guardian of the faith of his flock. And he added that, if he did not submit by signing the profession of faith, the law of the Church would follow its course.

On the Feast of Saint Peter's Chair, which is celebrated in the Catholic Church on January 18, Cardinal Vaughan sent a circular letter to the Catholic clergy of the diocese of Westminster:

Notice of Inhibition of Sacraments
Archbishop's House, Westminster
Feast of St. Peter's Chair, 1900

Rev. Dear Father:
Dr. St. George Mivart, in his articles entitled "The Continuity of Catholicism" and "Some Recent Apologists," in the "Nineteenth Century" and the "Fortnightly Review" for January, 1900, has declared, or at least seemed to declare, that it is permissible for Catholics to hold certain heresies—regarding the virginal birth of Our Lord and the perpetual virginity of the Blessed Virgin; the gospel account of the resurrection and the immunity of the sacred body from corruption; the reality and transmission of original sin; the redemption as a real satisfaction for the sins of men; the everlasting punishment of the wicked; the inspiration and integrity of Holy Scripture; the right of the Catholic church to interpret the sense of Scripture with authority; her perpetual retention of her doctrines in the same sense; not to speak of other false propositions. As he has thereby rendered his orthodoxy suspect, and has, moreover, confirmed the suspicion by failing, after three notifications, to sign the annexed profession of faith when tendered to him by me, it now becomes my duty to take further action, and I hereby inhibit him from approaching the sacraments, and forbid my priests to administer them to him, until he shall have proved his orthodoxy to the satisfaction of his ordinary. Believe me to be, Rev. dear Father, your faithful and devoted servant,

Herbert Cardinal Vaughan,
Archbishop of Westminster[65]

This was not properly an excommunication, rather just a deprivation of the sacraments.

It is difficult to establish with precision when this circular reached the clergy of the diocese and was communicated to Mivart. Perhaps this did not take place before the following letters were exchanged.

The correspondence still continued, although for only a short time. On January 14 Mivart replied with a long letter in which he asked Vaughn to pronounce on a paragraph of the profession of faith that spoke of the absence of errors in the Bible, and he made clear the difficulties he had in order to accept it. On January 21 Vaughan expressed his sadness over Mivart's illness; he spoke to him of humility and prayer for acceptance of the Church's teaching; he recommended various readings for his difficulty over Scripture, and added that perhaps a conversation with the priest Dr. Clarke or the Jesuit Father Tyrell would be more useful to him, because both would be capable of taking charge of his difficulties and helping him.

On February 23 Mivart thanked Vaughan for his last letter and replied with a very long letter. He cited a passage of the recent encyclical *Providentissimus Deus* (1893) where the pope spoke of the absence of error in the Bible; he said that this had caused him a crisis, and then he had applied to a good friend and excellent theologian, whose reflections had much helped him; but added that lately he had reached the conclusion that Catholic doctrine and science are fatally opposed, so that no one with common sense could agree with the Catholic Church: he had concluded that orthodox Catholicism was untenable. Perhaps he should have kept quiet, he added, but he felt an obligation to enlighten the people whom he had previously influenced to accept what now he sees as an error. His mortal illness made it necessary for him to make a decision. Therefore he wrote his articles in the most provocative way possible, to provoke a pronouncement. At this point in the letter, Mivart pronounced his own sentence, when he wrote: "I categorically refuse to sign the profession of faith."

Vaughan still replied to Mivart on January 25, addressing the question of Scripture. Because Mivart had referred literally to a passage of the encyclical of 1893, Vaughan added other passages that complemented it and seemed necessary for a correct interpretation. Finally, Vaughan assured Mivart that he was happy to make any sacrifice if he could really help him, and that he would pray to God that His grace would prevail and return Mivart to the Church. On January 27 Mivart sent his last letter. He reproached the cardinal for having acted precipitously in sending the circular to his priests, although he was sure he had done so out of a sense of duty, and he reiterated his opinions. That was the end of the correspondence. Mivart died on April 1.

Mivart, Vaughan, and the Holy Office

But Mivart's problems and their repercussions among English Catholics continued to demand the Holy Office's attention. The complications of the Mivart case found an immediate echo in 1899. It was known that Cardinal Vaughan had reported the events and had even consulted regarding the possible withholding of sacraments.

Monsignor Rafael Merry del Val, who was not yet a cardinal and who was later to become cardinal secretary of state, intervened in the case. Although he was Spanish by nationality, he had been born and educated in England, where he had excellent connections and was highly respected. Merry del Val was the next to write a report on the Mivart case.[66] In this report, Merry del Val recalled that, before Mivart's scandalous retraction, whereby he withdrew the submission he had made earlier to the condemnation of his articles on "Happiness in Hell," the London press and the Anglicans wanted to know what the Holy See would do. Perhaps they hoped to show that the Catholic Church tolerated the same lack of discipline and diversity of beliefs as were tolerated among Protestants, to discourage conversions. In these circumstances, Merry del Val continues, Cardinal Vaughan wrote that he wanted to convene a committee of theologians in his confidence to study Mivart's writings and report to him, after which he would be disposed to send a circular to the clergy of his diocese announcing that Mivart could not receive the sacraments until he showed his repentance by retracting his errors. The cardinal asked if an energetic action in this direction would have the full support of the Holy See, because it was easy to predict protests in the English press. Merry del Val adds that Father David Fleming, consultor of the Holy Office and a close friend of Mivart for many years, had tried to convince him of his error and could give an exact report on him.

The consultor David Fleming did in fact provide a four-page report handwritten in Latin, titled, "On the State of Catholicism in England," dated November 17, 1899.[67] Although he reports on serious matters, his tone was calm. Fleming had firsthand acquaintance with the subjects of his report and had even taken direct action on various occasions, so his knowledge was not superficial. Among the people he mentions, he pays special attention to Mivart, whom he criticizes severely. After providing many personal details, Fleming concludes that, were Mivart to apostatize publicly, a serious scandal would result. But he adds that, as a matter of conscience, he fears greatly for the possible harm done to the Church, and he says that there is still time to head that off. Fleming observes that, inasmuch as Cardinal Vaughan is afraid to treat Mivart harshly for fear of the scandal it might cause, the result is that Mivart acts with impunity to the great prejudice of the faithful.

At their meeting of Wednesday, November 22, 1899, the cardinals of the Holy Office asked to listen to a report from Monsignor Merry del Val and then requested that the reports of both Fleming and Merry del Val be printed so they could be studied. Merry del Val was to submit his report on December 8.[68] A report of two typewritten pages from January 1900 appears to have been the work of Merry del Val, informing the Holy Office about two articles published by Mivart the same month.[69] First, it refers to "Some Recent Catholic Apologists" in the *Fortnightly Review*, in whose last section Mivart refers to his articles from 1892–93 on "Happiness in Hell," condemned by the Holy Office and the Index. From that time on, the report continues, Mivart made a dubious and evasive submission, but it was not believed appropriate to insist on obtaining a more explicit one, as he had written: "In August I wrote to Cardinal Steinhuber, S.J., Prefect of the Congregation of the Index, to tell him that, inasmuch as my article had been again placed on the Index (in a new edition of this publication), if I did not receive a response to some of my questions, I would feel obliged to withdraw my submission. The answer I received does not respond to those questions and so I withdraw my submission."

In reality, there was a misunderstanding behind these words: a submission results in a public unburdening of the author whose writing has been condemned, but that has nothing to do with withdrawing the writing from the *Index*.

The report continues by stating that Mivart was famous in England, that he shared many ideas with Protestants, that he was considered to be the most distinguished representative of Catholic writers and scientists, and that the Holy See had awarded him the honorific title of doctor of philosophy. And he adds:

> It now seems to me that it is a little less than incredulous, although he always seeks to pass as a Catholic and to be known as such. . . . After the condemnation of his articles on hell, he has always gone further in his liberal and certainly unorthodox ideas, always showing himself as hostile to the Holy See.
>
> . . . Lamentably Mivart enjoys some sympathy and support from a group of Catholic clergy and laymen of the liberal school. Finally in this month of January 1900, in *The Nineteenth Century*, Mivart has written a completely scandalous and blasphemous article, in which he set out to show that the Church has continuously advanced, modifying its dogmas in accord with the times, while still remaining the same institution, and among many other very grave errors he ends by insinuating that, in time, the Church could even recognize that Saint Joseph was the true father of Jesus. If the Supreme [the Holy Office] might wish to take this article into consideration, an exact translation could be made.

On Wednesday, January 10, 1900, the cardinals of the Holy Office made various decisions with respect to Mivart's case. First, they decided to tell Cardinal Vaughan that he follow the course he deems most appropriate: that is, they gave him complete discretion to decide whether to prohibit Mivart from receiving the sacraments. They also decided to instruct the Congregation of the Index that, in the next edition of the *Index of Prohibited Books,* when referring to the condemnation of Mivart's writings on "Happiness in Hell," that the qualification "The author submitted in a praiseworthy manner and repudiated the articles" be deleted; and that Monsignor Merry del Val should translate Mivart's recent article into Latin or Italian, so that the Holy Office could examine it. And finally, the Holy Office should consider the question of how the doctorate of philosophy conferred upon him by the pope might be withdrawn.[70]

In accordance with these decisions, the assessor of the Holy Office wrote to the cardinal prefect of the Index on January 14 to inform him of the decision relative to the *Index.* This is recorded in the Diary of the Congregation of the Index (that the clause "the author submitted in a praiseworthy manner, etc." should be deleted from the *Index).*[71] For his part, Cardinal Vaughan sent to his priests the circular dated January 18, prohibiting Mivart from receiving the sacraments. Merry del Val translated Mivart's article "The Continuity of Catholicism" into Italian: it is preserved in the dossier of the Holy Office and was delivered to the commissioner of the Holy Office on February 8.[72]

On March 24, 1900, the Holy Office printed a fourteen-page summary of Mivart's article, "The Continuity of Catholicism," as had been decided on January 10.[73] A footnote advised, "Mr. George Mivart died several months after the article denounced was published." Nevertheless, the process set in motion by the article had already begun and continued on its way.

On Tuesday, May 14, 1900, in a meeting of the consultors of the Holy Office the new case was discussed, and it was decided to wait, pending new reports by Fleming and Merry del Val. On January 18, 1901, David Fleming signed a handwritten page titled "On the Case of Mr. Mivart, Deceased."[74] He says nothing new has occurred to him, ever since Cardinal Vaughan excommunicated Mivart for his stubbornness with respect to his heretical opinions and for which reason he had also been denied Church burial. Moreover, Vaughan, together with all of the bishops of England, had written a letter condemning "liberal Catholicism." In his last two articles, Mivart repeated things he had already said in others, adding some new problems. He had also written in favor of "Americanism" and its advocates. Fleming concludes that, even though the biblical question could be examined, and because the pope had already spoken with respect to Americanism, in his opinion nothing remained to be done, except to commend Cardinal Vaughan and the other bishops for their strength.

On January 21, 1901, in another handwritten page (seemingly written by someone from the Commissary of the Holy Office), it says that David Fleming's note was read and that, without waiting for Merry del Val's report, it is proposed that no further action be taken (*satis provisum*) or some similar wording. On Saturday, April 27, the Particular Congregation (of the consultors) agreed with the Commissary's proposal, namely, *provisum*. The Holy Office's task was completed.

Evolutionism and the
Condemnation of Mivart

The Mivart case is frequently presented as if its focus had been evolutionism, given that Mivart stood out and was especially well known as a proponent of the compatibility between evolutionism and Christianity. Many who learned that Mivart had been condemned by the Church concluded or took for granted that the Church had condemned evolution.

There were two separate official actions by the Church that affected Mivart. The first was the 1893 listing on the *Index of Prohibited Books* of his three articles "Happiness in Hell," which had nothing to do either with evolution or science. Mivart himself expressly conceived of this case as quite distinct from his previous battles. He distinguished a first campaign in defense of the compatibility of evolution and Christianity, which he counted as a victory, because the Church did not act. The second battle had to do with biblical criticism, and here he asserted that the Bible contains errors; Mivart also reckoned this a victory because, again, he was not officially censured by the Church. "Happiness in Hell" was really a third battle, related to the previous ones, but different from them. Here, Mivart intended to show that Catholic doctrine was not opposed to ethics. In this case he directly invaded theological territory on a nonscientific topic and clashed frontally with the authorities, who for the first time took official action.

The second official action was the prohibition from receiving sacraments in 1900, a few months before he died. It was provoked by Mivart, because he himself asserted that he had deliberately adopted a provocative tone in his final writings so that the authorities would have to act. His last writings contained a serious challenge to the doctrine and authority of the Church in more than one aspect. Science had something to do with it all, especially the biblical question, because Mivart had concluded that the Bible and Catholic doctrine could not be reconciled with science. But other contemporary Catholics, as serious as Mivart and more expert in theology, reached the opposite conclusion. Without attempting to pass judgment on Mivart's final stance, we can say that his attitude was not solely or principally determined by scientific motives and, more concretely, that evolution did not occupy a determining role in it.

It is significant that the Franciscan David Fleming, consultor of the Holy Office who was so harsh in his report on Mivart in the final phase of the case in 1899–1900, three years before, in 1896, had published a favorable review of John Zahm's *Evolution and Dogma*, a book that followed Mivart's line, in the *Dublin Review*. Clearly the reasons that led Fleming to criticize Mivart so severely and to propose that serious measures be taken against him had nothing to do with either evolution or science, neither of which are mentioned in Fleming's reports.

Contrary to what he had said in public over many years, Mivart concluded that there was an opposition between official Catholicism and science, but the Church authorities never acted against him because of his scientific views. Whatever his state of health might have been when he published his last articles, he expressed in them accusations against the Roman authorities, in a passionate tone, which on occasions departed from reality. For example, he claims that it had been demanded in Rome that Leroy publicly retract his convictions under the threat of a listing on the *Index;* Leroy retracted, but Mivart says later that his work was still listed on the *Index* afterward. He says he learned this from an unimpeachable source.[75] But all this is false. Mivart supplies more erroneous or confused information in the same context, conflating evolutionism with Americanism, and both with the Galileo case. He regards Rome's actions as a new Galileo case, but he interprets Galileo's case with a mixture of real facts, historical errors, and a passionate attack on the Roman curia.[76]

Cardinal Vaughan has been accused of acting precipitously, and the Roman authorities of supporting him.[77] Perhaps greater flexibility on the part of the authorities would have avoided a conclusion that can be considered lamentable or inevitable, according to one's point of view.[78] In any case, the archival data show that the Holy Office acted with its characteristic rhythm, following all the predictable steps, studying the issues in each phase with the aid of reports expressly prepared by experts. The rapid denouement of the controversy in January 1900 was deliberately provoked by Mivart, as he himself stated, when he published a strongly worded and provocative critique of Catholicism that made it impossible for the Church to look the other way. Mivart displayed no sadness at the turn of events: on the contrary, he declared he had been freed.

England and the Index

The procedures of the Congregation of the Holy Office and of the Index cannot be characterized by haste or improvisation. The rules were scrupulously obeyed. One confirmation of that comes from archival evidence referring to the force of the *Index of Prohibited Books* in England.

When Mivart, on August 29, 1899, wrote to Cardinal Steinhuber, prefect of the Congregation of the Index, asking for the reasons behind the condemnation of his articles on hell, the latter replied that the matter was not before the Congregation of the Index but rather the Holy Office, because it was there that the denunciation was addressed. In his letter, Mivart asserted that the Index had no force in England. Steinhuber replied that such was not the case. His response was not improvised. In the Archive of the Index there is an eighteen-page study titled *On the Doubt as to Whether the Rules of the Index Should Be Observed in England and How,* dated February 8, 1896,[79] written by the Jesuit consultor Franz Xaver Wernz,[80] with two appendices: the first is a four-page letter from Herbert Cardinal Vaughan, archbishop of Westminster, dated October 8, 1895, and the second is an two-page memorandum, transmitted by Cardinal Vaughan, reproducing the opinion of an English bishop (he does not say who), as he expressed it in a private letter to a priest and in a letter to his diocesans.[81]

Wernz's study was based on a report by the bishop of Nottingham (England) on the state of his diocese. The report had been sent to the Congregation for the Propagation of the Faith, which at the time had jurisdiction over the English Church. That congregation sent the report to the Congregation of the Index.[82] The bishop raised the question of the observance of the rules of the Index in England. He said that throughout England, since time immemorial, there existed the custom of not mentioning or teaching or observing the laws of the Congregation of the Index. He said he did not know whether this was an abuse or rather a tolerated custom. He asked Cardinal Vaughan, who thought that it was inopportune to insist on the laws of the Index in England. According to Vaughan, though, it was still suitable for believers to abstain from reading whatever was recognized as a danger to the faith. Moreover, it would be useful if bishops could grant necessary permissions without having to recur to the Holy See; and it would be further convenient if the laws of the Index could be simplified. Another unidentified English bishop proposed that the Church not insist on the obligations of the Index and gave some nuanced advice concerning how to proceed when confronted with questions or doubts raised by the faithful.

The consultor, Wernz, related in his report that he had lived in England for eight years and that ideas about the Index were colored by fantasies about the Inquisition and torture. But he concluded that there is no reason to deny that the laws of Index have the same validity in England as they have in any other country. He compares the situation of England with that of the United States, Holland, and Germany, where these rules are in force. He goes on to say that few books in English are on the *Index,* and that in England very few books are read that are not in English. Nor is there any reason to fear a possible scandal that the British press might sow, because such scandals only last until the next one comes along. Still, insofar as enforcing the Index's rules, Wernz advises not doing so with a lot of noise, or only as the act of a single bishop, but rather that all fifteen or sixteen English bishops agree on a practical and uniform course of action; for example, discussing the matter in seminaries and in meetings of parish priests, and letting them set the example for their parishioners. He further advises to grant English bishops the power to give permission to read prohibited books without having to apply to the Holy See, as the bishops of the United States, Germany, and Austria had been granted.

Therefore, Mivart's statement about the nonobligatory nature of the laws of the Index in England had some basis in fact, although it was inexact. That is the gist of the Congregation's response.

The Excommunication That Never Was
and Other Inaccuracies

It is not quite right to say that Mivart was excommunicated. Cardinal Vaughn's decision was titled "Notice of the Inhibition of Sacraments" and it stated: "I prohibit him from receiving the sacraments, and I prohibit my priests to administer them to him, until he demonstrates his orthodoxy to the satisfaction of his bishop."

Franz Xaver Wernz, whom we have met as a consultor of the Index and, later on, as general of the Jesuits, had written a treatise on canon law that includes an explanation of excommunication, its types, and its effects, which was published in the period that interests us (it is prior to the 1917 Code of Canon Law).[83] Of course, the deprivation of sacraments was one of the principal effects of excommunication, but is not the whole story. It was an ecclesiastical censure, not only a punishment, like deprivation of sacraments alone. What Cardinal Vaughan decided on was not separation from the communion of Catholics, which meant prohibiting participation in any sacred activity and even extended to external social activities. It might be thought that Mivart had been dealt the so-called *minor excommunication,* which also implied denial of sacraments. But full excommunication deprived one, for example, of the everyday assistance of the Church (for example, prayers for the dead), which did not occur in Mivart's case. In such matters, the Church imposed the strictest interpretation only when all the conditions stipulated by canon law were met. Inasmuch as, in addition, Cardinal Vaughan never used the term "excommunication" in this instance, it is incorrect to say that he excommunicated Mivart.

Perhaps we're splitting hairs because, for a layman to be deprived of the sacraments is perhaps the most important effect of excommunication. This explains that in the report prepared for internal Vatican use only by David Fleming (January 18, 1901, cited above) titled "On the Case of Mr. Mivart, Deceased," excommunication is mentioned, even though the usage was improper.

The *Dictionary of National Biography* says that Mivart was excommunicated.[84] Gruber titles the last chapter of his biography of Mivart "Excommunication," perhaps because it would have been less clear and less sonorous to have titled it "Deprivation of the Sacraments." In his article on Mivart in the *New Catholic Encyclopedia* of 2002, however, Gruber does not mention the nonexistent excommunication and, with complete accuracy, says that Vaughan denied him the sacraments.[85]

Imprecisions in Gruber's book are important because it has become the habitual source of information about Mivart. Gruber attributes the condemnation of
Mivart's writings on hell to a decree of the Holy Office, and when he then reproduces the text, it is clear that the reference is to the Congregation of the Index. In
this case there are complications that excuse Gruber, but it is clear he took his information from a secondhand source, an article published in England, as he indicates.[86] Gruber is familiar with the English sources, especially correspondence,
but he makes very limited use of Catholic sources, which, in the case of Mivart,
are important. The contents of the decree in question were not secret and can be
found in Vatican publications.

Other imprecisions of Gruber affect the substance of the case more directly, as
when he refers to Mivart's submission to the Index decree of 1893 in a most
tragic tone.[87] But we have seen that Mivart's letters to Cardinal Vaughan and
Vaughn's to Cardinal Mazzella transmitting Mivart's letter are completely normal and give no evidence of any personal tragedy.

Perhaps the most distorted aspect of Gruber's book is his treatment of the encyclical *Providentissimus Deus* of 1893 on the interpretation of Scripture. He
quotes from the text to show that the encyclical insisted on an extreme literalism
that had devastating effects on science, but he omits other passages of the same
encyclical that would convey a different impression. He goes on to assert categorically that the new exegetical approach developed by Mivart that fostered the
reconciliation of science and religion became a kind of heresy. He also notes that
within a few years the doctrine of scriptural infallibility condemned the evolution of the human body as erroneous: he cites in evidence the retractions of
Leroy, Zahm, and Hedley.[88] While it is true that these three cases were adjudicated in the years following the encyclical in question, there is no evidence what-
soever that the negative results of those cases were in any way a consequence of
the encyclical. The case of Caverni was prior to the encyclical and its result was
even more negative. This point is important because Gruber's assertions legit-
imize the notion that the encyclical condemned evolution as applied to man's
body,[89] which is false: the encyclical addresses neither evolution nor the origin of
man.

When he explains the condemnation of Mivart's articles on hell, Gruber criticizes Cardinal Vaughan, whom he presents in sinister tones, as aligned with in-
transigent positions and excessively preoccupied with money. Mivart would then
be the victim of a change of direction in the Catholic hierarchy. Yet, as we have
seen, Vaughan wrote to the Holy Office, on May 2, 1893, to report he had spoken
with Mivart, who was disposed to obey and continue his service to the Church;
he adds that, although some have asked for Mivart's condemnation, he asks that
it be avoided if possible, that Mivart be given time, and he confirms that the Eng-

lish bishops, with two exceptions, agreed with him.[90] Gruber's presentation cannot be sustained.

According to Gruber, Mivart's final unraveling began when, in 1899, he found out that the new edition of the *Index of Prohibited Books* still included his articles on hell and, filled with anxiety, he wrote the prefect of the Index asking for an explanation, so that he might understand the Church's position. But this supposition makes no sense. Even if he did not know the answer, anyone could have explained to him that accepting the decree of the Index did not mean that his writings would be delisted but that his acceptance of the condemnation would be made known, as indeed happened. Gruber places this episode in a tragic context: "it was [he says] in 1899 when the forced retractions of the Catholic evolutionists were made known."[91] But to whom does this refer? One supposes Leroy, Zahm, Bonomelli, and Hedley. Nevertheless, as we have seen, Leroy's letter of retraction was published in *Le Monde* in Paris on March 4, 1895; Zahm's letter to his Italian translator saw the light in 1899, but it was not a letter of retraction. Bonomelli's retraction letter was published in 1898 and was completely spontaneous, even though designed to avoid possible problems. Nor can it be said that Hedley's letter was in any way forced.

There is no doubt that Leroy's retraction only became generally known when *La Civiltà Cattolica* published it at the end of 1898, but it makes no sense to locate it on that date just to make all the presumed retractions coincide. Rather, the coincidence of the statements of Zahm and Bonomelli is completely logical because Bonomelli's problems arose out of praise of Zahm's book. Hedley's comments were provoked by the articles in *La Civiltà Cattolica* around this time. Therefore, the coincidence does not indicate that anything out of the ordinary was afoot.

The Church and Evolution

Was There a Policy?

It would be an illusion to expect a simple conclusion or a short summary of the six cases we have examined. Each was the product of very different circumstances, and if the cases appear to have been similar, that is not owing to any preconceived plan that guided the Church authorities. Indeed, in assessing the role of the Church, one of the principle conclusions that we might extract from the archival documents is that, in three of the cases examined (Bonomelli, Hedley, and Mivart) there was no action taken against the authors, nor did the Roman authorities make any decision at all, and that in the other three cases in which the Vatican intervened (Caverni, Leroy, and Zahm), it did so in response to denunciations that reached the Congregation of the Index from the outside. The Holy Office played no role in any of these cases. Therefore, we can say that the cases examined were not generated by any policy of the Roman authorities with respect to evolutionism.

Nevertheless, even though the cases are very different, they are also related, and they were so viewed in the articles published by *La Civiltà Cattolica* between 1898 and 1902. It is instructive, then, to reflect on their similarities and differences.

Similarities and Differences

Caverni's book was placed on the *Index* in 1878. Of the six cases, this is the only one in which the Holy See condemned a book for its defense of evolutionism and put its decision into practice—possibly because Caverni had no one to defend him, inasmuch as he was a priest of the diocese of Florence and his own archbishop was implicated in the denunciation of the book. This condemnation took place long before the publication in 1893 of the encyclical *Providentissimus Deus*, on biblical interpretation, indicating that the later condemnation of other evolutionist books was not simply a consequence of this encyclical. On the other hand, Mivart's book was published in 1871 and was not the object of any ecclesiastical censure, neither at the time nor later. The varying circumstances of each author no doubt influenced the course of the case: Caverni was a priest and published his book in Italy, whereas Mivart was a scientist and a layman and published his book in England. As we have seen, there were doubts about the effectiveness of the Index in England, where Mivart's intellectual prestige reflected well on the embattled Catholic minority. By contrast, Caverni's views met immediate resistance in Italy.

Mivart's problems with Church authorities were not owing to evolutionism. They were provoked by Mivart's incursions into theology, which, in the final and definitive clash, were accompanied by a pungent critique of the Church. The reactions of each author to Church condemnation were also very different. The priest Caverni absorbed the censure of his book, which appears not to have caused him further problems, whereas Mivart never absorbed the condemnation of his writings on hell, and this brought on the definitive break. In neither case was there any preconceived attitude on the part of the authorities, who simply reacted to the problems as they presented themselves. These are the two modal cases: the express condemnation of Caverni's book, and the complete absence of any condemnation of Mivart's.

The cases of Leroy and Zahm are intermediate between these two extremes. Leroy's book only drew fire after its second edition appeared and was denounced by an anonymous Frenchman. The examination of the book in the Congregation of the Index was highly complex. What is clear is no hard and fast criterion had developed on this question: the first report proposed that no measure be taken. At the meetings of the Congregation it was thought desirable to examine the question in greater detail but, even though two new reports were generated, each more critical than the first, neither proposed an outright condemnation. The fourth report, solicited in the middle of the process, was the most negative. The decision to condemn the book was made by the cardinals of the Index, but Leroy had effective backers: he was a Dominican, and the Dominicans had for centuries played an important role in the defense of Catholic doctrine, frequently availing themselves of the Congregation of the Index, whose secretary was likewise a Dominican. When he was summoned to Rome, Leroy was lodged in the very same house where the secretary of the Congregation lived, and with the express permission of the pope he was permitted to read one of the reports produced for the scrutiny of his book. In deference to Leroy and to the Dominicans, the process of condemnation of the book was never completed: the corresponding decree was not published and, as a result, the book never appeared on the *Index*. In return, Leroy was asked to issue a public retraction, which he did in a letter published in the Parisian daily *Le Monde*.

In effect, Leroy retracted because he knew that his thesis, "examined here in

Rome by the competent authority, had been judged unsustainable, above all because what refers to the body of man, by reason of such great incompatibility not only with the texts of the Bible but also with the principles of sound philosophy." The "authority" in question is not identified, but it doesn't take much imagination to figure it out, although in theory the reference might be only to the authority of the Dominicans alone. Thus, the letter did not overly bow to authority, and it had one additional advantage: it articulated the reason for the condemnation, which would never have figured in the decree of condemnation in the event that the Congregation had seen fit to publish it. In the final analysis, everyone won. Leroy did not have his book condemned. The Congregation of the Index published nothing; rather Leroy himself revealed publicly, under his own name, that there had been an action of the authorities against his book, which was enough to alert other Catholics who shared Leroy's ideas.

In the case of Zahm the denunciation came from a well-known cleric, surely related to the "conservative" group that, in Rome and in the United States, fought evolutionism and "Americanism" at the same time. Zahm's prestige and the fact that his book was immediately translated into Italian precipitated the events. All that was needed to condemn the book was a single report—written by the Dominican who had already twice intervened in the Leroy case and whose view of evolution was completely unfavorable. The complications, in this case, developed after the condemnation of the book. Zahm also had effective backers: the general superior of his congregation knew his way around Rome, and over a period of several months the influential American clerics who supported Americanism made many approaches on Zahm's behalf, on all levels, even reaching the pope in person. Zahm had no wish to retract, because he saw retraction as a victory for his adversaries (who were also opponents of Americanism). Finally the personal letter from Zahm to his Italian translator was presented by *La Civiltà Cattolica* as the equivalent of retraction, because Zahm had written: "I have learned from an irreproachable source that the Holy See is opposed to any further distribution of *Evolution and Dogma* and, therefore, I beg you to use all of your influence to withdraw the book from the market."

Zahm was duly satisfied when he learned that he was not asked to retract and that the efforts of his adversaries had ceased. Was he even aware that his letter to the translator had been represented as a retraction? He could not have foreseen that, for decades to come, that letter would be used as evidence that the Holy See had opposed the evolutionism that he defended. Those who sought to impede the diffusion of evolutionary ideas were also satisfied, since for the first time there was some public indication that the Holy See opposed these ideas, without it having been necessary to publish a single official document.

Thus, the cases of Leroy and Zahm were similar: the condemnation of the book was not published (and, as a result, the book was not listed in the *Index*), in return for a retraction (Leroy) or for a letter that had the same force (Zahm).

Bishop Bonomelli's case is very different, because it never triggered consideration by the Congregation of the Index. The rapidity with which Bonomelli published his letter of retraction forestalled it. Bonomelli was sympathetic to evolution, but he pointed out its limitations and was not disposed to fomenting a new conflict with the Holy See over the subject: he had already had several and would incur more in the future, but over other issues in which he found himself more deeply involved. Once again, the authorities did not have to publish anything in order to get the retraction.

In the case of Bishop Hedley, the Holy See did not intervene, and we cannot even talk of a retraction, properly speaking. He took a few steps back, but nothing more. Thus, it makes little sense to present Hedley as a case of either condemnation or retraction. The most salient aspect of the case is that it provoked the two articles in *La Civiltà Cattolica* that, with both its accurate judgments and its limitations, have served until to the present as the principal source for determining the actions of the Holy See with respect to evolutionism.

A Conspiracy against Evolution?

The archival materials show that the Roman authorities limited themselves to respond to the denunciations that reached it. Still it seems like too great a coincidence that in the decade of the 1890s several retractions and other public actions came about in relation to evolution. It is a fact that merits our attention and requires an explanation.

In Mivart's case, it has been held that the encyclical *Providentissimus Deus*, on the interpretation of Scripture, published by Leo XIII in 1893 led to the final conflict. There is no doubt that Mivart did not fully accept its contents. But although the encyclical contains parts that emphasize the literal interpretation of the Bible, there are other parts that leave a certain freedom. In fact, this encyclical was used in internal reports of the Congregation of the Index with considerable moderation, and at no time did it seem to have been a definitive argument against evolutionism. The encyclical cannot be considered, as at times it has, a point of inflection in the attitude of the authorities of the Church with respect to evolution: in reality, it does not mention evolution nor does it at any time refer to the origin of man's body. This does not exclude the possibility that the encyclical played some role in cases posterior to 1893, but if so there is no evidence of any definitive influence; and it should not be forgotten that the condemnation of Caverni's book, the only one that actually wound up on the *Index*, occurred much earlier than the encyclical's appearance.

Barry Brundell asserts that at the end of the nineteenth century a change of mind took place among the Roman authorities, and he attributes it to the influence of the Jesuits of *La Civiltà Cattolica*. We agree with Brundell that these Jesuits unleashed a running campaign against evolutionism. But such activity already existed long before the decade of the 1890s. Quite a few years before, *La Civiltà Cattolica* reacted strongly against Caverni's book, and with this motive it published a very long series of articles against evolutionism, over various years. Moreover, contrary to what Brundell says, the Jesuits' opposition to evolutionism was not limited to Rome. The Jesuits of the *Études religieuses* in Paris certainly engaged in a vigorous polemic with Leroy over the publication of his book. But, there is no proof of the existence of the type of conspiracy that Brundell imagines, in which the Jesuits of the *Civiltà* would have direct influence in the decisions of the Congregation of the Index. This might well be unimaginable: the archive shows who wrote the reports and made the decisions, and traces of conspiracies have not been found.

The temporal coincidence of several cases is not difficult to explain. Bishops Bonomelli and Hedley were implicated in evolutionism because they wrote commentaries on Zahm's book: the three cases thus coincided in time. *La Civiltà Cattolica* took advantage of these circumstances to give greater publicity to Leroy's retraction, which had been written several years before and up to that point was little known: in this way, the four cases coincided.

The *La Civiltà* articles create the impression that in scant time events happened that had never happened, but this is not certain: the condemnation of Caverni's book was much before the rest, Leroy's case was anterior to the rest, and the reserved attitude in the face of evolution was a constant among Roman authorities over many years.

Consensus and Uncertainty

The archival documents make clear that, in the internal deliberations of the Congregation of the Index, there was an almost general consensus for rejecting evolutionism, above all when it is applied to the origin of the human body. It is patent that in some cases uncertainties existed. In the case of Caverni there was unanimity behind Zigliara's proposal, not only among the consultors in the Preparatory Congregation but also among the cardinals in the General Congregation. But this did not happen years later, when the books of Leroy and Zahm were examined.

In the case of Leroy the opinions emitted were not only different but almost mutually contradictory. In the first report, written a short time after the publication of the encyclical *Providentissimus Deus*, Domenichelli admitted that it was correct to say that in the first chapters of Genesis the Bible uses figurative language *in a human mode, to condescend* (as Dante would say) *to our understanding,* and that the question of whether the biblical Creation narrative ought to be interpreted literally or allegorically is open to free debate among theologians. Moreover, Domenichelli reminds the reader that the task of the Congregation of the Index is to judge the book not by the criteria of science but only according to the teachings of the Church, and he mentions that many Catholic thinkers and theologians have accepted evolution as compatible with the Catholic faith as long as certain limits are imposed, especially in upholding the direct creation of each human soul by God. Finally, with regard to the origin of man, for Domenichelli Leroy's interpretation is sufficient to refute that which the Council of Cologne sought to condemn, that is, the formation of man's body by "spontaneous transformations." Domenichelli proposes finally that no measures be taken against Leroy and that if, on the contrary, it seems opportune to act because the theory seems too dangerous, it would be more to the point to specify the errors and condemn them with the appropriate censure, as the Church had done in other cases.

Domenichelli's suggestion went nowhere. In the course of the years here considered, the Roman authorities never formulated any explicit condemnation of evolutionism as a doctrine. There prevailed, instead, a more subtle attitude that led, on the basis of more severe reports, to the condemnation of Leroy's book, but without the condemnation being published as a decree.

In the first General Congregation on the Leroy case, the cardinals debated frankly, and concluded that the problem should be studied in greater depth. Secretary Cicognani wrote in the Diary that a decision had been reached "after diverse arguments over the subject and a serious discussion." According to the memorandum preserved in the archives, it appears that Cicognani initially agreed with Domenichelli's proposal or at least hoped it would be approved.

But then two new reports were ordered. In his report, the consultor Fontana, only recently named bishop, saw no reason to condemn Leroy's position, but because it still struck him as dangerous, the author should be warned. Tripepi was harsher. His report was based on the arguments of Cardinal Mazzella, and he saw a serious danger in the diffusion of evolutionary ideas. Yet he also failed to reach any drastic conclusion. He thought that it was necessary to take some measure, due to the anxiety and scandal suffered by Catholics who read these new doctrines that are contrary to the natural meaning of the Bible, and to the danger that the Catholic world, preoccupied by modern science, might progressively surrender the truths of religion. Still, he proposed not that the book be prohibited but rather that the author be warned and that he retire the remaining copies of his book from sale.

From Cicognani's memorandum we know that "the debate over the work of Father Leroy in the Preparatory Congregation was heated." He refers to the second Preparatory Congregation, in which the consultor Buonpensiere must have played an important role, as is clear from the hardening of the conclusion (the book should be condemned and the author invited to retract) and from the use of Buonpensiere's arguments to justify it.

The report by Buonpensiere, the consultor who a few years later would also write the report on Zahm's book, was completely negative. According to Buonpensiere, evolutionism cannot even be called a *hypothesis:* it is a simple *desideratum* of materialism: "therefore, the presumptuousness of those who seek to make evolution equivalent to Christian dogma is foolhardy and anti-Christian." In the second General Congregation, a hardening of the position finally crystallized. All but one cardinal formulated the decision that won the day: the book would be proscribed but without publishing the decree, and a public retraction would be sought from Leroy.

When Zahm's case reached the Congregation of the Index a few years later and in spite of the fact that it was Buonpensiere who again presented a completely negative report, new questions arose. For one thing, Buonpensiere proposed expressly condemning "the following proposition, or another like it, that is: 'God formed Adam's body not immediately from the mud of the earth, but out of the body of an anthropomorphic brute, whom he had prepared in order to be produced by the forces of natural evolution from lower matter.'" This condemnation was not approved. On the contrary, a notable disparity of opinion emerged in the Preparatory Congregation.

The summary of deliberations of the Preparatory Congregation began: "Bearing in mind that . . . this doctrine [evolution] is not opposed to any defined dogma nor does it impugn the faith, three favored that the author only be warned." Five proposed that the work be condemned and the decree published, because this doctrine "does not agree with the Bible, nor with the doctrine of the Church and its Holy Fathers and Theologians." One proposed the same, taking into account that "the Holy See made clear its position on this doctrine when it condemned the work of Leroy." Finally, one said "that it should not be condemned now, while the Holy Office is asked to resolve the doubt: does transformism contradict divine revelation? Afterward let this Congregation proceed against the book."

With respect to the General Congregation, Cicognani wrote that "the most Eminent Cardinals, after a long and serious discussion, decreed the prohibition" of Zahm's book. Again we encounter a serious debate, which is only logical, bearing in mind that there was no unanimity as to the theological standing of evolution. Domenichelli proposed that such standing be determined. Buonpensiere proposed a concrete definition. An anonymous consultor proposed that the question be clarified by appeal to the Holy Office. But the Congregation of the Index did not consider it necessary to take any of these steps and made the decisions that seemed most prudent.

A Continuing Condemnation?

Brian W. Harrison has analyzed some of the cases that we have examined, using documents from the Archive of the Holy Office. On the attitude of the Roman authorities with respect to evolution, Harrison writes:

> There was in fact a consistent, if relatively quiet, rejection of human evolution on the part of the See of Peter throughout the last three decades of the nineteenth century. Apart from the censures just mentioned, the Holy Office consultors Domenichelli and Tripepi both affirm that the anti-Darwinian decision of the German bishops at Cologne in 1860 was "approved" by Rome, and presumably this was the case (although this approval was apparently given little publicity and its documentation may well be still awaiting rediscovery in other archives of the Congregation). . . . Rome, it is true, did not exactly embark on a vigorous anti-evolution crusade, as did some Protestant denominations. No papal bulls or encyclicals thundered against the novel ideas, but neither did any such "heavyweight" documents ever condemn Galileo or Copernicus. It is also true (as we saw Fr. Tripepi complain obliquely) that Rome remained silent about some Catholic authors who were propagating the same evolutionary hypotheses as Leroy. However, when specific works such as his landed on Holy Office desks, confronting the Congregation with the need to speak out for or against, the decision was invariably and unambiguously negative.[1]

Harrison's basic point is correct: the Roman authorities consistently rejected evolution when applied to the human body. He is also right when he states that there was no systemic or noisy campaign, and the authorities limited themselves to acting on the denunciations that reached them. Harrison is less accurate, however, on a number of basic issues. First, Domenichelli and Tripepi were not consultors of the Congregation of the Holy Office, but rather of the Congregation of the Index. Harrison systematically confuses the two: he says that Chalmel's denunciation of Leroy was received by the Holy Office; he presents his citations as taken from the Acts of Holy Office, and the meetings and decisions of the Congregation of the Index as those of the Holy Office.[2] He does not mention the Congregation of the Index even once, but attributes all references (absolutely all: persons, documents, meetings, decisions) to the Congregation the Holy Office, and nothing (absolutely nothing) to the Index. This is important. The function and rank of the Holy Office and its decisions were much more important than the mission, much more concrete and modest, of the Congregation of the Index, which was ordinarily limited to listing publications on the *Index of Prohibited Books,* without specifying the reasons. The Congregation of the Index had an important but very limited, role. Moreover, only in the case of Caverni did the procedure run its course, with the book placed on the *Index:* in the other cases not even a single document was published.

In his classic book of 1932, Ernest C. Messenger devoted an entire chapter to examining the cases that occupy us here, using the few documents available at the time. Messenger concludes that the cases of Leroy and Zahm—the two cases that triggered the intervention of the Congregation of the Holy See (he states prudently that it could have been either the Holy Office or that of the Index)— were the result of "private ecclesiastical acts," and draws the following conclusions:

> But nevertheless it remains true that, as the *Civiltà* itself allows, *the Holy See did not condemn the theory in question by a public act,* and moreover, it deliberately abstained from doing so, "for the best reasons." The steps taken were private ones, and *nothing whatever about any of these incidents has ever appeared in any official journal of the Holy See.*
>
> Secondly, the action taken was confined to *particular individuals.*
>
> These two points are of the utmost importance. When the Holy See wishes to teach the whole world, the whole world is informed thereof. And when the Holy See wishes definitely to condemn any particular teaching, it does so in an unmistakable way: a document condemning the teaching in question is addressed to the whole world, or at any rate is published in the official organ of the Holy See.[3]

A "private ecclesiastical act" is an official ecclesiastical action, directed at particular persons (not at all Christians) and about which no document is made public (that is, nothing is published in the official media of the Church). In the cases of Leroy and Zahm, this is what happened. In both cases the Congregation of the Index examined the book in question, decided to condemn it, communicated this decision privately to the authors via their religious superiors, and commuted the publication of the decree in exchange for the author's personal, public retraction (in Zahm's case, not quite a retraction). There were official acts of the Congregation of the Index, but not public acts. The Congregation wanted to slow evolutionism, but it did so indirectly, without public notice, thinking that to be in the best interest of all parties.

No public condemnation of evolution occurred therefore. Harrison is right when he says that Buonpensiere and Tripepi seemed to want one. We have seen that Buonpensiere, at the end of one of his reports, even proposed a concrete form in which to pursue one. He did not meet with success.

We have also seen that the only book to be listed in the *Index* was Caverni's, and the Congregation of the Index wanted this case to serve as an "indirect condemnation" of evolution, but the wish was not realized. Bearing in mind that the decree did not specify the reasons for the prohibition and that the book's title contained no reference to evolution, the condemnation was hardly noticed and, even today it has been supposed that Caverni was condemned, in good part, for his critique of the ecclesiastical world. Books were prohibited for very different reasons, frequently circumstantial: it is enough to recall the condemnation of the pamphlet in which Bishop Bonomelli proposed a solution very similar to the one eventually adopted to govern relations between the Vatican and the Italian state.

It is important to observe that, in the cases examined, virtually all the actions related to evolution were directed to the Congregation of the Index and not to the Holy Office, which was the authority directly concerned with doctrine. *La Civiltà Cattolica* led its readers to suppose that the actions taken originated with the Holy Office and so it has been transmitted in Catholic theology texts for generations, seriously distorting the reality.

The archival documents reveal that there were actions of the Holy See aimed at the concept of evolution, but also that care was taken not to give them an official, public stamp. From one perspective it could be said that the Holy See did not condemn evolution because it never issued any official public condemnation. From a different perspective it could be said that it did condemn evolution, because evolutionist books were condemned, but these had no official, public character. The attitude was mainly adverse, but it was moderated by considerations of charity for those involved and the institutions with whom they were affiliated, and probably also by prudence in anticipation of future advances in science. Such prudence left some space open for the negotiation of individual cases.

A Pragmatic Policy

Perhaps it can be said that the Holy See adopted a pragmatic policy with respect to evolution. No energetic or proactive intervention was necessary, inasmuch as Catholic theologians were nearly unanimous in their opposition to evolution, whose proponents were, for the most part, non-Catholics. Moreover, the most ardent supporters of evolution often presented a strongly combative air, hostile to religion, so that it was not necessary to expressly condemn their works: it was evident that they opposed the Church. They defended not only evolution as a biological theory, but also as an ideology that made claims for all human reality, including religion, and which was rejected as a matter of course by Catholics.

Evolutionism was viewed by many Catholic theologians as a materialist and agnostic ideology based on a scientific theory that had no serious foundation. This ideology seemed opposed to Christian doctrine on the Bible, on Creation and divine action in the world, and on human beings. There was a consensus among theologians about Catholic doctrine, including aspects that, without expressly being dogmas of faith, were held to be closely related to them, and evolution, as it was presented by its most ardent proponents, certainly clashed with dogmas of faith and with other positions generally held by theologians.

The literal interpretation of Scripture was one of these aspects. Throughout the nineteenth century, as discoveries relating to ancient civilizations were discussed and debated, there arose problems regarding the literal and historic meaning of various passages of the Bible. This gave rise to an ongoing debate known in the Catholic world as "the biblical question." In 1893 Leo XIII's encyclical *Providentissimus Deus* countered interpretations that seemed excessively "liberal," although, at the same time, it offered moderate guidelines with respect to new issues. Evolution presented a direct challenge to the Genesis Creation narrative, particularly with respect to the special intervention of God in the creation of the body of the first man. There certainly existed in the Catholic tradition elements sufficient to reconcile evolution with the Bible, and there was no lack of Catholic theologians who proposed to resolve the difficulties. Nevertheless, the more severe attitude generally prevailed in the nineteenth century, only to progressively soften in the course of the twentieth.

Evolution could be reconciled with Creation and divine activity in the world by introducing the traditional distinction between God as "Prime Cause" and created "secondary causes" that act in accordance with the laws that God impressed upon nature. This was the means adopted by those who said that evolution could be harmonized with Christianity. But the stricter perspectives also prevailed here. Special creation of species was perceived as more in line with Christian doctrine and with tradition. This was especially clear with respect to the origin of the human body. The Catholic position on this issue also softened in the course of the twentieth century.

The predominance of more traditional attitudes is probably related to the predictable consequences that evolution could have with respect to Catholic doctrine, which affect not only the issues just mentioned, but others as well—for example, the doctrines of the human soul and original sin. In the closing decades of the nineteenth century a consensus on these subjects prevailed among Catholic theologians. Those who confronted evolution could not refrain from warning that, if evolution were accepted, problems would arise in these other areas. It could have been relatively easy to harmonize evolution and Christianity by proposing that Scripture does not speak of science and by utilizing the distinction between primary and secondary causality. But it was more complex to harmonize evolution with the existence of an immortal, spiritual soul in each human being, and with the existence of original sin committed by a first human couple and transmitted by generation to all humankind. In all probability, these problems were all present in the frontal rejection of evolutionism in the nineteenth century. Catholic harmonizers accepted the Catholic doctrine on the soul and original sin, but it was obvious that evolution could cause problems in these areas.

It is no wonder that, in these circumstances, the position of Roman authorities was pragmatic. One might even think that they preferred to avoid condemnations, preferring to use less controversial measures, such as personal retractions that did not directly compromise the Vatican but which still were sufficient in those times to slow the diffusion of evolutionary ideas. In a period when an article in *La Civiltà Cattolica* could be taken as authoritative for theology, it was not necessary to adopt more severe measures that might, moreover, have seemed risky.

In Galileo's Shadow

The shadow of Galileo surely fell over the evolution controversy. The Roman authorities were aware that they ought not interfere with science. In Galileo's epoch modern science was just beginning. But since Galileo's condemnation in 1633, much had changed. The publication of Newton's *Principia* in 1687 signaled the definitive consolidation of a kind of science that, since then, has made spectacular advancements. Part of the achievement of the Scientific Revolution was to have instated an approach to science whereby change and the replacement of less accurate by more accurate paradigms had become expected. Especially in that context, and in view of the ever-present shadow of the Galileo affair, it was only prudent to avoid anything that looked like a clash with science.

To do this, the simplest tactic was to deny that evolutionism was an authentic science. This was repeatedly stated in the decades after 1859. In the battle that the Jesuits of *La Civiltà Cattolica* unleashed against evolutionism, this argument played a key role, until well into the twentieth century. Those Catholics who argued for the compatibility between evolution and Christianity were reproached for having favored above their faith a doctrine that, although it was presented as scientific, was nothing more than a concatenation of falsehoods lacking any serious basis.

Certainly, there were aspects of evolution that remained cloaked in mystery. Only in the twentieth century did the development of genetics permit a realistic analysis of variation and heredity, using resources that were lacking in the nineteenth century. The age of the Earth was another speculative point. It was not difficult to find scientists of some prestige who added their voices to critiques of evolution in general or of Darwinism (natural selection) in particular. Catholic theologians required that, before posing theological problems of great significance, a theory that seemingly lacked solid proof be analyzed with a critical spirit.

It is easy to confirm Galileo's shadow in the way evolution was assessed, not only by its critics but by its own partisans. As we have seen, Caverni cited in his favor ideas that Galileo had proposed for scriptural interpretation, emphasizing that it is not the intention of the Bible to teach us scientific truths but rather to show us the road to salvation. In the written report of the General Congregation in which Caverni's work was examined, it was expressly warned that shouts would be heard against the condemnation of Caverni, alluding to the example of Galileo and suggesting that the Congregation of the Index is not a competent tribunal to pass judgments on science. Leroy said that the same thing would happen to evolution as happened with Galileo: that after having initially alarmed the orthodox, truth would be distinguished from the exaggerations of both sides, and it would end with the road opened. On examining Leroy's book, Tripepi said he did not understand how Leroy could compare evolution with the Galileo case, because Galileo, although he made some errors in the way he presented his theory, found support in scientific arguments as well as in many texts, well interpreted, of Scripture. Buonpensiere commented that in the case of Galileo there had been no persistent condemnation of the system he defended, and he was condemned less for the substance of his argument than on the way he defended it. Zahm thought that what stopped the publication of the decree of condemnation of his book was the fear that it could be considered as a condemnation of evolution and his case might be compared with that of Galileo. Bonomelli said that, just as the ideas of Galileo opened new perspectives, the same would happen with evolution. In one sense or another, the comparison of the Galileo case with problems caused by evolutionism was in the back of the participants' minds.

An evident similarity between the two cases is that both were theories that were not as yet firmly established among scientists. In the case of Galileo, in the seventeenth century, the eleven theologians of the Holy Office whose opinion was sought began by saying that the theory (that the Earth moved) was absurd; once this proposition was introduced, it was easy to dismiss it from the theological point of view. In the case of nineteenth-century evolutionism, theologians presented a whole series of scientific objections, supported by references to, and statements by, scientists, to make the point that it was not logical to base an appeal to theology on such a weak foundation. In both cases, as the scientific arguments grew stronger, theological resistance decreased.

A notable difference between the two cases is that in the case of Galileo the Roman authorities proceeded expeditiously, both in 1616 and in 1633. In 1616 they ordered the suspension of Copernicus's book until required changes were made; these were approved four years later and consisted in a few minor changes in the text, to present the movement of the earth as a simple hypothesis. In 1633 they listed the *Dialogo* of Galileo in the *Index of Prohibited Books* and condemned its author to house arrest. In both instances the decisions were made quite quickly, and there was no calm, objective discussion of heliocentrism. By contrast, theological opposition to evolution lasted for several decades and, despite its ardor, it led to no public condemnation of evolution.

The most noteworthy episodes, those of Leroy and Zahm, were settled by a letter of retraction from Leroy and a letter from Zahm to his Italian translator. The Roman authorities had no difficulty in condemning the works of Leroy and Zahm, but decided not to publish the condemnations and were content that it become known, via the authors of these books, that the Holy See had opposed their circulation. In the case of Caverni, the decree prohibiting his book was published, but it had no effect, since it mentioned only the book's title, which did not mention evolution. The deliberations and votes that occurred in the cases studied show that opposition to evolution prevailed in the Congregation of the Index, but there were notable differences in the stances of individual members. It was generally recognized that the official Magisterium of the Church took no position on evolution, and it was even suggested that the Holy Office be asked for an opinion, as a leading doctrinal authority. But such an opinion was never sought, nor did the Holy Office pronounce.

Whatever other motives there may have been, the desire not to compromise the authority of the Church in an issue related to science was one of the reasons for the blandness of the measures adopted. For the rest, the analysis of the six cases clearly shows that the way events developed was quite complex: it is enough to recall the multiple deliberations of the Congregation of the Index in the case of Leroy, and the lengthy negotiations made both for and against Zahm, reaching the highest level of authority, and with results that hung in the balance for a long time.

NOTES

Introduction

1. John Paul II, Address to the Pontifical Academy of Sciences, October 22, 1996, p. 371, n. 3.

2. Ibid., p. 372. Instead of "new knowledge has led to recognize that the theory of evolution is more than a hypothesis," the English version of the text, as published in the English edition of *L'Osservatore Romano* and in the book we quote, says, "new knowledge has led to the recognition of more than one hypothesis in the theory of evolution." This seems to be a misunderstanding of the original French text: "de nouvelles connaissances conduisent à reconnaître dans la théorie de l'évolution plus qu'une hypothèse" (*L'Osservatore Romano*, October 24, 1996, p. 6). The Italian translation in the issue just quoted says: "conducono a non considerare più la teoria dell'evoluzione una mera ipotesi" (p. 7). In all cases, the final part of the paragraph refers to "this theory," in the singular. Father Robert Dempsey, editor of the English-language version of *L'Osservatore*, said on November 19, 1996, that the newspaper had published an overly literal translation of the French-language message that "obscures the real meaning of the text." He added that, "The Pope's real meaning was that it is now possible to recognize that the theory of evolution is more than a hypothesis." Moreover, the whole paragraph clearly shows the meaning of the phrase.

3. Messenger, *Evolution and Theology.*

4. See, for example, Ferngren, *Science and Religion.*

5. See, for example, Moore, *The Post-Darwinian Controversies;* Roberts, *Darwinism and the Divine in America.*

6. Lindberg and Numbers, *God and Nature.*

7. Brooke, *Science and Religion.*

8. Ibid., p. 5.

Chapter 1. The New Documents

1. For a historical and juridical account of both congregations, see Del Re, *La Curia romana*, pp. 89–101 ("S. Congregazione per la Dottrina della Fede") and 325–29 ("Congregazione dell'Indice").

2. On the history of the various editions of the *Index of Prohibited Books*, see Bujanda, *Index de Rome, 1557, 1559, 1564*, pp. 27–44 ("Introduction").

3. For an edition containing all the books listed on the Roman *Index* from 1600 to 1966, with a comprehensive introduction, see Bujanda and Richter, *Index librorum prohibitorum 1600–1966*. Bujanda has also published other studies on the editions of the *Index* published in Anvers, Spain, Louvain, Milan, Paris, Portugal, and Venice.

4. See the Apostolic Letter of Paul VI, titled *Integrae Servandae*. The specification on the Index is from a *notificatio* of the Congregation for the Doctrine of the Faith, dated June 14, 1966: *Acta Apostolicae Sedis* 58 (1966): 445: "eundem tamen non amplius vim legis ecclesiasticae habere cum adiectis censuris."

5. We use the term "Congregation" to designate both the organism or Congregation of the Index, as well as its meetings, the Preparatory and General Congregations: in each instance we will make clear to which we refer.

6. "Propaganda Fide," also called simply "Propaganda."In the period we discuss, the Church of the United States was a dependency of this congregation.

7. We will cite this archive as ACDF (Archives of the Congregation for the Doctrine of the Faith), then indicate whether the Holy Office (S. Oficio) or the Index (Index).

8. Darwin, in his biography of his grandfather, noted that *Zoonomia* "was honoured by the Pope by being placed in the 'Index Expurgatorius'"; *Charles Darwin's The Life of Erasmus Darwin*, p. 35. Note that when the president of the Royal Society stated that the *Origin of Species* had been expressly excluded from the grounds on which the Copley Medal was awarded to Darwin in 1864, T. H. Huxley complained that such an exclusion was the equivalent of placing the work on an *index expurgatorius*; see *The Correspondence of Charles Darwin*, vol. 12: *1864*, pp. 447, 452.

9. *Index librorum prohibitorum*, p. 7.

10. One of the two bishops was a Benedictine monk.

11. ACDF, Index, Protocolli, 1894–96, fol. 128, p. 1.

12. Jacolliot, *Christna et le Christ*, p. 1.

13. For some of the political baggage acquired by evolutionism in England and France, see Desmond, *The Politics of Evolution*.

14. Scheeben, *Handbuch der katholischen Dogmatik*, book III: *Schöpfungslehre*, pp. 160–61.

15. Mazzella, *De Deo creante*, pp. 343–74.

16. Charles Darwin, *Autobiography*, in Darwin and Huxley, *Autobiographies*, p. 71.

17. Bowler, *The Eclipse of Darwinism*, pp. 3–19.

18. Huxley, *Evolution*, p. 23.

19. Ibid., pp. 23–24.

20. We cite the documents of this council in the 1961 reprint edition, which reproduces exactly an edition of 1923: Mansi, *Sacrorum Conciliorum collectio*, cols. 25–166.

21. Ibid., cols. 25–46.

22. The delay was probably due to the fact that Caterini was named prefect of the Congregation of the Council on September 26, 1860. Therefore, when the acts of the council reached Rome, he was not yet prefect. He must have begun his review shortly after starting work in his new office.

23. Mansi, *Sacrorum Conciliorum collectio*, cols. 67–74.

24. Ibid., col. 72.

25. Wernz, *Ius decretalium*, 2:1091 (emphasis in original text).

26. Mansi, *Sacrorum Conciliorum collectio*, col. 91.

27. Pesch, *Praelectiones dogmaticae quas in Collegio Ditton-Hall habebat*, 3:58–59. In the notes, Pesch cites Leroy's book and letter as well as the book by Zahm.

28. Tanquerey, *Synopsis Theologiae Dogmaticae Specialis*, 2:504–5.

29. Beraza, *Tractatus de Deo creante*, 467–76.

30. Boyer, *Tractatus de Deo creante et elevante*, p. 186.

31. Rahner, "De Deo creante et elevante et de peccato originali," p. 76.

32. Parente, *De creatione universali*, p. 73.

33. Ruiz de la Peña, *Imagen de Dios*, p. 251.

34. Alszeghy, "El evolucionismo y el Magisterio de la Iglesia," pp. 367–68.

35. Ramon Juste, "La teología católica y el problema de la evolución humana," pp. 393–414.

36. Messenger, *Evolution and Theology*, pp. 232–39.

37. Brandi, "Evoluzione e Domma. Erronee informazioni di un inglese," pp. 75–77.

38. Ibid., p. 77.

39. Bricarelli, "La dissoluzione dell'evoluzione," pp. 668–86.

40. See De Rosa, *La Civiltà Cattolica*.

41. Zahm, *Evolution and Dogma*, pp. 352–67.

42. Brandi, "Evoluzione e Domma. Erronee informazioni di un inglese," p. 76.

43. Brundell, "Catholic Church Politics and Evolution Theory, 1894–1902," pp. 81–95.

44. Harrison, "Early Vatican Responses to Evolutionist Theology."

Chapter 2. An Ineffective Decree

1. *Acta Sanctae Sedis* 11 (1878): 204 (Rome: Typis Polyglottae Officinae S. C. De Propaganda Fide; reprint, New York and London: Johnson Reprint Corporation, 1968).

2. Pagnini, *Profilo di Raffaello Caverni*, p. 43.

3. Cappelletti and di Trocchio, "Caverni, Raffaello," *Dizionario biografico degli italiani*, 23:86.

4. For biographical details, see the works of Pagnini and of Cappelletti and di Trocchio, cited in nn. 2 and 3, and also, Betti, *Raffaello Caverni, 1837–1900*.

5. Caverni, *Storia del metodo sperimentale in Italia*, 6 vols. Five volumes were published in Caverni's lifetime, and a sixth, incomplete, after his death.

6. See an interesting appreciation of Caverni by Castagnetti and Camerota, "Raffaello Caverni and His *History of the Experimental Method in Italy*," pp. 327–39.

7. Garin, *Scienza e vita civile nel Rinascimento italiano*, p. 60.

8. On evolution in Florence, see Landucci, *Darwinismo a Firenze*; and in Italy generally, Pancaldi, *Darwin in Italy*. The Italian translation of *Origin of Species* was published in 1864 as *Sull'origine delle specie per elezione naturale* (Modena: Zanichelli), translated by Giovanni Canestrini and Leonardo Salimbeni. On Canestrini, who subsequently translated most of Darwin's books into Italian, see Minelli and Casolati, *Giovanni Canestrini: Zoologist and Darwinist*.

9. Caverni, *De' nuovi studi della Filosofia*. In 1878 the *Rivista Universale* changed its name to *La Rassegna Nazionale*.

10. Ibid., pp. 24–28.

11. Ibid., p. 172.

12. Salis Seewis, review of *De' nuovi studi della Filosofia. Discorsi di Raffaello Caverni a un giovane studente*, *La Civiltà Cattolica*: (I) 10th ser., 4 (1877): 570–80; (II) 10th ser., 5 (1878): 65–76.

13. Ibid. (II), p. 66.

14. Ibid., p. 67.

15. The articles, in forty-one sections, were published in *La Civiltà Cattolica*, 10th ser., 5 (1878): 52–64, 160–73, 288–97, 527–39; 10th ser., 6 (1878): 17–34, 269–78, 685–96; 10th ser., 7 (1878): 166–76, 432–43, 674–91; 10th ser., 8 (1878): 158–71, 397–410, 670–82; 10th ser., 9 (1879): 158–70, 324–34, 556–69; 10th ser., 10 (1879): 35–45, 291–301, 542–55; 10th ser., 11 (1879): 19–28, 174–82, 284–94, 579–89; 10th ser., 12 (1879): 33–48, 291–300, 548–59; 11th ser., 1 (1880): 142–54, 411–23; 11th ser., 2 (1880): 34–44, 272–84, 560–71; 11th ser., 3 (1880): 40–56, 273–83, 538–52, 680–95; 11th ser., 4 (1880): 38–51, 159–71. All published references to this series of articles are incomplete, including those in the indexes of *La Civiltà Cattolica*.

16. Caterini, *Dell'origine dell'uomo secondo il trasformismo*. See the review in *La Civiltà Cattolica*, 12th ser., 6 (1884): 73–76.

17. Pietro Caterini, "La scienza e l'uomo bestia," XXXIV, *La Civiltà Cattolica*, 11th ser., 2 (1880): 274–77.

18. Pietro Caterini, "La scienza e l'uomo bestia," XXXV, ibid., pp. 564–65.

19. Pietro Caterini, "D'alcuni principii filosofici rispetto al trasformismo," XXXVII, *La Civiltà Cattolica*, 11th ser., 3 (1880): 279–80.

20. Pietro Caterini, "Come entrino la fede e la teologia nella questione trasformistica," XXXIX, ibid., p. 681.

21. Pietro Caterini, "Come entrino la fede e la teologia nella questione trasformistica," XLI, *La Civiltà Cattolica*, 11th ser., 4 (1880): 171.

22. ACDF, Index, Protocolli, 1875–78, fol. 342.

23. ACDF, Index, Atti e Documenti della S. C., 1878–85, fol. 4.

24. Salis Seewis, reviews of *Du Darwinisme, ou l'Homme Singe*, by C. James, and *Le Darwinisme. Ce qu'il y a de vrai et de faux dans cette théorie*, by E. de Hartmann, pp. 449–58.

25. Zigliara, *Summa philosophica in usum scholarum*, 2:148–53.

26. Zigliara, *Propaedeutica ad Sacram theologiam in usum scholarum*, pp. 27–28.

27. ACDF, Index, Protocolli, 1878–81, fol. 71, pp. 18–19.

28. Darwin reacted with indignation at Charles Lyell's penchant for referring to "Lamarck's theory improved by yourself." Darwin to Lyell: "Lastly you refer repeatedly to my view as a modification of Lamarck's doctrine of development and progression. . . . I believe this way of putting the case is very injurious to its acceptance, as it implies necessary progression . . .; I must add that Henrietta, who is a first rate critic & to whom I have *not said a word* about Lamarck, last night said, 'Is it fair that Sir C. Lyell always calls your theory a modification of Lamarck's?'" (Darwin to Lyell, March 12–13, 1863, *The Correspondence of Charles Darwin*, vol. 11: *1863*, p. 223).

29. In scholastic philosophy "estimative" refers to *vis* or *facultas aestimativa*, which is the equivalent in animals of the cogitative sense in humans—what today we call instinct.

30. ACDF, Index, Protocolli, 1878–81, fol. 71, p. 18.

31. Ibid., pp. 18–19.

32. Ibid., fol. 66. See also ACDF, Index, Diari, vol. XX, p. 202.

33. ACDF, Index, Diari, vol. XX, p. 203.

34. ACDF, Index, Protocolli, 1878–81, fol. 73.

35. Probably Segismund Wiese, whose drama *Jesus* was placed on the *Index* by the decree of August 8, 1898.

36. ACDF, Index, Diari, vol. XX, p. 204. A copy of the decree can be found in ACDF, Index, Protocolli, 1878–81, fol. 76.

37. The letter is reproduced in Pagnini, *Profilo di Raffaello Caverni*, pp. 40–41.

38. Ibid., pp. 41–43.

39. Caverni, *Dell'antichità dell'uomo secondo la scienza moderna*.

Chapter 3. Retraction in Paris

1. A necrological note published in *Annales dominicaines*, 1905, p. 289, gives the date as January 15, 1828. *Archivum Fratrum Praedicatorum* 31 (1961): 360, gives January 28, and January 25 appears in another document in the Dominican archives.

2. To add "Marie" before "Dalmace" was a Dominican custom, to display the veneration of the Virgin Mary. We are grateful to Fr. Michel Albaric, archivist of the Dominican Province of France, for copies of the documents of August 28, 1851, and August 28, 1852.

3. Hinnebusch, *The Dominicans*, pp. 151, 155–59; Bonvin, *Lacordaire-Jandel*.

4. Duval, "Le rapport du P. Lacordaire au Chapitre de la Province de France (Septembre 1854)," pp. 326–64. In this list Leroy appears as no. 58 (p. 360).

5. *Annales dominicaines*, 1905, p. 290.

6. Leroy, *L'évolution des espèces organiques*.

7. Leroy, *L'évolution restreinte aux espèces organiques*. Hereafter references are cited in text.

8. Conry, *L'introduction du darwinisme en France au XIXᵉ siécle*, p. 474.

9. Mansi, *Sacrorum Conciliorum collectio*, col. 91.

10. On Bonniot, see Patrick Tort, "Bonniot, Joseph de," in Tort, *Dictionnaire du darwinisme et de l'évolution*, p. 377.

11. Bonniot, "Le transformisme et l'athéisme," pp. 428–40.

12. Ibid., p. 428. In another article, Bonniot defends the fixity of species and states that evolution is absurd: Bonniot, "Essai philosophique sur le transformisme," pp. 337–68. In his book, Leroy had criticized Bonniot's ideas on species, saying that they were not demonstrated, and that the whole problem revolves around this issue (Leroy, *L'évolution restreinte aux espèces organiques*, pp. 85–93).

13. Bonniot, *La bête, question actuelle*, a collection of articles; and *La bête comparée à l'Homme*.

14. On Brucker, see Patrick Tort, "Brücker, Joseph," in Tort, *Dictionnaire du darwinisme*

et de l'évolution, p. 445. Tort writes "Brücker." We use the form "Brucker," which is how it appeared in *Études.* The variation is probably due to the Alsatian origin of Brucker.

15. Brucker, "Les jours de la création et le transformisme," pp. 567–92; "L'origine de l'homme d'après la Bible et le transformisme," pp. 28–50. The quotation from the second article is from p. 43.

16. Brucker, "L'origine de l'homme," p. 43.

17. Brucker, "Bulletin Scripturaire," pp. 488–97.

18. Quoted in ibid., p. 492.

19. Ibid., pp. 496–97.

20. Gardeil, "L'évolutionnisme et les principes de S. Thomas," p. 30.

21. Portalié, "Le R. P. Frins et la *Revue Thomiste,*" pp. 58–59.

22. Leroy, "Correspondance au R. P. directeur de la *Revue Thomiste,*" pp. 532–35.

23. Ibid., pp. 534–35. Leroy uses the term *daltonisme,* meaning color-blindness, after the chemist, John Dalton (1766–1844), who described it.

24. Dierckx, *El hombre-mono y los precursores de Adán.* The original was published in 1894 (*L'homme-singe et les précurseurs d'Adam en face de la science et de la théologie*).

25. Ibid., pp. 136–37.

26. Ibid., pp. 139–40.

27. Ibid., pp. 146–48.

28. Letter of Ch. Chalmel to the Congregation of the Index, June 20, 1894: ACDF, Index, Protocolli, 1894–96, fol. 71.

29. The title "Officier d'Académie" is one of the honorary titles ("les palmes académiques") created by Napoleon in 1808 as awards in the field of teaching and research. It was intended mainly for high school teachers and staff, but from 1866 it might be granted also to other people who have made important contributions to the educational world.

30. At the time, the archivist and *minutante* was Antonio Gandolfi. From 1898 on, the *Annuario Pontificio* records a second official, Salvatore de Angelis, with the post of *protocollista.*

31. ACDF, Index, Atti e Documenti della S. C., 1886–97, fol. 254.

32. Ibid., fol. 255.

33. ACDF, Index, Diari, vol. XXII, fol. 8r.

34. Ibid., fol. 3v.

35. ACDF, Index, Protocolli, 1894–96, fol. 88, p. 2.

36. Domenichelli himself says so at the beginning of his report.

37. ACDF, Index, Protocolli, 1894–96, fols. 86 and 128. The archive contains two copies of the report; the second was used in a later stage of the case. Hereafter references to folio 86 are cited in text.

38. Cf. Brundell, "Catholic Church politics and evolution theory," 87.

39. Domenichelli provides a brief quotation from *In Secund. Sent.,* d. 12, a. 1, q. 11.

40. The idea mentioned by Domenichelli refers to Saint Augustine, *De Genesis ad litteram,* lib. I, 18–19.

41. ACDF, Index, Diari, vol. XXII, fol. 6r: Et quoad 3ᵐ omnes unanimiter fuerunt in voto: Dilata; et scribat alter, reassumptis ad rem facientibus. Also in ACDF, Index, Protocolli, 1894–96, fol. 82.

42. ACDF, Index, Diari, vol. XXII, fol. 6r. Pierotti was named a cardinal in 1896 and continued in the Congregation of the Index as cardinal member.

43. Ibid., fol. 6r–v.

44. ACDF, Index, Protocolli, 1894–94, fol. 88.

45. ACDF, Index, Diari, vol. XXII, fol. 7r.

46. ACDF, Index, Protocolli, 1894–96, fol. 88.

47. ACDF, Index, Diari, vol. XXII, fol. 6v.

48. ACDF, Index, Protocolli, 1894–96, fols. 123 and 124 (the archive preserved two copies of the report). Hereafter references to folio 123 are cited in text.

49. Fontana cites Cuvier, Linnaeus, Agassiz, Quatrefages, Favre, Andrassy, Brucker, Jousset, Lavaud, Contejeau, and Bonniot.

50. Leroy, *L'évolution restreinte aux espèces organiques*, p. 248.

51. ACDF, Index, Protocolli, 1894–96, fols. 125 and 126 (two copies). Hereafter references to folio 125 are cited in text. The report is divided into seventeen sections numbered at the margin. One can distinguish three principal parts, each covering one of the three questions formulated by the General Congregation, plus a short introduction (§1) and a few concluding points (§§15–17). The only peculiarity of organization is that he discusses the three points in reverse order, and with very uneven coverage. First he considers the "Doctrine of the Author on the Formation of the First Man" (§§2–8, pp. 2–31). Then he turns to the "System of Evolution of Organic Species, as Proposed by the Author" (§§9–13, pp. 31–44), and he ends with the "Exegetical Criteria That the Author Believes Should Be Followed in the Interpretation of Genesis" (§14, pp. 44–49). The order of the answers, and their decreasing length provides a clue to Tripepi's strategy.

52. The reference is to Mazzella, *De Deo creante*, §496, p. 344.

53. Jules Fabre d'Envieu (1821–1901) was professor of sacred history at the Sorbonne.

54. John Gmeiner, born in Bavaria in 1847, later emigrated to the United States, where he was professor of theology in seminaries in Milwaukee (Wisconsin) and St. Paul (Minnesota). Gmeiner was an early Catholic partisan of evolution, "the most progressive of the liberals before Zahm" with respect to evolution, and he advocated a "minimalist interpretation of ecclesial authority" in matters where science and theology were at odds; Appleby, "*Church and Age Unite!*" pp. 23–25. Gmeiner called for "a believer's interpretation of Darwinism"; Appleby, "Exposing Darwin's 'Hidden Agenda': Roman Catholic Responses to Evolution, 1875–1925," p. 185.

55. Leo XIII's encyclical *Providentissimus Deus*, on biblical studies (November 18, 1893).

56. Cf. Mazzella, *De Deo creante*, §§513–14, pp. 352–53.

57. The text closely followsMazzella's wording in *De Deo creante*, §514, p. 353.

58. Several pages later Tripepi again mentions the article in which Leroy tackles the question of the formation of Eve's body, "seeking to defend himself from Brucker's criticisms, Leroy was obliged to proclaim the immediate act of God in the formation of the body of Eve" (ACDF, Index, Protocolli, 1894, p. 30) Tripepi refers to a response to Brucker that Leroy published in *La Science Catholique* (February 1892). But, in fact, Leroy had stated that no believer could doubt that Eve's body was created from that of Adam (Leroy, *L'évolution restreinte aux espèces organiques*, pp. 156–57).

59. Cf. Mazzella, *De Deo creante*, §§517–19, pp. 356–64.

60. The texts cited appear in Mazzella, *De Deo creante*, pp. 351, 360–61, 362, 363.

61. W. Devivier, *Cours d'apologétique chrétienne*, p. 73.

62. Leo XIII, encyclical *Providentissimus Deus*, p. 282.

63. Ibid., pp. 286–87.

64. Alfred Firmin Loisy (1857–1940) was the principal exponent of biblical modernism in France.

65. Leo XIII, encyclical *Providentissimus Deus*, pp. 288–89.

66. According to d'Hulst in a letter of December 7, 1893: "I have learned that Cardinal Mazzella made an authentic scene before the Pope in order to put me on the *Index* by name, and that the Holy Father energetically declined to do so, and for this I am grateful to him with the affection of a son"; cited by Baudrillart, *Vie de Mgr. d'Hulst*, 2:172–73. Francesco Beretta has identified Countess Frignet as the recipient of the letter; Beretta, *Monseigneur d'Hulst et la science chrétienne*, p. 417, where there is another letter by d'Hulst with a text similar to the one cited.

67. Guglielmo D'Ambrogi (1833–95) was named bishop of Tivoli two months later, in March 1895, but he died several months later.

68. ACDF, Index, Diari, vol. XXII, fol. 9r. The master general of the Dominicans then was Andreas Frühwirth (1845–1933).

69. Barry Brundell refers to this memorandum as if it had been written by Vannutelli: "Catholic Church Politics and Evolution Theory," 88. But any possible doubt is resolved merely by reading the heading: "Memorandum for the Most Eminent Cardinal Prefect" (*Pro-memoria per l'Emo. Card. Prefetto*). The error could have been caused by another error of the archive itself, because in the index of the volume in which this document is found the heading is written, wrongly: "Pro-memoir of the Most Eminent Cardinal Prefect" (*Pro-memoria dell'Emo. Card. Prefetto*). Cf. ACDF, Index, Protocolli, 1894–96, volume index.

70. ACDF, Index, Protocolli, 1894–96, fol. 127.

71. Ibid., fol. 119.

72. Ibid., fols. 117–18. Folio 117 contains only the title of the report (the verso is blank). The next page was mistakenly numbered 118, as if it were a new document.

73. Ibid., p. 3. In reality, the Preparatory Congregation had taken place on Thursday the 17th.

74. Ibid. (emphasis in the original text).

75. Ibid., p. 4 (emphasis in the original text).

76. *On the Origin of Species: A Facsimile of the First Edition*, pp. 245, 248, 252. Darwin's fullest discussion of hybrid sterility is in *The Variation of Plants and Animals under Domestication*, 2: chaps. 15–19.

77. "Notes on the Causes of Cross and Hybrid Sterility," in *The Correspondence of Charles Darwin*, vol. 10: *1862*, pp. 700–711. The notion that new species arise through hybridization is an old one. Linnaeus thought so, and so did Cesalpino.

78. From the Darwin's "Big Species Book" (1856), quoted by Harvey, "Fertility or Sterility?" p. 58.

79. Hull, *Science as a Process*, p. 103.

80. Cited by Zahm, *Evolution and Dogma*, p. 184.

81. Ibid., pp. 188–91.

82. ACDF, Index, Protocolli, 1894–96, fols. 117–18, p. 4.

83. Ibid., p. 8. Buonpensiere drafted the proposed phrasing in Latin: "*Leroy Opus aut proscribendum, vel si mitius supprimendum, et interim Auctor moneatur*" (emphasis in the original).

84. Brundell, "Catholic Church Politics and Evolution Theory," p. 88.

85. ACDF, Index, Diari, vol. XXII, fol. 9r.

86. ACDF, Index, Protocolli, 1894–96, fol. 131.

87. Ibid., fol. 128.

88. Ibid.

89. ACDF, Index, Diari, vol. XXII, fol. 9v.

90. ACDF, Index, Protocolli, 1894–96, fols. 132 and 133.

91. ACDF, Index, Diari, vol. XXII, fol. 10r.

92. In a letter from Leroy to Cardinal Steinhuber, February 2, 1897: ACDF, Index, Protocolli, 1897–99, fol. 53.

93. There is a copy of the letter, as it was published in *Le Monde,* in the Archive of the Congregation of the Index: ACDF, Index, Protocolli, 1894–96, fol. 134.

94. ACDF, Index, Protocolli, 1894–96, fols. 132 and 133. A complete issue of the newspaper was archived among other miscellaneous documents: cf. ACDF, Index, Atti e Documenti della S. C., 1886–97, fol. 202.

95. ACDF, Index, Diari, vol. XXII, fol. 10r.

96. ACDF, Index, Protocolli, 1894–96, fol. 169.

97. At the head of a copy of Tripepi's report, Cicognani wrote: "By permission authorized by the Holy Father, this has been given to Father Leroy to read, with the obligation to return it. February 20, 1895. Father Secretary": ACDF, Index, Protocolli, 1894–96, fol. 126, p. 1. A scrap of paper was pasted over the name of the consultor, so that Leroy could not identify it.

98. ACDF, Index, Atti e Documenti della S. C., 1886–97, fol. 262.

99. Brucker, *Questions actuelles d'Écriture Sainte,* pp. vii–viii.

100. Ibid., pp. ix–x.

101. ACDF, Index, Protocolli, 1897–99, fol. 53.

102. Ibid., fol. 61.

103. ACDF, Index, Diari, vol. XXII, fol. 27r, March 26, 1897.

104. The index of the archive wrongly indicates that it was addressed to Cicognani.

105. ACDF, Index, Protocolli, 1897–99, fol. 52.

106. Ibid., fol. 54.

107. Ferrata first wrote that publication "ought not be permitted," but then substituted "cannot."

108. ACDF, Index, Diari, vol. XXII, fol. 29r, June 19, 1897.

109. ACDF, Index, Protocolli, 1897–99, fol. 55.

110. Ibid., pp. 5–6. Buonpensiere adds parenthetically: "I do not quite remember under what title of the *Index* the works of Darwin are to be found. I think under the name of an Italian translator." Buonpensiere was wrong. No book of Charles Darwin was listed on the *Index.* He may have confused Charles with his grandfather, Erasmus, whose *Zoonomia* was listed because it was considered materialist.

111. Marselli, *Origini dell'umanità; Le grandi razze dell'umanità.*

112. See the report of the consultor, Eusebio de Monte Santo, Capuchin, for the Preparatory Congregation of June 18, 1881; ACDF, Index, Protocolli, 1878–81, fol. 201.

113. ACDF, Index, Protocolli, 1882–84, fol. 5.

114. "This is the same power that will be communicated to the soul of each individual on the last day, to resuscitate our ashes in an instant: *in ictu oculi,* as the Apostle puts it" (Buonpensiere report, ACDF, Index, Protocolli, 1897–99, fol. 55, p. 47).

115. ACDF, Index, Diari, vol. XXII, fol. 30v, August 14, 1897.

116. ACDF, Index, Protocolli, 1897–99, fol. 61.

117. ACDF, Index, Diari, vol. XXII, 1902, p. 98, January 7, 1902.

118. Leroy, *Lettre à M. l'Abbé A. Farges.* We are grateful to Fr. Michel Albaric, archivist of the Dominican Province of France, for a photocopy of this publication.

119. According to Saint Thomas, the human body has no existence distinct from that of the soul, that is, it does not exist without the soul. Hence the human body does not exist as such until the complete human being comes into existence, which presupposes the infusion of the human soul.

120. Marie-Dalmace Leroy, review of *L'origine des espèces,* by J. Guibert, pp. 735–41.

121. ACDF, Index, Diari, vol. XXII, 1901, p. 93, November 21, 1901.

122. ACDF, Index, Protocolli, 1900–1902, fol. 197.

123. ACDF, Index, Diari, vol. XXII, 1901, p. 97, December 22, 1901.

124. Ibid., 1902, p. 98, January 7, 1902.

125. ACDF, Index, Protocolli, 1900–1902, fol. 198.

126. ACDF, Index, Diari, vol. XXII, 1902, p. 99, January 13, 1902.

127. ACDF, Index, Protocolli, 1900–1902, fol. 199.

128. González de Arintero, *La evolución,* Introducción general, vol. 1. On Arintero, see Huerga, "La evolución," pp. 127–53, and Alonso Lobo, *Padre Arintero. Un maestro di vita spirituale* and *El Padre Arintero, precursor clarividente del Vaticano II.*

129. Cited by González de Arintero, *La evolución,* p. 93.

130. Ibid.

131. Ibid., pp. 100–101, 104–6.

132. Arintero accepted all of what today we call microevolution, by observing that that the standard Thomist conception of species is not the Linnaean species but the Linnaean class.

133. González de Arintero, *La evolución,* p. 162.

134. Ibid., p. 161.

135. Brandi, "Evoluzione e domma," p. 48.

136. Arnould, "Le singe, le dominicain et les jésuites," pp. 255–64. See also Arnould, *Darwin, Teilhard de Chardin et Cie.*

137. Harrison, "Early Vatican Responses to Evolutionist Theology."

138. Ibid., pp. 7, 9 (nn. 40, 56, with the corresponding texts), referring to Zubizarreta, *Theologia dogmatico-scholastica,* 2:479, nn. 5, 6.

139. Zubizarreta refers to Brandi's article in *La Civiltà Cattolica,* "Evoluzione e domma," where Leroy's letter is reproduced.

140. For a listing of all the works placed on the Roman *Index of Prohibited Books* since

1600, with the date of the corresponding decree indicated, see Bujanda and Richter, *Index librorum prohibitorum 1600–1966*.

141. Harrison, "Early Vatican Responses to Evolutionist Theology," p. 9.

142. Brundell, "Catholic Church Politics and Evolution Theory," pp. 81–95.

143. Ibid., pp. 87–88. In note 19, he refers to "the complete documentation" that is to be found in the Archives of the Congregation for the Doctrine of the Faith. But, in fact, he does not include all the documentation. There are various documental series in the Archive of the Index. In Brundell's account there are three kinds of omissions: documents in the two volumes of Protocolli that he cites; the Protocolli of 1900–1902, where there are other documents referring to Leroy; and the Diari, which he does not even mention.

Chapter 4. Americanism and Evolutionism

1. The congregation was founded in France, and its official name is Congrégation de Sainte-Croix. Thus we here use the form Congregation of Holy Cross (not Congregation of *the* Holy Cross).

2. Zahm, *Sound and Music*.

3. John A. Zahm, *Catholic Science and Catholic Scientists* (Philadelphia: H. L. Kilner, 1893); *Science catholique et savants catholiques* (Paris: Lethielleux, 1895); *Scienza cattolica e scienziati cattolici* (Genova: Fassicomo, 1896).

4. Zahm, *Moses and Modern Science*.

5. John A. Zahm, *Bible, Science, and Faith* (Baltimore: Murphy, 1894); *Bible, science et foi* (Paris: Lethielleux, 1894); *Bibbia, scienza e fede* (Siena: Presso l'Ufficio della Biblioteca del clero, 1895); *Biblia, ciencia y fe* (Madrid: La España Moderna, 1912).

6. Zahm, *Scientific Theory and Catholic Doctrine*.

7. Zahm, *Science and the Church*.

8. John A. Zahm, *Evolution and Dogma* (Chicago: D. H. McBride, 1896). Hereafter references are cited in text. Translations include *Evoluzione e dogma* (Siena: Presso l'Ufficio della Biblioteca del clero, 1896); *L'évolution et le dogme* (Paris: Lethielleux, 1897); and *La evolución y el dogma* (Madrid: Sociedad Editorial Española, 1905).

9. Weber, *Notre Dame's John Zahm*, pp. 87–88, 94.

10. J. A. Zahm to A. F. Zahm, July 22, 1896, University of Notre Dame Archives (hereafter cited as UNDA), Albert F. Zahm Collection, box 4, folder 7. Albert Francis Zahm (1862–1954) was professor of engineering at Catholic University of America from 1895 to 1908. A pioneer in aeronautical engineering, he installed the first wind tunnel at an American university.

11. Salis Seewis, "La generazione spontanea e la filosofia antica," pp. 142–52; "Sant'Agostino e la generazione spontanea primitiva," pp. 421–38; "S. Tommaso e la generazione spontanea primitiva," pp. 676–91; "Le origini della vita sulla terra secondo il Suarez," pp. 168–76.

12. Darwin had left design intact but did away with a designer; the "invisible hand" of natural selection accounts for Paley's perfect adaptation. Thus Frank Burch Brown has characterized Darwin's theory a "theology without religion"; "The Evolution of Darwin's Theism," p. 18. Julian Huxley had noted the same some decades earlier: "Late nineteenth-

century Darwinism came to resemble the early nineteenth-century school of Natural Theology, Paley *redivivus*, one might say, but philosophically upside-down, with Natural Selection instead of a Divine Artificer as the Deus ex machina." Huxley, *Evolution: The Modern Synthesis*, p. 23.

13. Marquis de Nadaillac, review of *L'évolution et le Dogme*, pp. 229–46.

14. F. David (pseudonym of David Fleming), review of *Evolution and Dogma*, pp. 245–55.

15. Salis Seewis, "Evoluzione e Domma pel Padre J. A. Zahm," pp. 201–4.

16. Pisani, "Le Congrès de Fribourg," pp. 119–25.

17. Beretta, "Monseigneur d'Hulst, les Congrès Scientifiques Internationaux des Catholiques et la question biblique," pp. 79–80.

18. Leo XIII's letter was published in *Revue des Sciences Ecclésiastiques* 56 (1887): 379–81.

19. See, for example, Beretta, "Monseigneur d'Hulst, les Congrès Scientifiques Internationaux des Catholiques et la question biblique," pp. 114–17.

20. Pisani, "Les Congrès Scientifiques Internationaux des Catholiques," p. 115.

21. Zahm, "Évolution et téléologie," pp. 403–19.

22. ACDF, S. Oficio, Censurae Librorum 1896–97, 19: Malta—Intorno all'opera del P. J. A. Zahm C.S.C. (America) sull'Evoluzione e Dogma.

23. Ibid.

24. Ibid.

25. ACDF, S. Oficio, Decreta S. O., 1897, Feria V loco IV Die 6 Maii 1897.

26. ACDF, S. Oficio, Minutari, 1897, p. 246.

27. Parocchi, Vannutelli, Verga, Mazella and Aloisi-Masella. Cf. ACDF, S. Oficio. Decreta S. O., 1897, Feria V loco IV Die 6 Maii 1897.

28. ACDF, Index, Protocolli, 1897–99, fol. 179.

29. Cf. Mazzella, *De Deo creante*, pp. 154–66 (on evolution in general) and 343–74 (on the origin of man).

30. Salis Seewis, "La generazione spontanea e la filosofia antica," pp. 142–52; "Sant'Agostino e la generazione spontanea primitiva," pp. 421–38; "S. Tommaso e la generazione spontanea primitiva," pp. 676–91; "Le origini della vita sulla terra secondo il Suarez," pp. 168–76.

31. ACDF, Index, Protocolli, 1897–99, fol. 180. Hereafter references are cited in text.

32. Buonpensiere says he will provide further illustrations of Zahm's misperception of ancient thinkers in sections 19, 20, 26, 27, 28, and 35 of his report.

33. Quotations are from Zahm, *Evolution and Dogma*, pp. 196, 201.

34. The Italian edition that Buonpensiere used has translation errors, some of which affect the passages he cites, for example, pp. 10, 11, 12 of the report (121, 122, 201 of the book, respectively). In some cases, the parts that Buonpensiere liked were poorly translated. In the copy of the report cited here, these errors are marked. Another consultor (see subsequent discussion in the text) had been charged with comparing the Italian text with the English original. His report was favorable, and he included in his report the English passages that had been poorly translated into Italian.

35. ACDF, Index, Protocolli, 1897–99, fol. 181.

36. ACDF, Index, Diari, vol. XXII, fol. 36r, May 29, 1898, and fol. 36v, May 31, 1898.

37. Ibid., fol. 38r, August 5, 1898.

38. ACDF, Index, Protocolli, 1897–99, fol. 191–92.

39. ACDF, Index, Diari, vol. XXII, fol. 39r, August 5, 1898.

40. ACDF, Index, Protocolli, 1897–99, fol. 191.

41. ACDF, Index, Diari, vol. XXII, fol. 39v, September 1, 1898.

42. Ibid.

43. Three other cardinal members, Aloisi-Masella, Mazella, and Satolli, were not in Rome and did not participate in this meeting.

44. ACDF, Index, Protocolli, 1897–99, fol. 193.

45. ACDF, Index, Diari, vol. XXII, fol. 39v, September 3 and 5, 1898; Protocolli, 1897–99, fol. 194.

46. ACDF, Index, Diari, vol. XXII, fol. 41r, September 9, 1898.

47. ACDF, Index, Protocolli, 1897–99, fol. 277.

48. Ibid., fol. 179.

49. ACDF, Index, Diari, vol. XXII, fol. 42r, November 4, 1898.

50. Ibid., fol. 48r, February 3, 1899.

51. See O'Connell, *John Ireland and the American Catholic Church.*

52. From 1875 he was coadjutor bishop (thus, with the right of succession); in 1884 he became titular bishop when the previous bishop died.

53. See Fogarty, *The Vatican and the Americanist Crisis.*

54. See Hales, *The Catholic Church in the Modern World,* pp. 168–69.

55. Elliott, *The Life of Father Hecker.*

56. Elliott, *La Vie du Père Hecker.*

57. ACDF, Index, Protocolli, 1897–99, fol. 140.

58. Ibid, fol. 142, pp. 44–48.

59. Ibid., fol. 143, pp. 31–32.

60. Ibid., fol. 155.

61. Ibid., fol. 156.

62. J. A. Zahm to A. F. Zahm, June 24, 1896, UNDA, Albert F. Zahm Collection, box 4, folder 7.

63. J. A. Zahm to A. F. Zahm, September 13, 1896, and October 2, 1896, ibid.

64. Alfonso Maria Galea (1861–1941) was a merchant, educator, and politician from Malta, known for his charitable work for religious institutions.

65. J. A. Zahm to A. F. Zahm, August 23, 1896, UNDA, Albert F. Zahm Collection, box 4, folder 7.

66. J. A. Zahm to A. F. Zahm, November 11, 1896, ibid.

67. J. A. Zahm to A. F. Zahm, December 6, 1896, ibid.

68. S. Vannutelli to D. J. O'Connell, September 10, 1897; Diocese of Richmond Archives (copy at UNDA, MDRI, M8 reel 7).

69. Indiana Province Archives of the Congregation of Holy Cross (hereafter cited as IPA), J. A. Zahm Collection, box 1, folder 6.

70. Note from S. Vannutelli to D. J. O'Connell, July 31, 1897, in ibid.

71. J. A. Zahm to A. F. Zahm, August 1, 1897, UNDA, Albert F. Zahm Collection, box 4, folder 7.

72. Kraus, "The Catholic Congress at Fribourg," p. 1.

73. Ibid. The complete text is found in Fogarty, *The Vatican and the Americanist Crisis*, pp. 319–26.

74. J. Ireland to J. A. Zahm, St. Paul, September 11, 1897, IPA, J. A. Zahm Collection, box 1, folder 6.

75. J. J. Keane to D. J. O'Connell, Washington, October 15, [1897], Diocese of Richmond Archives (copy at UNDA, MDRI, M18 reel 7).

76. J. L. Spalding to J. A. Zahm, Peoria, December 28, 1897, IPA, J. A. Zahm Collection, box 1, folder 6.

77. Fogarty, *The Vatican and the American Hierarchy*, pp. 156–57.

78. Cf. ACDF, Index, Protocolli, 1897–99, fol. 140.

79. The date of the letter of denunciation is unclear: there is a "5," preceded by a "2" and a "1," with no clear indication which number is meant or whether both have been struck. So the day could be November 5, 15 or 25. In the Congregation of the Index the denunciation was noted on November 25, with the indication that the prefect had entrusted the book to the secretary to be assigned to a consultor; so it is possible that several days had already passed since the denunciation was received (ACDF, Index, Diari, vol. XXII, fol. 32r). At the end of E. Buonpensiere's report, Zardetti's denunciation is also reproduced, with a date of November 5 (ACDF, Index, Protocolli, 1897–99, fol. 180, p. 54).

80. ACDF, Index, Protocolli, 1897–99, fol. 179, p. 6.

81. J. A. Zahm to D. J. O'Connell, February 5, 1898, UNDA, J. A. Zahm Collection, box 2, folder 17.

82. J. J. Keane to J. A. Zahm, January 20, 1898, IPA, J. A. Zahm Collection, box 1, folder 7.

83. John A. Zahm, "Evolution and Teleology," *Appleton's Popular Science Monthly* 52 (April 1898): 815–24; "Évolution et téléologie," *Revue des Questions Scientifiques* 43 (April 1898): 403–19.

84. Cf. ACDF, Index, Protocolli, 1897–99, fol. 142: "If we compare this letter [of denunciation] with Martel's articles perhaps we will find some of the motives for thinking, with some basis, that the letter was inspired by them, the more so since in France everyone is aware of the agreement of the Bishop of Nancy with the newspaper, *La Vérité*."

85. McAvoy, *The Americanist Heresy*, pp. 166–73.

86. D. J. O'Connell to J. A. Zahm, July 10, 1898, UNDA, J. A. Zahm Collection, box 1, folder 12.

87. Klein, *Americanism*, pp. 138–39.

88. McAvoy, *The Americanist Heresy*, p. 173.

89. Fogarty, *The Vatican and the American Hierarchy*, pp. 165–66.

90. Born in Goldach, Switzerland, August 29, 1847; died in Milwaukee, August 4, 1930.

91. Fogarty, *The Vatican and the American Hierarchy*, p. 167.

92. D. J. O'Connell to J. A. Zahm, July 10, 1898, UNDA, J. A. Zahm Collection, box 1, folder 12.

93. The letters were written on July 11, 14, and 16: Fogarty, *The Vatican and the American Hierarchy*, pp. 170–71.

94. S. M. Brandi to M. A. Corrigan, July 11, 1898, UNDA, Archdiocese of New York Collection, I–1-i (2) 1894–98.

95. S. M. Brandi to M. A. Corrigan, October 12, 1898, ibid.

96. Fogarty, *The Vatican and the American Hierarchy*, pp. 171, 175.

97. Ibid., p. 175, n. 88.

98. ACDF, Index, Diari, vol. XXII, fol. 48v.

99. Born in Cerasa (Pesaro, Italy), March 2, 1860; died in Senigallia, September 8, 1938. He was professor of theology of the Roman Seminary, consultor of the Congregation of the Index and the Congregation of Studies, and pro-secretary of the Pontifical Academy of Theology. On April 19, 1900, he was elected bishop of Senigallia (Marche, Italy).

100. Born in Vaucouleurs (diocese of Verdun), February 28, 1863; died in Rome, May 20, 1936. After joining the order of the Servants of Mary, he studied in England, where he was ordained. He was rector of the College of Saint Alexis, professor of the Urbanian Pontifical Atheneum, and consultor of the Congregation of Studies. He became a cardinal in 1927, and the following year was named prefect of the Congregation of Religious.

101. The pope had indicated to the prefect, Steinhuber, that after the case, which he had reserved, was resolved, the relevant files would be deposited at the Index.

102. ACDF, Index, Protocolli, 1897–99, fol. 159 (whoever numbered the fascicle first wrote "160," then corrected it to "159."

103. Ibid., fol. 161.

104. Ibid., fol. 159, p. 1.

105. Ibid., p. 4. The expression of Saint Vincent of Lerins (the original Latin is "Retenta est antiquitas explosa novitas") refers to the decision of the Pope Saint Stephen to annul the decrees of a synod of Carthage, which prescribed the rebaptism of converted heretics. Therefore, in its original context it had no normative and general meaning, but rather was descriptive of a particular case.

106. ACDF, Index, Protocolli, 1897–99, fol. 158. Unsigned, no date. On the last page another hand (probably Thomas Esser O. P., successor of Cicognani as secretary of the Index) has noted in pencil: "This is the writing of Rev. M. Lepidi, Master of the Apostolic Palace."

107. Ibid., fol. 157. Same annotation the previous document.

108. ACDF, Index, Diari, vol. XXII, fol. 49r.

109. Ibid., fol. 49v.

110. ACDF, Index, Protocolli, 1897–99, fol. 162–64. Among them is a letter from the superior general of the Paulists, William Deshon, and another from Félix Klein.

111. Leo XIII, letter *Testem benevolentiae* to Cardinal James Gibbons, archbishop of Baltimore, concerning some opinions known by the name of Americanism (January 22, 1899), *Acta Sanctae Sedis* 31 (1898–99): 470–71.

112. Ibid., p. 479.

113. Cf. Fogarty, *The Vatican and the Americanist Crisis*, p. 286.

114. ACDF, Index, Diari, vol. XXII, fol. 41r.

115. Weber, *Notre Dame's John Zahm*, p. 107.

116. Ibid., p. 144.

117. Cf. J. A. Zahm to D. J. O'Connell, September 28, 1898, UNDA, J. A. Zahm Collection, box 2, folder 17; J. J. Keane to D. J. O'Connell, Washington, October 7, 1898, Diocese of Richmond Archives (copy held at UNDA, MDRI, M18 reel 7).

118. J. Ireland to D. J. O'Connell, St. Paul, October 27, 1898, Diocese of Richmond Archives (copy held at UNDA, MDRI, M18 reel 7).

119. J. J. Keane to J. A. Zahm, Washington, September 28, 1898, UNDA, J. A. Zahm Collection, box 1, folder 12.

120. J. A. Zahm to D. J. O'Connell, September 28, 1898, UNDA, J. A. Zahm Collection, box 2, folder 17.

121. Ibid.

122. D. J. O'Connell to J. A. Zahm, "Tritone," October 15, 1898, UNDA, J. A. Zahm Collection, box 1, folder 12. "Tritone" was O'Connell's residence, in the Palazzo Torlonia, Via del Tritone 60, Rome.

123. Telegram, D. J. O'Connell to J. A. Zahm, Genazzano, no date, no signature, UNDA, J. A. Zahm Collection, box 1, folder 12.

124. G. Français to J. A. Zahm, Neuilly, September 29, 1898, IPA, J. A. Zahm Collection, box 1, folder 7.

125. D. J. O'Connell to J. A. Zahm, October 31, 1898, UNDA, J. A. Zahm Collection, box 1, folder 12.

126. J. A. Zahm to D. J. O'Connell, October 31, 1898, UNDA, J. A. Zahm Collection, box 2, folder 17.

127. J. Ireland to D. J. O'Connell, St. Paul, October 27, 1898, ibid.

128. J. A. Zahm to D. J. O'Connell, Notre Dame, October 20, 1898, UNDA, J. A. Zahm Collection, box 1, folder 17.

129. S. Parravicino to D. J. O'Connell, November 10, 1898, Diocese of Richmond Archives (copy held at UNDA, MDRI, M18 reel 7).

130. Cf. G. Français to J. A. Zahm, Neuilly, November 10, 1898, UNDA, J. A. Zahm Collection, box 1, folder 12.

131. D. J. O'Connell to J. A. Zahm, Rome, October 31, 1898, ibid.

132. G. Français to J. A. Zahm, Neuilly, November 10, 1898, UNDA, J. A. Zahm Collection, box 1, folder 12. Original in French. There is an English translation in the same folder, which we have used, with corrections.

133. ACDF, Index, Diari, vol. XXII, fol. 42r.

134. D. J. O'Connell to J. A. Zahm, November 7, 1898, UNDA, J. A. Zahm Collection, box 1, folder 12.

135. Telegram, D. J. O'Connell to J. A. Zahm, November 7, 1898, ibid.

136. D. Fleming to J. A. Zahm, October 20, 1898, ibid.

137. J. J. Keane to J. A. Zahm, November 9, 1898, ibid.

138. Telegram, D. J. O'Connell to J. A. Zahm, November 11, 1898, ibid.

139. D. J. O'Connell to S. Parravicino, November 12, 1898, UNDA, Americanism Collection, M21, reel 1.

140. J. Legrand to Français, November 11, 1898, UNDA, J. A. Zahm Collection, box 1, folder 12. We have made minor corrections to the English translation in the same folder.

141. D. J. O'Connell to J. A. Zahm, November 27, 1898, ibid.

142. Ibid.

143. J. J. Keane to J. A. Zahm, December 10, 1898, ibid.

144. J. Ireland to J. A. Zahm, December 13, 1898, ibid.

145. J. A. Zahm to D. J. O'Connell, December 19, 1898, UNDA, J. A. Zahm Collection, box 2, folder 17.

146. S. M. Brandi to M. A. Corrigan, January 2, 1899, UNDA, Archdiocese of New York Collection, I–1-i (2) 1898–1910.

147. Brandi, "Evoluzione e domma," pp. 34–49. Brandi always wrote to Corrigan in English.

148. G. Français to J. A. Zahm, January 12, 1899, UNDA, J. A. Zahm Collection, box 1, folder 12.

149. G. Français to J. A. Zahm, March 10, 1899, ibid.

150. G. Français to J. A. Zahm, April 18 and April 24, 1899, ibid.

151. S. M. Brandi to M. A. Corrigan, March 28, 1899, UNDA, Archdiocese of New York Collection, I–1-i (2) 1898–1910.

152. ACDF, Index, Diari, vol. XXII, fol. 48r.

153. J. A. Zahm to D. J. O'Connell, March 30, 1899, UNDA, J. A. Zahm Collection, box 2, folder 17.

154. J. A. Zahm to J. Ireland, March 31, 1899, ibid. Early in 1899, Ireland made a trip through Italy and Europe. Cf. O'Connell, *John Ireland and the American Catholic Church*, pp. 461–66.

155. James A. Burns, Diary, April 5, 1899, IPA, James Burns Personal Papers, accession no. 1991/27, blocknotes 2.

156. D. J. O'Connell to J. A. Zahm, April 12, 1899, UNDA, J. A. Zahm Collection, box 1, folder 12.

157. Cf. Cenci, *Il Cardinale Merry del Val*, p. 6.

158. J. A. Burns, Diary, May 23, 1899, IPA, James Burns Personal Papers, accession no. 1991/27, blocknotes 2.

159. Zahm, *L'évolution et le dogme.*

160. M. Cicognani to G. Français, April 25, 1899, UNDA, J. A. Zahm Collection, box 1, folder 12. In this archive are preserved an English translation and several copies in Italian, with quite a few errors which we have corrected following the draft manuscript by Cicognani: ACDF, Index, Protocolli, 1897–99, fol. 275.

161. Cf. ACDF, Index, Diari, vol. XXII, fol. 51r.

162. G. Français to M. Cicognani, Neuilly-Seine, April 29, 1899, ACDF, Index, Protocolli, 1897–99, fol. 276.

163. J. A. Zahm to Flageollet, May 16, 1899, ibid., fol. 273 (copy sent to Sacred Congregation of the Index).

164. J. A. Zahm to M. Cicognani, May 16, 1899, ibid., fol. 274.

165. P. Lethielleux to G. Français, Paris, May 20, 1899, UNDA, J. A. Zahm Collection, box 1, folder 12.

166. G. Français to J. A. Zahm, Le Fleix (Dordogne), May 27, 1899, ibid.

167. G. Français to J. A. Zahm, Ernée (Mayenne), June 3, 1899, ibid.

168. G. Français to J. A. Zahm, Neuilly-Seine, July 22, 1899, ibid.

169. G. Français to J. A. Zahm, Neuilly-Seine, October 24, 1899, ibid.

170. G. Français to J. A. Zahm, Neuilly-Seine, November 27, 1899, ibid.

171. G. Français to J. A. Zahm, Neuilly-Seine, April 24, 1899, ibid. Frederick Linnbarn, C. S. C. (1864–1915) was named bishop of Dacca in 1909.

172. J. A. Burns, Diary, June 2, 1899, IPA, James Burns Personal Papers, accession no. 1991/27, blocknotes 2.

173. Telegram from D. J. O'Connell to J. A. Zahm, Rome, June 6, 1899, UNDA, J. A. Zahm Collection, box 1, folder 12.

174. J. A. Zahm to D. J. O'Connell, June 21–24, 1899, UNDA, J. A. Zahm Collection, box 2, folder 17.

175. Sebastiano Martinelli, born in Borgo Sant'Anna (Lucca, Italy), August 20, 1848; died in Rome, July 4, 1918. He was prior general of the Order of Saint Augustine. Named apostolic delegate in the United States, April 18, 1896, he was created cardinal by Leo XIII in 1910.

176. Frederick Z. Rooker (1861–1907) had been vice-rector of the American College in Rome, when Denis O'Connell was rector. In 1895 he was appointed secretary of the Apostolic Delegation in Washington. In 1903 he as named bishop of Jaro (Philippines), where he died in 1907.

177. Weber, *Notre Dame's John Zahm*, p. 121.

178. Ibid., pp. 121–22.

179. Sacred Congregation of the Index to Français, September 10, 1898, UNDA, J. A. Zahm Collection, box 1, folder 12.

180. J. A. Zahm to A. M. Galea, May 16, 1899; *La Civiltà Cattolica*, 17th ser., 7 (1899): 125.

181. Declaration of Alfonso M. Galea, May 31, 1899, ibid.

182. J. A. Zahm to D. J. O'Connell, June 21–24, 1899, UNDA, J. A. Zahm Collection, box 2, folder 17.

183. Cf. Weber, *Notre Dame's John Zahm*, pp. 122–23.

184. Born in New York, January 28, 1835; died, March 15, 1905. He was a grandson of Elizabeth Ann Seton. In 1897 he published a short work of popular science, *A Glimpse of Organic Life, Past and Present*.

185. W. Seton to J. A. Zahm, Huntington, New York, July 3, 1899, UNDA, J. A. Zahm Collection, box 1, folder 12.

186. G. Français to J. A. Zahm, Neuilly-Seine, July 22, 1899, ibid.

187. J. A. Zahm to D. J. O'Connell, "En route to Washington," October 6, 1899, UNDA, J. A. Zahm Collection, box 2, folder 17.

188. L. Bufalini to A. M. Galea, Siena, June 23, 1899, IPA, J. A. Zahm Collection, box 1, folder 8.

189. Hedley, "Physical Science and Faith," pp. 241–61.

190. Hedley, "Dr. Mivart on Faith and Science," pp. 401–19; Mivart, "Letter from Dr. Mivart on the Bishop of Newport's Article on Our Last Number," pp. 180–87; Hedley, "The Bishop of Newport's Rejoinder," pp. 188–89.

191. Hedley, "Physical Science and Faith," pp. 243–44.

192. Ibid., p. 253.

193. Brandi, "Evoluzione e domma," pp. 34–49.

194. Ibid., p. 46.

195. Brandi, "Evoluzione e Domma. Erronee informazioni di un inglese," pp. 75–77.

196. González de Arintero, *La evolución*, Introducción general, pp. 82 (Genesis), 96 (species), 98–100 (Galileo), 119 (Albertus Magnus), 154–56 (probability of evolution).

197. Ibid, p. 161.

198. Ibid., pp. 162–63, 165.

199. Brundell, "Catholic Church Politics and Evolution Theory," pp. 89–90.

Chapter 5. Condemned for Evolutionism?

1. Popular missions were organized programs of public preaching, designed to reinvigorate the faith of the Catholic masses. For a description of a mid-nineteenth-century mission in London, "the object of which was to give an impetus to the work of education by the revival of religion in the families of the children who frequented them," see Bowden, *The Life and Letters of Frederick William Faber*, pp. 369–70.

2. On the life and work of Bonomelli, see Malgeri, "Bonomelli, Geremia," 12:298–303.

3. "Cronaca contemporanea," *La Civiltà Cattolica* 65, 3 (1914): 632–33.

4. Bonomelli, *Roma e l'Italia e la realtà delle cose*. This work by placed on the *Index of Prohibited Books* by decree of the Congregation of the Index, April 13, 1889.

5. See Gallina, *Il problema religioso nel Risorgimento*.

6. ACDF, S. Oficio, St. St. L-2-b, Cremona 62, argomento: Bonomelli.

7. Gallina, *Il problema religioso nel Risorgimento*, p. 38.

8. ACDF, S. Oficio, St. St. L-2-b, Cremona 62, argomento: Bonomelli: in a note dated January 3, 1893, it is written that these documents refer to a decision not made by the Holy Office.

9. Monsabré, *Esposizione del Dogma Cattolico*, 18 vols.; *Introduzione al Dogma Cattolico*, 4 vols.

10. Bonomelli to S. Parravicino, October 4, 1898, cited in Ornella Confessore, *L'Americanismo cattolico in Italia*, p. 58.

11. Fogazzaro, *L'origine dell'uomo e il sentimento religioso*. Fogazzaro was quite well read both in philosophy and science. He was an ardent harmonizer of evolution and Catholicism, in which the word of God functions as an internal force guiding the process of evolution. See Mazhar, *Catholic Attitudes to Evolution in Nineteenth-Century Italian Literature*, pp. 201–44.

12. G. Bonomelli to A. Fogazzaro, Cremona, March 17, 1893, in Marcora, *Corrispondenza Fogazzaro-Bonomelli*, pp. 143–44.

13. G. Bonomelli to A. Fogazzaro, Cremona, March 31, 1893, ibid., p. 145.

14. G. Bonomelli to A. Fogazzaro, Cremona, July 9, 1893, ibid., p. 149.

15. Salis Seewis, review of *L'origine dell'uomo e il sentimento religioso*, by A. Fogazzaro, pp. 199–211, 324–39.

16. Ibid., p. 339.

17. G. Bonomelli to A. Fogazzaro, Cremona, April 4, 1893, in Marcora, *Corrispondenza Fogazzaro-Bonomelli*, p. 146.

18. G. Bonomelli to A. Fogazzaro, Cremona, May 7, 1893, in ibid., pp. 148–49.

19. G. Bonomelli to A. Fogazzaro, Cremona, July 9, 1893, in ibid., pp. 149–50.

20. A. Fogazzaro to G. Bonomelli, Montegalda, October 26, 1897, in ibid., pp. 34–35.

21. G. Bonomelli to A. Fogazzaro, Cremona, November 1, 1897, in ibid., p. 168.

22. Bonomelli, *Seguiamo la ragione*; "Appendice importante" appears on pp. 201–13.

23. D. J. O'Connell to S. Parravicino, November 12, 1898, in Confessore, *L'Americanismo cattolico in Italia*, p. 171 (our translation).

24. G. Bonomelli to A. Fogazzaro, Masino, August 4, 1898, in Marcora, *Corrispondenza Fogazzaro-Bonomelli*, p. 173. As G. Gallina rightly observes (*Il problema religioso nel Risorgimento*, p. 92), Bonomelli refers to Zahm's book, *Evolution and Dogma*, and not, as stated by the editor of the correspondence (Marcora, *Corrispondenza Fogazzaro-Bonomelli*, p. 174), to Zahm's 1897 Fribourg lecture.

25. This information is found in two letters from Cardinal Antonio Agliardi to Bonomelli, dated September 26 and October 20, 1898 (*Biblioteca Ambrosiana, Archivio Bonomelliano*, Milan), cited by Gallina, *Il problema religioso nel Risorgimento*, pp. 92–93.

26. *Lega Lombarda* (Milan), no. 287, Tuesday–Wednesday, October 25–26, 1898, p. 1. A copy is preserved in ACDF, S. Oficio, St. St. L-2-b, Cremona 62, argomento: Bonomelli.

27. That is, the Preparatory Congregation of August 5, 1898; the General Congregation of September 1; approval of the pope, September 3, 1898.

28. Cardinal Antonio Agliardi to Bonomelli, October 28, 1898 (*Biblioteca Ambrosiana, Archivio Bonomelliano*, Milan), cited by Gallina, *Il problema religioso nel Risorgimento*, p. 93.

29. G. Bonomelli to A. Fogazzaro, Cremona, November 1, 1898, in Marcora, *Corrispondenza Fogazzaro-Bonomelli*, pp. 176–77.

30. S. Parravicino to D. J. O'Connell, November 10, 1898, in Confessore, *L'Americanismo cattolico in Italia*, p. 156 (our translation).

31. G. Bonomelli to A. Fogazzaro, Cremona, November 6, 1898, in Marcora, *Corrispondenza Fogazzaro-Bonomelli*, pp. 178–79.

32. John C. Hedley, "Physical Science and Faith," *Dublin Review* 123 (July–October 1898): 241–61.

33. Unsigned review of "Physical Science and Faith," by John C. Hedley, published in the *Tablet*, October 29, 1898, p. 690.

34. "Le idee di un Vescovo sull'Evoluzione," *La Rassegna Nazionale* 103 (November 16, 1898): 418, signed with the pseudonym "Theologus."

35. S. Parravicino to D. J. O'Connell, November 10, 1898, in Confessore, *L'Americanismo cattolico in Italia*, p. 156 (our translation).

36. Brandi, "Evoluzione e domma," pp. 34–49.

37. G. Bonomelli to A. Fogazzaro, Cremona, November 23, 1898, in Marcora, *Corrispondenza Fogazzaro-Bonomelli*, p. 180.

38. G. Bonomelli to A. Fogazzaro, Cremona, November 28 1898, in ibid., p. 181.

39. G. Bonomelli to A. Fogazzaro, Cremona, January 13, 1899, in ibid., p. 183.

40. G. Bonomelli to A. Fogazzaro, Cremona, March 4, 1899, in ibid., pp. 184–85.

41. "Cronaca contemporanea," *La Civiltà Cattolica*, 17th ser., 4 (1898): 362–63.

Chapter 6. "The Erroneous Information of an Englishman"

1. Gruber, *A Conscience in Conflict*, p. 188.

2. Hedley, letter to the editor ("Physical Science and Faith"), *Tablet*, January 14, 1899, p. 59.

3. Brandi, "Evoluzione e Domma. Erronee informazioni di un inglese," pp. 75–77.

4. Ward, "Bishop Hedley," pp. 1–12.

5. Unsigned review of "Physical Science and Faith," by John C. Hedley, *Tablet*, October 29, 1898, p. 690.

6. Parravicino (under the pseudonym "Theologus"), "Le idee di un Vescovo sull'Evoluzione," pp. 418–20.

7. Hedley, "Physical Science and Faith," pp. 241–61. Hereafter references are cited in text.

8. S. Parravicino to D. J. O'Connell, November 10, 1898, Diocese of Richmond Archives (copy held at UNDA, MDRI, M18 reel 7).

9. S. Parravicino to D. J. O'Connell, November 10, 1898, in Confessore, *L'Americanismo cattolico in Italia*, p. 156.

10. Brandi, "Evoluzione e domma." Hereafter references are cited in text.

11. Letters from Cardinal Antonio Agliardi to Geremia Bonomelli, September 20 and October 20, 1898 (*Biblioteca Ambrosiana, Archivio Bonomelliano*, Milan), cited by Gallina, *Il problema religioso nel Risorgimento*, pp. 92–93.

12. Hedley, letter to the editor ("Physical Science and Faith"), *Tablet*, January 14, 1899, p. 59.

13. Cited by Jones, *England and the Holy See*, pp. 198–99. We have cited this passage from Messenger, *Evolution and Theology*, pp. 235–36.

14. Brandi, "Evoluzione e domma," pp. 47–49.

15. Brandi, "Evoluzione e domma. Erronee informazioni di un inglese."

Chapter 7. Happiness in Hell

1. On Mivart's relationship with Darwin and Darwinism, see Vorzimmer, *Charles Darwin*, pp. 225–51; Hull, *Darwin and His Critics*, pp. 351–415; and Richards, *Darwin*, pp. 225–30, 353–63. There is a lucid discussion of "how Darwin resolved Mivart's challenge of incipient stages" (that is, how can so many organs be altered all together and adaptively to produce a modified descendant) in Gould, *The Structure of Evolutionary Theory*, pp. 1218–24. On p. 1220, however, Gould repeats the cliché that Mivart was excommunicated because of his evolutionist views.

2. We have taken many details about Mivart from Gruber, *A Conscience in Conflict*.

3. Richards, *Darwin*, p. 355.

4. From Wright's review of *The Genesis of Species*, North American Review 113 (1871): 63–103; quoted by Hull, *Darwin and His Critics*, p. 398.

5. Darwin to J. D. Hooker, September 16, 1871; quoted by Hull, *Darwin and His Critics*, p. 353.

6. Mivart, *On the Genesis of Species*, p. 116; cited by Vorzimmer, *Charles Darwin*, p. 231.

7. Vorzimmer, *Charles Darwin*, p. 232.

8. Vorzimmer concludes (ibid., p. 248): "Darwin never succeed in reestablishing either the necessity or the sufficiency of the thesis of natural selection." On this score, see Hull, *Darwin and His Critics*, p. 412.

9. Mivart, *On the Genesis of Species*.

10. Ibid., pp. 1, 24–31.

11. Ibid., pp. 268–70.

12. Ibid., pp. 279–83. In a famous riposte to "Mr. Darwin's Critics" (*Contemporary Review* 18 [1871]: 443–76), Thomas H. Huxley analyzed passages from Suárez in mock scholastic style and concluded, *contra* Mivart, that Suárez held to a literal interpretation of Genesis. See Richards, *Darwin*, pp. 227–28.

13. Mivart, *On the Genesis of Species*, p. 291.

14. Hull, "Darwinism as a Historical Entity," p. 797.

15. Mivart, *On the Genesis of Species*, p. 305.

16. Ibid., pp. 306–7.

17. Mivart, *Lessons from Nature as Manifested in Mind and Matter*, esp. chap. 6, pp. 128–91; *On Truth: A Systematic Inquiry*.

18. *On Truth: A Systematic Inquiry*, p. 527.

19. Gruber, *A Conscience in Conflict*, p. 111.

20. "Notices of Books," *On the Genesis of Species*, by St. George Mivart, F.R.S. (London: Macmillan & Co., 1871), *Dublin Review*, n.s., 16 (January–April 1871): 482–86.

21. "Evolution and Faith," *Dublin Review*, n.s., 17 (July–October 1871): 1–40.

22. Ibid., p. 24.

23. Ibid., p. 39.

24. "Notices of Books. *The Contemporary Review* for November, 1871, and January, 1872. Papers by Mr. Huxley and Mr. St. George Mivart," *Dublin Review*, n.s., 18 (January–April 1872): 195–200.

25. Gruber, *A Conscience in Conflict*, esp. pp. 164–65, 188–89, 197–98, 213, 227–28.

26. Hedley, "Dr. Mivart on Faith and Science," pp. 401–19.

27. Mivart, "Letter from Dr. Mivart on the Bishop of Newport's Article in Our Last Number," pp. 180–87.

28. Ibid., p. 182.

29. Hedley, "The Bishop of Newport's Rejoinder," pp. 188–89.

30. Murphy, "Dr. Mivart on Faith and Science," pp. 400–411.

31. Ibid., p. 407.

32. Ibid., p. 408.

33. Ibid., pp. 409–10.

34. Ibid., pp. 410–11.

35. ACDF, Index, Diari, vol. XXII, fol. 39r, August 5, 1898.

36. ACDF, Index, Protocolli, 1897–99, fol. 180, pp. 45, 51, 53.

37. Mivart, "Happiness in Hell," pp. 899–919.

38. Bagshawe, *A Pastoral Letter by Edward, Bishop of Nottingham*.

39. Clarke, "Happiness in Hell: A Reply," pp. 83–92; Mivart, "The Happiness in Hell: A Rejoinder," pp. 320–38; "Last Words on the Happiness in Hell: A Rejoinder," pp. 637–51.

40. Mivart, "Happiness in Hell," pp. 918–19.

41. We cite the documents contained in this folder from Censura Librorum as "ACDF, S. Oficio, C. L. 1893, Mivart," adding the identification of the document in question in each case.

42. ACDF, S. Oficio, C. L. 1893, Mivart: Letter from the Bishop of Nottingham to the Congregation for the Propagation of the Faith, February 24, 1893.

43. Ibid., Relazione del P. M. Ludovico Hickey, March 1893.

44. Ibid., Relazione 2ª del P. M. Ludovico Hickey, June 1893.

45. Not to be confused with another priest, a friend of Mivart, who is also usually referred to as R. F. Clarke. This is Robert F. Clarke, and he was not a Jesuit but a secular priest. Mivart calls attention to this point in his second article, "The Happiness in Hell: A Rejoinder," p. 321.

46. ACDF, S. Oficio, C. L. 1893, Mivart: R. F. Clarke to Cardinal Vaughan, May 12, 1893; also found in Relazione 2ª, p. 16.

47. ACDF, S. Oficio, C. L. 1893, Mivart: Relazione 2ª, pp. 18–19.

48. Ibid., pp. 20–21.

49. ACDF, S. Oficio, C. L. 1893, Mivart: Relazione 1ª, p. 12.

50. ACDF, S. Oficio, C. L. 1893, Mivart: Relazione 2ª, p. 17.

51. Ibid., pp. 16–17. The original letter is also found in this folder.

52. ACDF, S. Oficio, C. L. 1893, Mivart: folio "Feria II die 3 julii 1893."

53. ACDF, S. Oficio, Decreta S. O., 1893, July 19, 1893, pp. 193, 195, 198.

54. ACDF, Index, Protocolli, 1891–94, fol. 104.

55. Ibid., fol. 105.

56. Ibid., fol. 111; *Acta Sanctae Sedis*, 26 (1893–94): 59–60.

57. Gruber, *A Conscience in Conflict*, p. 185.

58. Ibid.

59. ACDF, Index, Protocolli, 1891–94, fol. 107.

60. Ibid.

61. ACDF, Index, Protocolli, 1894–96, fol. 31; *Acta Sanctae Sedis* 26 (1893–94): 703–4.

62. Ibid., fol. 297 A.

63. Mivart, "The Continuity of Catholicism," pp. 51–72; "Some Recent Catholic Apologists," pp. 24–44.

64. The letters mentioned here, besides being published in the *Times* on Mivart's initiative, were collected and published, together with Cardinal Vaughan's circular and Mivart's two articles of 1900, as *Under the Ban.*

65. Ibid., pp. 31–32.

66. ACDF, S. Oficio, C. L. 1893, Mivart: R. Merry del Val, "Domanda del Cardinale Vaughan."

67. Ibid.: David Fleming, "De Statu Rei Catholicae in Anglia," November 17, 1899.

68. Ibid.: manuscripts of November 22 and December 8, 1899.

69. Ibid.: typewritten report beginning: "Nella importante Rivista inglese."

70. Ibid.: two manuscripts on the meeting of January 10, 1900.

71. ACDF, Index, Protocolli, 1900–1902, fol. 2; Diari, vol. XXII, folio 58, January 14, 1900, 58r.

72. ACDF, S. Oficio, C. L. 1893, Mivart: "La continuità del cattolicismo." Articolo scritto dal Dottor St George Mivart nella Rivista *The Nineteenth Century* (Gennaio 1900). Traduzione.

73. Ibid.: "Sinossi di un articolo di Giorgio Mivart," April 1900.

74. Ibid.: D. Fleming, "De Causa D. Mivart iam defuncti," April 27, 1901.

75. Mivart, "Some Recent Catholic Apologists," in *Under the Ban*, pp. 52–53.

76. Ibid., p. 51.

77. Root, "The Final Apostasy of St. George Jackson Mivart," pp. 15–17.

78. In the article cited, Root seems to consider it lamentable. The biographer Gruber seems to consider it inevitable, because apparently he takes it for granted that science is impossible to harmonize with Christianity.

79. ACDF, Index, Protocolli, 1894–96, fol. 249.

80. Some years later, on September 8, 1906, Wernz was elected general of the Jesuits. A native of Wurtemberg, he was an expert in canon law.

81. ACDF, Index, Protocolli, 1894–96, fol. 243.

82. Ibid., fols. 241 and 242.

83. Wernz, *Ius decretalium*, 6:188–206.

84. J. M. R., "Mivart, St. George Jackson," in *The Dictionary of National Biography*, p. 1053.

85. Gruber, "Mivart, George Jackson, St.," 9:747.

86. Gruber, *A Conscience in Conflict*, pp. 184–85, 246.

87. Ibid., pp. 185–86, 246.

88. Ibid., pp. 186–88.

89. So states John Hedley Brooke in his valuable book *Science and Religion*, p. 309.

90. ACDF, S. Oficio, C. L. 1893, Mivart: Relazione 2ᵃ, pp. 16–17.

91. Gruber, *A Conscience in Conflict*, pp. 191–92.

Chapter 8. The Church and Evolution

1. Harrison, "Early Vatican Responses to Evolutionist Theology," 9–10.

2. Ibid., pp. 2, 6, 7, 8, 9, 10, 11.

3. Messenger, *Evolution and Theology*, p. 237. In a note explaining the reference to "*any official journal of the Holy See*," Messenger adds "such as the *Acta*. The *Civiltà* is not an official organ in the canonical sense."

BIBLIOGRAPHY

Alonso Lobo, Arturo. *El Padre Arintero, precursor clarividente del Vaticano II.* Salamanca: San Esteban, 1970.

———. *Padre Arintero. Un maestro di vita spirituale.* Rome: San Sisto Vecchio, 1975.

Alszeghy, Zoltan. "El evolucionismo y el Magisterio de la Iglesia." *Concilium,* no. 26 (June 1967): 366–73.

Appleby, R. Scott. *"Church and Age Unite!" The Modernist Impulse in American Catholicism.* Notre Dame, Ind.: University of Notre Dame Press, 1992.

———. "Exposing Darwin's 'Hidden Agenda': Roman Catholic Responses to Evolution, 1875–1925." In *Disseminating Darwinism: The Role of Place, Race, Religion, and Gender,* ed. Ronald L. Numbers and John Stenhouse, 173–207. Cambridge: Cambridge University Press, 1999.

Arnould, Jacques. *Darwin, Teilhard de Chardin et Cie. L'Église et l'évolution.* Paris: Desclée, 1996.

———. "Le singe, le dominicain et les jésuites. Dalmace Leroy et la question de l'origine de l'homme au cours de l'année 1893." *Mémoire dominicaine* 13, no. 2 (1998): 255–64.

Bagshawe, Edward G. *A Pastoral Letter by Edward, Bishop of Nottingham, on Hell, and the State of the Condemned.* Nottingham, 1892.

Baudrillart, Alfred. *Vie de Mgr. d'Hulst.* 3rd ed. 2 vols. Paris: J. De Gigord, 1925.

Beraza, Blas. *Tractatus de Deo creante.* Bilbao: Elexpuru, 1921.

Beretta, Francesco. *Monseigneur d'Hulst et la science chrétienne. Portrait d'un intellectuel.* Paris: Beauchesne, 1996.

———. "Monseigneur d'Hulst, les Congrès Scientifiques Internationaux des Catholiques et la question biblique: La liberté de la science chrétienne au service du renouvellement de la théologie." In *Monseigneur d'Hulst fondateur de l'Institut catholique de Paris,* ed. Claude Bressolette, 75–135. Paris: Beauchesne, 1998.

Betti, Umberto, ed. *Raffaello Caverni, 1837–1900. Antologia di scritti.* Florence: Giampiero Pagnini, 1991.

Bonniot, Joseph de. "Essai philosophique sur le transformisme." *Études* 26 (March 1889): 337–68.

———. "Le transformisme et l'athéisme." *Études* 17 (September 1873): 428–40.

———. *La bête comparée à l'Homme.* Paris: 1889.

————. *La bête, question actuelle.* Tours: Alfred Mame, 1874.

Bonomelli, Geremia. *Roma e l'Italia e la realtà delle cose.* Florence: Cellini, 1889.

————. *Seguiamo la ragione.* Milan: Cogliati, 1898.

Bonvin, Bernard. *Lacordaire-Jandel. La restauration de l'Ordre dominicain en France après la Révolution; suivi de l'édition originale et annotée du Mémoire Jandel.* Paris: Éditions du Cerf, 1989.

Bowden, John Edward. *The Life and Letters of Frederick William Faber.* Baltimore: John Murphy, 1869.

Bowler, Peter J. *The Eclipse of Darwinism: Anti-Darwinian Evolution Theories in the Decades around 1900.* Baltimore: Johns Hopkins University Press, 1983.

Boyer, Charles. *Tractatus de Deo creante et elevante.* 3rd ed. Rome: Gregorian University, 1940.

Brandi, Salvatore M. "Evoluzione e Domma." *La Civiltà Cattolica,* 17th ser., 5 (1898): 34–49.

————. "Evoluzione e Domma. Erronee informazioni di un inglese." *La Civiltà Cattolica,* 18th ser., 6 (1902): 75–77.

Bressolette, Claude, ed. *Monseigneur d'Hulst fondateur de l'Institut catholique de Paris. Actes du colloque d'Hulst tenu à l'Institut Catholique de Paris les 21 et 22 novembre 1996.* Paris: Beauchesne, 1998.

Bricarelli, Carlo. "La dissoluzione dell'evoluzione." *La Civiltà Cattolica,* 17th ser., 6 (1899): 668–86.

Brooke, John Hedley. *Science and Religion: Some Historical Perspectives.* Cambridge: Cambridge University Press, 1991.

Brown, Frank Burch. "The Evolution of Darwin's Theism." *Journal of the History of Biology* 19 (1986): 1–45.

Brucker, Joseph. "Bulletin Scripturaire." *Études* 28 (September–December 1891): 488–97.

————. "Les jours de la création et le transformisme." *Études* 26 (April 1889): 567–92.

————. "L'origine de l'homme d'après la Bible et le transformisme." *Études* 26 (May 1889): 28–50.

————. *Questions actuelles d'Écriture Sainte.* Paris: Victor Retaux, 1895.

Brundell, Barry. "Catholic Church Politics and Evolution Theory, 1894–1902." *British Journal for the History of Science* 34 (2001): 81–95.

Bujanda, Jesús Martínez de. *Index de Rome, 1557, 1559, 1564. Les premiers index romains et l'index du Concile de Trente.* Québec: Éditions de l'Université de Sherbrooke; Genève: Librairie Droz, 1990.

Bujanda, Jesús Martínez de, and Marcella Richter. *Index librorum prohibitorum 1600–1966.* Québec: Éditions de l'Université de Sherbrooke; Montréal: Médiaspaul; Genève: Librairie Droz, 2002.

Cappelletti, Vincenzo, and Federico di Trocchio. "Caverni, Raffaello." In *Dizionario biografico degli italiani,* 23:85–88. Rome: Società Grafica Romana, 1979.

Castagnetti, Giuseppe, and Michele Camerota. "Raffaello Caverni and His *History of the Experimental Method in Italy.*" In *Galileo in Context,* ed. Jürgen Renn, 327–39. Cambridge: Cambridge University Press, 2001.

Caterini, Pietro. *Dell'origine dell'uomo secondo il trasformismo. Esame scientifico, filosofico, teologico.* Prato: Giachetti, 1884.

Caverni, Raffaello. *De' nuovi studi della Filosofia. Discorsi a un giovane studente.* Florence: Carnesecchi, 1877.

———. *Dell'antichità dell'uomo secondo la scienza moderna.* Florence: Cellini, 1881.

———. *Storia del metodo sperimentale in Italia.* 6 vols. Florence: Civelli, 1891–1910; reprint, Bologna: Forni, 1970, and New York: Johnson Reprint Corporation, 1972.

Cenci, Pio. *Il Cardinale Merry del Val, Segretario di Stato di San Pio X Papa.* Rome and Turin: Roberto Berruti, 1933.

Clarke, Richard F. "Happiness in Hell: A Reply." *Nineteenth Century* 33 (1893): 83–92.

Confessore, Ornella. *L'Americanismo cattolico in Italia.* Rome: Studium, 1984.

Congregation for the Doctrine of the Faith. *Notificatio* (June 14, 1966). *Acta Apostolicae Sedis* 58 (1966): 445.

Conry, Yvette. *L'introduction du darwinisme en France au XIXe siécle.* Paris: J. Vrin, 1974.

Darwin, Charles. *Charles Darwin's The Life of Erasmus Darwin.* Ed. Desmond King-Hele. Cambridge: Cambridge University Press, 2003.

———. *The Correspondence of Charles Darwin.* Vol. 10: *1862.* Cambridge: Cambridge University Press, 1997.

———. *The Correspondence of Charles Darwin.* Vol. 11: *1863.* Cambridge: Cambridge University Press, 1999.

———. *The Correspondence of Charles Darwin.* Vol. 12: *1864.* Cambridge: Cambridge University Press, 2001.

———. *On the Origin of Species: A Facsimile of the First Edition.* Ed. Ernst Mayr. Cambridge, Mass.: Harvard University Press, 1964.

———. *Sull'origine delle specie per elezione naturale, ovvero Conservazione delle razze perfezionate nella lotta per l'esistenza.* Trans. Giovanni Canestrini and Leonardo Salimbeni. Modena: Zanichelli 1864.

———. *The Variation of Plants and Animals under Domestication.* 2nd rev. ed. 2 vols. London: John Murray, 1875.

Darwin, Charles, and Thomas Henry Huxley. *Autobiographies.* Ed. Gavin de Beer. Oxford: Oxford University Press, 1983.

de Nadaillac, Jean-François-Albert du Pouget, Marquis. Review of *L'évolution et le dogme,* by John A. Zahm. *Revue des Questions Scientifiques* 40 (July 1896): 229–46.

De Rosa, Giuseppe. *La Civiltà Cattolica. 150 anni al servizio della Chiesa, 1850–1999.* Rome: La Civiltà Cattolica, 1999.

Del Re, Niccolò. *La Curia romana. Lineamenti storico-giuridici.* 3rd ed. Rome: Edizioni di Storia e Letteratura, 1970.

The Dictionary of National Biography. Ed. Leslie Stephen and Sidney Lee. Oxford: Oxford University Press, 1973.

Desmond, Adrian. *The Politics of Evolution: Morphology, Medicine, and Reform in Radical London.* Chicago: University of Chicago Press, 1989.

Dierckx, F. *El hombre-mono y los precursores de Adán ante la ciencia y la teología.* Madrid: Gregorio del Amo, 1895.

————. *L'homme-singe et les précurseurs d'Adam en face de la science et de la théologie.* Brussels: Societé Belge de Librairie, 1894.

Duval, André. "Le rapport du P. Lacordaire au Chapitre de la Province de France (Septembre 1854)." *Archivum Fratrum Praedicatorum* 31 (1961): 326–64.

Elliot, Walter. *The Life of Father Hecker.* New York: Columbus Press, 1891.

————. *La Vie du Père Hecker.* Paris: Librairie Victor Lecoffre, 1897.

Ferngren, Gary B., ed. *Science and Religion: A Historical Introduction.* Baltimore: Johns Hopkins University Press, 2002.

Fleming, David. Review of *Evolution and Dogma,* by John A. Zahm. *Dublin Review* 119 (July–October 1896): 245–55.

Fogarty, Gerald P. *The Vatican and the American Hierarchy from 1870 to 1965.* Päpste und Papsttum, vol. 21. Stuttgart: Anton Hiersemann, 1982.

————. *The Vatican and the Americanist Crisis: Denis O'Connell, American Agent in Rome, 1885–1903.* Miscellanea Historiae Pontificiae, vol. 36. Rome: Università Gregoriana Editrice, 1974.

Fogazzaro, Antonio. *L'origine dell'uomo e il sentimento religioso.* Milan: Galli, 1893.

Gallina, Giuseppe. *Il problema religioso nel Risorgimento e il pensiero di Geremia Bonomelli.* Miscellanea Historiae Pontificiae, vol. 35. Rome: Università Gregoriana Editrice, 1974.

Gardeil, Ambroise. "L'évolutionnisme et les principes de S. Thomas." *Revue Thomiste* 1 (1893): 27–45.

Garin, Eugenio. *Scienza e vita civile nel Rinascimento italiano.* Bari: Laterza, 1965.

Glick, Thomas F., ed. *The Comparative Reception of Darwinism.* 2nd ed. Chicago: University of Chicago Press, 1988.

González de Arintero, Juan. *La evolución y la filosofía cristiana.* Introducción general and vol. 1: *La evolución y la mutabilidad de las especies orgánicas.* Madrid: Gregorio del Amo, 1898.

Gould, Stephen Jay. *The Structure of Evolutionary Theory.* Cambridge, Mass.: Harvard University Press, 2002.

Gruber, Jacob W. *A Conscience in Conflict: The Life of St. George Jackson Mivart.* New York: Temple University Publications by Columbia University Press, 1960; reprint, Westport, Conn.: Greenwood Press, 1980.

————. "Mivart, George Jackson, St." In *New Catholic Encyclopedia,* 2nd ed., 9: 746–47. Detroit: Thomson/Gale; Washington, D.C.: Catholic University of America, 2003.

Hales, Edward Elton Young. *The Catholic Church in the Modern World: A Survey from the French Revolution to the Present.* Garden City, N.Y.: Hanover House, 1958.

Harrison, Brian W. "Early Vatican Responses to Evolutionist Theology." *Living Tradition,* no. 93 (May 2001), 11 pp.

Harvey, Joy. "Fertility or Sterility? Darwin, Naudin and the Problem of Experimental Hybridity." *Endeavor* 27 (June 2003): 57–62.

Hedley, John C. "The Bishop of Newport's Rejoinder." *Dublin Review,* n.s., 19 (January–April 1888): 188–89.

————. "Dr. Mivart on Faith and Science." *Dublin Review,* n.s., 18 (July–October 1887) 401–19.

————. Letter to the editor. *Tablet*, January 14, 1899: 59.

————. "Physical Science and Faith." *Dublin Review* 123 (July–October 1898): 241–61.

Hinnebusch, William A. *The Dominicans: A Short History.* New York: Alba House, 1975.

Huerga, Alvaro. "La evolución: Clave y riesgo de la aventura intelectual arinteriana." *Studium* 7 (1967): 127–53.

Hull, David L. *Darwin and His Critics: The Reception of Darwin's Theory of Evolution by the Scientific Community.* Cambridge, Mass.: Harvard University Press, 1973.

————. "Darwinism as a Historical Entity: A Historiographical Proposal." In *The Darwinian Heritage*, ed. David Kohn, 773–812. Princeton: Princeton University Press, 1985.

————. *Science as a Process: An Evolutionary Account of the Social and Conceptual Development of Science.* Chicago: University of Chicago Press, 1988.

Huxley, Julian. *Evolution: The Modern Synthesis.* New York: Harper, 1943.

Index librorum prohibitorum. Rome: Tipografia Vaticana, 1900.

Jacolliot, Louis. *Christna et le Christ.* New ed. Paris: Flammarion, 1913.

John Paul II. Address to the Pontifical Academy of Sciences (October 22, 1996). In *Papal Addresses to the Pontifical Academy of Sciences 1917–2002*, 370–74. Vatican City: Pontifical Academy of Sciences, 2003.

Jones, S. *England and the Holy See: An Essay towards Reunion.* London: Longmans, 1902.

Juste, Ramón. "La teología católica y el problema de la evolución humana. Un siglo de historia eclesiástica." *Revista española de teología* 25 (1965): 393–414.

Klein, Félix. *Americanism: A Phantom Heresy.* Atchinson, Kans.: Aquin Book Shop, 1987.

Kohn, David, ed. *The Darwinian Heritage.* Princeton: Princeton University Press, 1985.

Kraus, Franz Xavier ("Spectator"). "The Catholic Congress at Fribourg." *Catholic Citizen* 27, no. 48 (September 18, 1897): 1 and 4.

Landucci, Giovanni. *Darwinismo a Firenze. Tra scienza e ideologia (1860–1900).* Florence: Olschki, 1977.

Leo XIII. Encyclical *Providentissimus Deus*, on Biblical studies (November 18, 1893). In *Acta Sanctae Sedis* 26 (1893–94): 269–92 (Rome: Ex Typographia Polyglotta S. Congr. De Propaganda Fide; reprint, New York and London: Johnson Reprint Corporation 1969).

————. Letter addressed to the Congrès Scientifiques Internationaux des Catholiques (May 20, 1887). *Revue des Sciences Ecclésiastiques* 56 (1887): 379–81.

————. Letter *Testem benevolentiae* to Cardinal James Gibbons, Archbishop of Baltimore, concerning some opinions known by the name of Americanism (January 22, 1899). In *Acta Sanctae Saedis* 31 (1898–99): 470–71 (Rome: Ex Typographia Polyglotta S. Congr. De Propaganda Fide; reprint, New York and London: Johnson Reprint Corporation 1969).

"Le P. Dalmace Leroy." *Annales dominicaines* (1905): 289–91.

Leroy, Marie-Dalmace. "Correspondance au R. P. directeur de la *Revue Thomiste.*" *Revue Thomiste* 1 (1893): 532–35.

————. *L'évolution des espèces organiques.* Paris: Perrin, 1887.

————. *L'évolution restreinte aux espèces organiques.* Paris and Lyons: Delhomme et Briguet, 1891.

————. *Lettre à M. l'Abbé A. Farges.* Paris: October 1898.

————. Review of *L'origine des espèces*, by J. Guibert. *Revue Thomiste* 7 (1899): 735–41.

Lindberg, David C., and Ronald L. Numbers, eds. *God and Nature: Historical Essays on the Encounter between Christianity and Science*. Berkeley: University of California Press, 1986.

Malgeri, Francesco. "Bonomelli, Geremia." In *Dizionario biografico degli italiani*, 12:298–303. Rome: Società Grafica Romana, 1970.

Mansi, Ioannes Dominicus. *Sacrorum Conciliorum nova et amplissima collectio*. Vol. 48. Graz: Akademische Druck-u. Verlagsanstalt, 1961.

Marcora, Carlo, ed. *Corrispondenza Fogazzaro-Bonomelli*. Milan: Vita e Pensiero, 1968.

Marselli, Niccola. *Le grandi razze dell'umanità*. Turin and Rome: Ermanno Loescher, 1880.

————. *Origini dell'umanità*. Turin and Rome: Ermanno Loescher, 1879.

Mazhar, Noor Giovanni. *Catholic Attitudes to Evolution in Nineteenth-Century Italian Literature*. Venice: Istituto Veneto di Scienze, Lettere ed Arti, 1995.

Mazzella, Camillo. *De Deo creante*. 4th ed. Rome: Forzani, 1896.

McAvoy, Thomas Timothy. *The Americanist Heresy in Roman Catholicism, 1895–1900*. Notre Dame, Ind.: Notre Dame University Press, 1961.

Messenger, Ernst Charles. *Evolution and Theology: The Problem of Man's Origin*. New York: Macmillan, 1932.

Minelli, Alessandro, and Sandra Casolati, eds. *Giovanni Canestrini: Zoologist and Darwinist*. Venice: Instituto Veneto di Scienze, 2001.

Mivart, St. George Jackson. "The Continuity of Catholicism." *Nineteenth Century* 47 (1900): 51–72.

————. "Happiness in Hell." *Nineteenth Century* 32 (1892): 899–919.

————. "The Happiness in Hell: A Rejoinder." *Nineteenth Century* 33 (1893): 320–38.

————. "Last Words on the Happiness in Hell: A Rejoinder." *Nineteenth Century* 33 (1893): 637–51.

————. *Lessons from Nature as Manifested in Mind and Matter*. New York: Appleton 1876.

————. "Letter from Dr. Mivart on the Bishop of Newport's Article on Our Last Number." *Dublin Review*, n.s., 19 (January–April 1888): 180–87.

————. *On the Genesis of Species*. New York: Appleton, 1871.

————. *On Truth: A Systematic Inquiry*. London: Kegan Paul, 1889.

————. "Some Recent Catholic Apologists." *Fortnightly Review* 67 (1900): 24–44.

Monsabré, Jacques-Marie-Louis. *Esposizione del Dogma Cattolico*. 2nd ed. 18 vols. Turin: Marietti; Cremona: E. Maffezzoni, 1893–95.

————. *Introduzione al Dogma Cattolico*. 4 vols. Turin: Marietti; Cremona: E. Maffezzoni, 1890–95.

Moore, James R. *The Post-Darwinian Controversies: A Study of the Protestant Struggle to Come to Terms with Darwin in Great Britain and America, 1870–1900*. Cambridge: Cambridge University Press, 1979.

Murphy, Jeremiah. "Dr. Mivart on Faith and Science." *Dublin Review*, n.s., 19 (January–April 1888): 400–411.

Numbers, Ronald L., and John Stenhouse, eds. *Disseminating Darwinism: The Role of Place, Race, Religion, and Gender*. Cambridge: Cambridge University Press, 1999.

O'Connell, Marvin R. *John Ireland and the American Catholic Church.* St. Paul: Minnesota Historical Society Press, 1988.

Pagnini, Sara. *Profilo di Raffaello Caverni (1837–1900). Con appendice documentaria.* Florence: Pagnini e Martinelli, 2001.

Pancaldi, Giuliano. *Darwin in Italy: Science across Cultural Frontiers* Trans. Ruey Brodine Morelli. Bloomington: Indiana University Press, 1991.

Papal Addresses to the Pontifical Academy of Sciences, 1917–2002. Vatican City: Pontifical Academy of Sciences, 2003.

Parente, Pietro. *De creatione universali: de angelorum hominisque elevatione et lapsu.* Collectio theologica romana, vol. 4. 4th ed. Turin: Marietti, 1959.

Parravicino, Sabina ("Theologus"). "Le idee di un Vescovo sull'Evoluzione." *La Rassegna Nazionale* 104 (November 16, 1898): 418–20.

Paul VI. Apostolic Letter *Integrae Servandae. Acta Apostolicae Sedis* 57 (1965): 952–55.

Pesch, Christian. *Praelectiones dogmaticae quas in Collegio Ditton-Hall habebat.* Vol. 3: *De Deo creante et elevante. De Deo fine ultimo.* 3rd ed. Freiburg: Herder, 1908.

Pisani, P. "Le Congrès de Fribourg." *Revue du Clergé Français* 3 (1897): 119–25.

———. "Les Congrès Scientifiques Internationaux des Catholiques." *Revue du Clergé Français* 4 (1898): 109–15.

Portalié, E. "Le R. P. Frins et la *Revue Thomiste.*" *Études* 30, no. 59 (May–August 1893): 37–64.

Rahner, Karl. "De Deo creante et elevante et de peccato originali." Typescript source notes. Innsbruck, 1953.

Richards, Robert J. *Darwin and the Emergence of Evolutionary Theories of Mind and Behavior.* Chicago: University of Chicago Press, 1987.

Roberts, Jon H. *Darwinism and the Divine in America: Protestant Intellectuals and Organic Evolution, 1859–1900.* Notre Dame, Ind.: University of Notre Dame Press, 2001.

Root, John D. "The Final Apostasy of St. George Jackson Mivart." *Catholic Historical Review* 71 (1985): 1–25.

Ruiz de la Peña, Juan Luis. *Imagen de Dios. Antropología teológica fundamental.* 3rd ed. Santander: Sal Terrae, 1996.

Salis Seewis, Francesco. "Evoluzione e Dogma pel Padre J. A. Zahm." *La Civiltà Cattolica,* 16th ser., 9 (1897): 201–4.

———. "La generazione spontanea e la filosofia antica." *La Civiltà Cattolica,* 16th ser., 11 (1897): 142–52.

———. "Le origini della vita sulla terra secondo il Suarez." *La Civiltà Cattolica,* 16th ser., 12 (1897): 168–76.

———. Review of *De' nuovi studi della Filosofia. Discorsi di Raffaello Caverni a un giovane studente.* (I) *La Civiltà Cattolica* 10th ser., 4 (1877): 570–80; (II) *La Civiltà Cattolica* 10th ser., 5 (1878): 65–76.

———. Review of *L'origine dell'uomo e il sentimento religioso,* by A. Fogazzaro. *La Civiltà Cattolica,* 15th ser., 8 (1893): 199–211 and 324–39.

———. Reviews of *Du Darwinisme, ou l'Homme Singe,* by C. James, and *Le Darwinisme. Ce qu'il y a de vrai et de faux dans cette théorie,* by E. de Hartmann. *La Civiltà Cattolica,* 10th ser., 2 (1877): 449–58.

————. "S. Tommaso e la generazione spontanea primitiva." *La Civiltà Cattolica,* 16th ser., 11 (1897): 676–91.

————."Sant'Agostino e la generazione spontanea primitiva." *La Civiltà Cattolica,* 16th ser., 11 (1897): 421–38.

Scheeben, Matthias Joseph. *Handbuch der katholischen Dogmatik.* Book III: *Schöpfungslehre.* Ed. Wilhelm Breuning and Franz Lakner. In *Gesammelte Schriften,* vol. 5. 3rd ed. Freiburg: Herder, 1961.

Tanquerey, Adolphe. *Synopsis Theologiae Dogmaticae Specialis* Vol. 2: *de Fide, de Deo Uno et trino, de Deo Creante et Elevante, de Verbo Incarnato.* 13th ed. Rome, Tournai, and Paris: Desclée, 1911.

Tort, Patrick, ed. *Dictionnaire du darwinisme et de l'évolution.* Paris: Presses Universitaires de France, 1996.

Under the Ban: Correspondence between Dr. St. George Mivart and Herbert Cardinal Vaughan. New York: Tucker, 1900.

Vorzimmer, Peter J. *Charles Darwin: The Years of Controversy; The Origin of Species and Its Critics, 1859–1882.* Philadelphia: Temple University Press, 1970.

Ward, Wilfrid. "Bishop Hedley." *Dublin Review* 158 (January–April 1916): 1–12.

Weber, Ralph Edward. *Notre Dame's John Zahm: American Catholic Apologist and Educator.* Notre Dame, Ind.: University of Notre Dame Press, 1961.

Wernz, Franz Xaver. *Ius decretalium ad usum praelectionum in scholis textus canonici sive iuris decretalium.* Vol. 2: *Ius constitutionis Ecclesiae Catolicae.* Rome: S. C. de Propaganda Fide, 1899.

————. *Ius decretalium ad usum praelectionum in scholis textus canonici sive iuris decretalium.* Vol. 6: *Ius poenale Ecclesiae Catholicae.* Prati: Giachetti, 1913.

Zahm, John Augustine. *Bibbia, scienza e fede.* Siena: Presso l'Ufficio della Biblioteca del clero, 1895.

————. *Bible, science et foi.* Paris: Lethielleux, 1894.

————. *Bible, Science, and Faith.* Baltimore: Murphy, 1894.

————. *Biblia, ciencia y fe.* Madrid: La España Moderna, 1912.

————. *Catholic Science and Catholic Scientists.* Philadelphia, Pa.: H. L. Kilner, 1893.

————. *Evolution and Dogma.* Chicago: D. H. McBride, 1896.

————. *Evoluzione e dogma.* Siena: Presso l'Ufficio della Biblioteca del clero, 1896.

————. *L'évolution et le dogme.* Paris: Lethielleux, 1897.

————. *La evolución y el dogma.* Madrid: Sociedad Editorial Española, 1905.

————. "Evolution and Teleology." *Appleton's Popular Science Monthly* 52 (1898): 815–24.

————. "Évolution et téléologie." *Revue des Questions Scientifiques* 43 (April 1898): 403–19.

————. *Moses and Modern Science.* Philadelphia: D. J. Gallagher, 1894.

————. *Science and the Church.* Chicago: D. H. McBride, 1896.

————. *Science catholique et savants catholiques.* Paris: Lethielleux, 1895.

————. *Scientific Theory and Catholic Doctrine.* Chicago: D. H. McBride, 1896.

————. *Scienza cattolica e scienziati cattolici.* Genova: Fassicomo, 1896.

————. *Sound and Music.* Chicago: McClurg, 1892.

Zigliara, Tommaso. *Propaedeutica ad Sacram theologiam in usum scholarum.* 4th ed. Rome: S. C. de Propaganda Fide, 1897.

———. *Summa philosophica in usum scholarum.* 8th ed. Paris and Lyons: Delhomme et Briguet, 1891.

Zubizarreta, Valentín. *Theologia dogmatico-scholastica ad mentem S. Thomae Aquinatis.* Vol. 2: *De Deo uno, de Deo trino et de Deo creatore.* 3rd ed. Bilbao: Eléxpuru Hermanos, 1926.